Cellular Physiology and Neurophysiology

Look for these other *Mosby Physiology Monograph Series* titles:

▪   ▪   ▪   ▪   ▪   ▪   ▪   ▪   ▪   ▪   ▪   ▪   ▪   ▪   ▪

BLANKENSHIP: *Neurophysiology* (978-0-323-01899-9)

CLOUTIER: *Respiratory Physiology* (978-0-323-03628-3)

HUDNALL: *Hematologic Physiology and Pathophysiology* (978-0-323-04311-3)

JOHNSON: *Gastrointestinal Physiology, 7th edition* (978-0-323-03391-6)

KOEPPEN & STANTON: *Renal Physiology, 4th edition* (978-0-323-03447-0)

LEVY & PAPPANO: *Cardiovascular Physiology, 9th edition* (978-0-323-03446-3)

PORTERFIELD & WHITE: *Endocrine Physiology, 3rd edition* (978-0-323-03666-5)

Cellular Physiology and Neurophysiology

SECOND EDITION

By

MORDECAI P. BLAUSTEIN, MD
Professor, Departments of Physiology and Medicine
Director, Maryland Center for Heart Hypertension and Kidney Disease
University of Maryland School of Medicine
Baltimore, Maryland

JOSEPH P. Y. KAO, PhD
Professor
Center for Biomedical Engineering and Technology
and
Department of Physiology
University of Maryland School of Medicine
Baltimore, Maryland

DONALD R. MATTESON, PhD
Associate Professor
Department of Physiology
University of Maryland School of Medicine
Baltimore, Maryland

ELSEVIER
MOSBY

MOSBY

1600 John F. Kennedy Blvd.
Ste 1800
Philadelphia, PA 19103-2899

CELLULAR PHYSIOLOGY AND NEUROPHYSIOLOGY
Copyright © 2012 by Mosby, an imprint of Elsevier Inc.
Copyright © 2004 by Mosby, Inc., an affiliate of Elsevier Inc.

ISBN: 978-0-3230-5709-7

Cartoon in Chapter 1 reproduced with the permission of *The New Yorker*.

Notices

Knowledge and best practice in this field are constantly changing. As new research and experience broaden our understanding, changes in research methods, professional practices, or medical treatment may become necessary.

Practitioners and researchers must always rely on their own experience and knowledge in evaluating and using any information, methods, compounds, or experiments described herein. In using such information or methods they should be mindful of their own safety and the safety of others, including parties for whom they have a professional responsibility.

With respect to any drug or pharmaceutical products identified, readers are advised to check the most current information provided (i) on procedures featured or (ii) by the manufacturer of each product to be administered, to verify the recommended dose or formula, the method and duration of administration, and contraindications. It is the responsibility of practitioners, relying on their own experience and knowledge of their patients, to make diagnoses, to determine dosages and the best treatment for each individual patient, and to take all appropriate safety precautions.

To the fullest extent of the law, neither the Publisher nor the authors, contributors, or editors, assume any liability for any injury and/or damage to persons or property as a matter of products liability, negligence or otherwise, or from any use or operation of any methods, products, instructions, or ideas contained in the material herein.

Library of Congress Cataloging-in-Publication Data

Cellular physiology and neurophysiology / edited by Mordecai P. Blaustein, Joseph P.Y. Kao, and Donald R. Matteson.—2nd ed.
    p. ; cm.—(Mosby physiology monograph series)
  Rev. ed. of: Cellular physiology / Mordecai P. Blaustein, Joseph P.Y. Kao, Donald R. Matteson. c2004.
  Includes bibliographical references and index.
  ISBN 978-0-323-05709-7 (pbk. : alk. paper)
  I. Blaustein, Mordecai P. II. Kao, Joseph P. Y. III. Matteson, Donald R. IV. Blaustein, Mordecai P. Cellular physiology. V. Series: Mosby physiology monograph series.
  [DNLM: 1. Cell Physiological Phenomena. 2. Biological Transport—physiology. 3. Muscle Contraction—physiology. 4. Nervous System Physiological Processes. QU 375]
  571.6—dc23
                                                                    2011036478

Acquisitions Editor: Bill Schmitt
Developmental Editor: Margaret Nelson
Publishing Services Manager: Peggy Fagen/Hemamalini Rajendrababu
Project Manager: Divya Krish
Designer: Steven Stave

Printed in United States

Last digit is the print number: 9 8 7 6 5 4 3

PREFACE

Knowledge of cellular and molecular physiology is fundamental to understanding tissue and organ function as well as integrative systems physiology. Pathological mechanisms and the actions of therapeutic agents can best be appreciated at the molecular and cellular level. Moreover, a solid grasp of the scientific basis of modern molecular medicine and functional genomics clearly requires an education with this level of sophistication.

The explicit objective of *Cellular Physiology and Neurophysiology* is to help medical and graduate students bridge the divide between basic biochemistry and molecular and cell biology on the one hand and organ and systems physiology on the other. The emphasis throughout is on the functional relevance of the concepts to physiology. Our aim at every stage is to provide an intuitive approach to quantitative thinking. The essential mathematical derivations are presented in boxes for those who wish to verify the more intuitive descriptions presented in the body of the text. Physical and chemical concepts are introduced wherever necessary to assist students with the learning process, to demonstrate the importance of the principles, and to validate their ties to clinical medicine. Applications of many of the fundamental concepts are illustrated with examples from systems physiology, pharmacology, and pathophysiology. Because physiology is fundamentally a science founded on actual measurement, we strive to use original published data to illuminate key concepts.

The book is organized into five major sections, each comprising two or more chapters. Each chapter begins with a list of learning objectives and ends with a set of study problems. Many of these problems are designed to integrate concepts from multiple chapters or sections; the answers are presented in Appendix E. Throughout the book key concepts and new terms are highlighted. A set of multiple-choice review questions and answers is contained in Appendix F. A review of basic mathematical techniques and a summary of elementary circuit theory, which are useful for understanding the material in the text, are included in Appendixes B and D respectively. For convenience Appendix A contains a list of abbreviations symbols and numerical constants.

We thank our many students and our teaching colleagues whose critical questions and insightful comments over the years have helped us refine and improve the presentation of this fundamental and fascinating material. Nothing pleases a teacher more than a student whose expression indicates that the teacher's explanation has clarified a difficult concept that just a few moments earlier was completely obscure.

Mordecai P. Blaustein
Joseph P. Y. Kao
Donald R. Matteson

ACKNOWLEDGMENTS

■　■　■　■　■　■　■　■　■　■　■

We thank Professors Clara Franzini-Armstrong and John E. Heuser for providing original electron micrographs, and Jin Zhang for an original figure. We are indebted to the following colleagues for their very helpful comments and suggestions on preliminary versions of various sections of the book: Professors Mark Donowitz and Luis Reuss (Chapters 10 and 11); Professors Thomas W. Abrams, Bradley E. Alger, Bruce K. Krueger, Scott M. Thompson, and Daniel Weinreich (Section IV); Professors Martin F. Schneider and David M. Warshaw (Section V); and Professor Toby Chai (Chapter 16). We also thank the *New Yorker* for permission to reproduce the cartoon in Chapter 1.

CONTENTS

SECTION II
Ion Channels and Excitable Membranes

ION CHANNEL DIVERSITY 87

SECTION III
Solute Transport

CHAPTER 9

ELECTROCHEMICAL POTENTIAL ENERGY AND TRANSPORT PROCESSES 103

CHAPTER 10

PASSIVE SOLUTE TRANSPORT 113

CHAPTER **11**

ACTIVE TRANSPORT 133

SECTION IV
Physiology of Synaptic Transmission

SYNAPTIC PHYSIOLOGY I 155

SYNAPTIC PHYSIOLOGY II 181

SECTION V
Molecular Motors and Muscle Contraction

CHAPTER **14**

MOLECULAR MOTORS AND THE MECHANISM OF MUSCLE CONTRACTION....211

CHAPTER **15**

EXCITATION-CONTRACTION COUPLING IN MUSCLE 229

CHAPTER **16**

MECHANICS OF MUSCLE CONTRACTION

Cellular Physiology and Neurophysiology

Section I

Fundamental Physicochemical Concepts

1

INTRODUCTION: HOMEOSTASIS AND CELLULAR PHYSIOLOGY

OBJECTIVES

1. Understand the need to maintain the constancy of the internal environment of the body and the concept of homeostasis.

2. Understand the hierarchical view of the body as an ensemble of distinct compartments.

3. Understand the composition and structure of the lipid bilayer membranes that encompass cells and organelles.

4. Understand why the protein-mediated transport processes that regulate the flow of water and solutes across biomembranes are essential to all physiological functions.

HOMEOSTASIS ENABLES THE BODY TO SURVIVE IN DIVERSE ENVIRONMENTS

Humans are independent, free-living animals who can move about and survive in vastly diverse physical environments. Thus we find humans inhabiting habitats ranging from the frozen tundra of Siberia and the mountains of Nepal* to the jungles of the Amazon and the deserts of the Middle East. Nevertheless, the elemental constituents of the body are cells, whose survival and function are possible only within a narrow range of physical and chemical conditions, such as temperature, oxygen concentration, osmolarity, and pH.

Therefore the whole body can survive under diverse external conditions only by maintaining the conditions around its constituent cells within narrow limits. In this sense the body has an **internal environment,** which is maintained constant to ensure survival and proper biological functioning of the body's cellular constituents. The process whereby the body maintains constancy of this internal environment is referred to as **homeostasis.**[†] When homeostatic mechanisms are severely impaired, as in a patient in an intensive care

---

* The adaptability of humans can be surprising: humans can survive on Mount Everest, which, at 29,028 feet above sea level, is at the cruising altitude of jet airplanes. At the summit the temperature is approximately −40° Celsius (same as −40° Fahrenheit), the thin atmosphere supplies only approximately one third of the oxygen at sea level, and the relative humidity is zero.

[†] The concept of the internal environment was first advanced by the 19th-century pioneer of physiology, Claude Bernard, who discussed it in his book, *Introduction à l'étude de la médecine expérimentale* in 1865. Bernard's often-quoted dictum is: "The constancy of the internal environment is the prerequisite for a free life." ("*La fixeté du milieu intérieur est la condition de la vie libre.*" from *Leçons sur les phénomènes de la vie communs aux animaux et aux végétaux*, 1878.) The term "homeostasis" was introduced by Walter B. Cannon in his physiology text, *The Wisdom of the Body* (1932).

unit, artificial life support systems become necessary for maintaining the internal environment.

Achieving homeostasis requires various component physiological systems in the body to function coordinately. The musculoskeletal system enables the body to be motile and to acquire food and water. The gastrointestinal system extracts nutrients (sources of both chemical energy, such as sugars, and essential minerals, such as sodium, potassium, and calcium) from food. The respiratory (pulmonary) system absorbs oxygen, which is required in oxidative metabolic processes that "burn" food to release energy. The circulatory system transports nutrients and oxygen to cells while carrying metabolic waste away from cells. Metabolic waste products are eliminated from the body by the renal and respiratory systems. The complex operations of all the component systems of the body are coordinated and regulated through **biochemical signals** released by the endocrine system and disseminated by the circulation, as well as through electrical signals generated by the nervous system.

THE BODY IS AN ENSEMBLE OF FUNCTIONALLY AND SPATIALLY DISTINCT COMPARTMENTS

The organization of the body may be viewed hierarchically (Figure 1-1). The various systems of the body not only constitute functionally distinct entities, but also comprise spatially and structurally distinct compartments. Thus the lungs, the kidneys, the various endocrine glands, the blood, and so on are distinct compartments within the body. Each compartment has its own local environment that is maintained homeostatically to permit optimal performance of different physiological functions.

Compartmentation is an organizing principle that applies not just to macroscopic structures in the body, but to the constituent cells as well. Each cell is a compartment distinct from the extracellular environment and separated from that environment by a membrane (the *plasma membrane*). The intracellular space of each cell is further divided into subcellular compartments (cytosol, mitochondria, endoplasmic reticulum, etc.). Each of these subcellular compartments is encompassed within its own membrane, and each has a different microscopic internal environment to allow

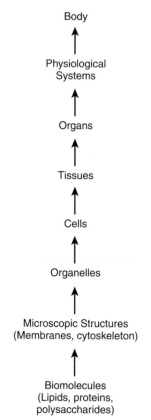

FIGURE 1-1 ■ Hierarchical view of the organization of the body. *(Modified from Eckert R, Randall D: Animal physiology, ed 2, San Francisco, 1983, WH Freeman.)*

different cellular functions to be carried out optimally (e.g., protein synthesis in the cytosol and oxidative metabolism in the mitochondria).

The Biological Membranes That Surround Cells and Subcellular Organelles Are Lipid Bilayers

As noted previously, cells and subcellular compartments are separated from the surrounding environment by biomembranes. Certain specific membrane proteins are inserted into these **lipid bilayer membranes.** Many of these proteins are **transmembrane proteins** that mediate the transport of various solutes or water across the bilayers. Ion channels and ion pumps are examples of such transport proteins. Other transmembrane proteins have signaling functions and transmit information from one side of the membrane to the other. Receptors for neurotransmitters, peptide hormones, and growth factors are examples of signaling proteins.

Biomembranes Are Formed Primarily from Phospholipids but May Also Contain Cholesterol and Sphingolipids

Most of the lipids that make up biomembranes are *phospholipids*. These **amphiphilic** (or **amphipathic**) **phospholipids** consist of a **hydrophilic** (water-loving), or **polar**, phosphate-containing head group attached to two **hydrophobic** (water-fearing), or **nonpolar**, fatty acid chains. The phospholipids assemble into a sheet or *leaflet*. The polar head groups pack together to form the hydrophilic surface of the leaflet, and the nonpolar hydrocarbon fatty acid chains pack together to form the hydrophobic surface of the leaflet. Two leaflets combine at their hydrophobic surfaces to form a bilayer membrane.

The bilayer presents its two hydrophilic surfaces to the aqueous environment, whereas the hydrophobic fatty acid chains remain sequestered within the interior of the membrane (Figure 1-2). The individual lipid molecules within the bilayer are free to move and are not rigidly packed. Therefore the lipid bilayer membrane behaves in part like a two-dimensional fluid and is frequently referred to as a **fluid mosaic.**

Biomembranes typically also contain other lipids such as cholesterol and sphingolipids. For example, in animals, biomembranes usually contain significant amounts of cholesterol, a nonphospholipid whose presence alters the fluidity of the membrane.

Biomembranes Are Not Uniform Structures

Different biomembranes vary in their lipid composition. For example, the plasma membrane is rich in cholesterol but contains almost no cardiolipin (a structurally complex phospholipid); the reverse is true for the mitochondrial membranes. Even the lipid compositions of the two leaflets constituting a single bilayer membrane can differ. For example, whereas phosphatidyl choline is most abundant in the outer leaflet of the plasma membrane, phosphatidyl serine is found almost exclusively in the inner leaflet. Such asymmetry can be maintained because flip-flop of lipid molecules from one leaflet to the other occurs naturally at an extremely slow rate.

Some cytoskeletal proteins bind to membrane proteins. These interactions enable the cytoskeleton to confer structural integrity on the membrane. Just as important, such interactions, by grouping and "tethering" membrane proteins, also organize membrane proteins into functional **membrane microdomains.** Such microdomains are compositionally and functionally different from other regions of the membrane. Thus it should be apparent that most biomembranes are not uniform either in composition or in architecture but are highly

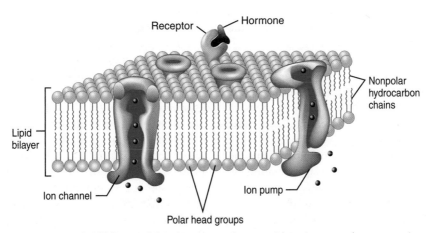

FIGURE 1-2 ■ Lipid bilayer of the plasma membrane, with various membrane proteins that serve transport and signaling functions. The locations of the polar head groups and nonpolar hydrocarbon chains of the phospholipids in the bilayer are shown. Also represented are a hormone receptor, an ion channel, and an ion pump.

organized structures with different microdomains serving different functions.

TRANSPORT PROCESSES ARE ESSENTIAL TO PHYSIOLOGICAL FUNCTION

Each compartment within the body, whether microscopic or macroscopic, has the optimal biochemical composition to enable a different set of physiological processes to take place. However, those very physiological processes tend to alter the composition within the compartments. In this light, homeostasis within each compartment implies that transport processes must operate continuously to adjust and maintain the internal environment of each compartment, including microscopic compartments such as those within subcellular organelles. Therefore transport mechanisms are central to homeostasis. Moreover, coordinated regulation of the physiological functions that occur in distinct compartments implies communication, that is, the transmission and reception of signals, between different compartments. At the subcellular level this is achieved through the generation and movement of biochemical signals, including **second messengers** such as inositol trisphosphate (IP$_3$), cyclic adenosine monophosphate (cAMP), or calcium ions (Ca^{2+}).

As noted earlier, extracellular (or intercellular) communication is mediated by biochemical signals as well as by electrical signals. Many biochemical signals (e.g., hormones and growth factors) are secreted by specialized cells and are disseminated through the circulation to distant targets. Other biochemical signals (e.g., neurotransmitters; see Section IV) mediate local intercellular communication. The electrical signals are generated and propagated through the transport of certain ions across the membranes of "excitable" cells (see Chapters 5 to 7). By their nature, the signaling mechanisms themselves alter the composition of the cells from which they originate. Thus the composition of those cells, too, must be continually restored. Therefore transport processes are also fundamental to the coordinated regulation of physiological processes in the body. Indeed, when membrane transport processes go awry, as may occur with mutations in transport proteins, homeostatic mechanisms are disrupted and physiology is adversely affected (this is

referred to as **pathophysiology**). Examples of pathophysiological mechanisms are presented throughout this book.

CELLULAR PHYSIOLOGY FOCUSES ON MEMBRANE-MEDIATED PROCESSES AND ON MUSCLE FUNCTION

The foregoing description implies that homeostasis and its regulation depend on transport and signaling processes that occur at or through biological membranes. For this reason such **membrane-mediated processes** are essential to physiology and are a central theme of this text (see Chapters 2 to 13). Of these membrane-mediated processes, passive diffusion and osmosis are fundamental physical processes that can occur *directly* through any lipid bilayer membrane and are the topics of Chapters 2 and 3, respectively. Most of the membrane-mediated processes can occur only through the agency of diverse protein machinery (e.g., ion channels, solute transporters, and transport ATPases or "pumps") residing in cellular membranes. These membrane protein–dependent processes are the subject of Chapters 4 to 13. A schematic representation of a cellular (plasma) membrane and some of the transport and signaling processes it mediates is shown in Figure 1-2.

Although processes mediated by cellular membranes are fundamental to physiological function, they take place on a microscopic scale. The maintenance of life also requires action on a macroscopic scale. Thus acquisition of food and water requires body mobility; nutrient extraction requires maceration of food and its passage through the gastrointestinal tract; intake of oxygen and expulsion of carbon dioxide require expansion and contraction of air sacs in the lungs; and distribution of nutrients and dissemination of endocrine signals to various tissues require rapid transport of material through circulation. All these processes require movement on a macroscopic scale. The evolutionary solution to the problem of large-scale movements is *muscle*. For this reason the cellular mechanisms underlying muscle function constitute the other major theme of this text (see Section V). The subject of cellular physiology comprises the two major themes described previously.

SUMMARY

1. To survive under extremely diverse conditions, the body must be able to maintain a constant internal environment. This process is referred to as *homeostasis*.

2. Homeostasis requires the coordination and regulation of numerous complex activities in all the component systems of the body.

3. The body can be viewed in terms of a hierarchical organization in which compartmentation is a major organizing principle.

4. Cells and subcellular organelles are compartments that are encompassed within biomembranes, which are essentially lipid bilayer membranes.

5. Biomembranes are composed primarily of phospholipids and integral membrane proteins; the membranes may also contain other lipids such as cholesterol and sphingolipids.

6. Most of the integral membrane proteins span the membrane (i.e., they are transmembrane proteins) and are involved in signaling or in the transport of water and solutes across the membrane. These processes are essential for homeostasis.

7. Biomembranes are usually nonuniform structures: the inner and outer leaflets often have different composition. Many integral membrane proteins bind to elements of the cytoskeleton and may be organized into microdomains with specialized functions.

8. The transport processes mediated by integral membrane proteins such as channels, carriers, and pumps in cell and organelle membranes are essential for physiological function.

9. The maintenance of life also depends on movement on a macroscopic scale. Such movements are mediated by muscle.

KEY WORDS AND CONCEPTS

- Internal environment
- Homeostasis
- Biochemical signals
- Compartmentation
- Lipid bilayer membranes
- Transmembrane proteins
- Amphiphilic (or amphipathic) phospholipids
- Hydrophilic (polar)
- Hydrophobic (nonpolar)
- Fluid mosaic
- Membrane microdomains
- Second messengers
- Pathophysiology
- Membrane-mediated processes

BIBLIOGRAPHY

Alberts B, Johnson A, Lewis J, et al: *Molecular biology of the cell*, ed 7, New York, NY, 2007, Garland Science.

Bernard C: *An introduction to the study of experimental medicine* (translated by H.C. Greene, from the French: *Introduction à l'étude de la médecine expérimentale*, Paris, 1865, JB Baillière), New York, NY, 1957, Dover.

Bernard C: *Leçons sur les phénomènes de la vie communs aux animaux et aux végétaux*, vol I, Paris, France, 1878, JB Baillière.

Cannon WB: *The wisdom of the body*, New York, NY, 1932, WW Norton.

Eckert R, Randall D: *Animal physiology*, ed 2, San Francisco, CA, 1983, WH Freeman.

Gennis RB: *Biomembranes*, New York, NY, 1989, Springer-Verlag.

Vance DE, Vance JE: *Biochemistry of lipids and membranes*, Menlo Park, CA, 1985, Benjamin Cummings.

"In order to be free I had to make certain adjustments."

2

DIFFUSION AND PERMEABILITY

OBJECTIVES

1. Understand that diffusion is the migration of molecules *down* a concentration gradient.
2. Understand that diffusion is the result of the purely *random* movement of molecules.
3. Define the concepts of *flux* and membrane *permeability* and the relationship between them.

DIFFUSION IS THE MIGRATION OF MOLECULES DOWN A CONCENTRATION GRADIENT

Experience tells us that molecules always move spontaneously from a region where they are more concentrated to a region where they are less concentrated. As a result, concentration differences between regions become gradually reduced as the movement proceeds. **Diffusion** always transports molecules from a region of high concentration to a region of low concentration because the underlying molecular movements are completely *random*. That is, any given molecule has no preference for moving in any particular direction. The effect is easy to illustrate. Imagine two adjacent regions of comparable volume in a solution (Figure 2-1). There are 5200 molecules in the left-hand region and 5000 molecules in the right-hand region. For simplicity, assume that the molecules may move only to the left or to the right. Because the movements are random, at any given moment approximately half of all molecules would move to the right and approximately half would move to the left. This means that, on average, roughly 2600 would leave the left side and enter the right side, whereas 2500 would leave the right

and enter the left. Therefore a *net* movement of approximately 100 molecules would occur across the boundary going from left to right. This net transfer of molecules caused by **random movements** is indeed from a region of higher concentration into a region of lower concentration.

FICK'S FIRST LAW OF DIFFUSION SUMMARIZES OUR INTUITIVE UNDERSTANDING OF DIFFUSION

The preceding discussion indicates that the larger the difference in the number of molecules between adjacent compartments, the greater the net movement of molecules from one compartment into the next. In other words, the *rate* at which molecules move from one region to the next depends on the concentration difference between the two regions. The following definitions can be used to obtain a more explicit and quantitative representation of this observation:

1. **Concentration gradient** is the change of concentration, ΔC, with distance, Δx (i.e., $\Delta C/\Delta x$).
2. **Flux** (symbol J) is the amount of material passing through a certain cross-sectional area in a certain amount of time.

7

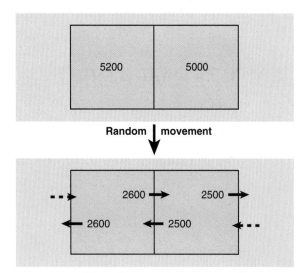

FIGURE 2-1 ■ Two adjacent compartments of comparable volume in a solution. The left compartment contains 5200 molecules, and the right compartment contains 5000 molecules. If the molecules can only move randomly to the left or to the right, approximately half of all molecules would move to the right and approximately half would move to the left. This means that, on average, roughly 2600 would leave the left side and enter the right side, whereas 2500 would leave the right and enter the left.

With these definitions, the earlier observation can be simply restated as "flux is proportional to concentration gradient," or

$$J \propto \frac{\Delta C}{\Delta x} \qquad [1]$$

By inserting a proportionality constant, D, we can write the foregoing expression as an equation:

$$J = -D \frac{\Delta C}{\Delta x} \qquad [2]$$

The proportionality constant, D, is referred to as the **diffusion coefficient** or **diffusion constant**. The minus sign accounts for the fact that the diffusional flux, or movement of molecules, is always *down* the concentration gradient (i.e., flux is from a region of high concentration to a region of low concentration). The graphs in Figure 2-2 illustrate this sign convention.

Equation [2] applies to the case in which the concentration gradient is linear, that is, a change in concentration, ΔC, for a given change in distance, Δx. For cases in which the concentration gradient may not be linear, the equation can be generalized by replacing the linear concentration gradient, $\Delta C/\Delta x$, with the

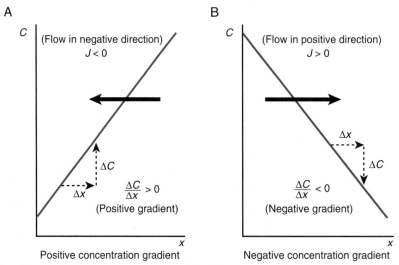

FIGURE 2-2 ■ The direction (sign) of the concentration gradients is opposite to the direction (sign) of the flux. **A,** A positive concentration gradient: the concentration increases as we move in the positive direction along the x-axis ($\Delta C/\Delta x > 0$). The flux being driven by this positive gradient is in the negative direction. The concentration increases from left to right, but the flux is going from right to left. **B,** A negative concentration gradient: the concentration decreases as we move in the positive direction along the x-axis ($\Delta C/\Delta x < 0$). The flux being driven by this negative gradient is in the positive direction. The concentration increases from right to left, but the flux is going from left to right.

more general expression for concentration gradient, dC/dx (a derivative). The diffusion equation now takes the form

$$J = -D \frac{dC}{dx} \qquad [3]$$

This equation is also known as **Fick's First Law of Diffusion**. It is named after Adolf Fick, a German physician who first analyzed this problem in 1855.

To complete the discussion of Fick's First Law, we should examine the dimensions (or units) associated with each parameter appearing in Equation [3]. Because flux, J, is the quantity of molecules passing through unit area per unit time, it has the dimensions of "moles per square centimeter per second" ($= [mol/cm^2]/sec = mol \cdot cm^{-2} \cdot sec^{-1}$). Similarly, the concentration gradient, dC/dx, being the rate of change of concentration with distance, has dimensions of "moles per cubic centimeter per centimeter" ($= [mol/cm^3]/cm = mol \cdot cm^{-4}$). For all the units to work out correctly in Equation [3], the diffusion coefficient, D, must have dimensions of cm^2/sec ($= cm^2 \cdot sec^{-1}$).

ESSENTIAL ASPECTS OF DIFFUSION ARE REVEALED BY QUANTITATIVE EXAMINATION OF RANDOM, MICROSCOPIC MOVEMENTS OF MOLECULES

Random Movements Result in Meandering

The most important characteristics of diffusion can be appreciated just by considering the simplest case of random molecular motion—that of a single molecule moving randomly along a single dimension. The situation is presented graphically in Figure 2-3.

The molecule is initially (at Time $= 0$) at some location that for convenience we simply refer to as 0 on the distance scale. During every time increment, Δt, the molecule can take a step of size δ either to the left or to the right. A typical series of 20 random steps is shown in Figure 2-3. Two features are immediately apparent from the figure. First, when a molecule is moving randomly, it does not make very good progress in any particular direction; it tends to meander back and forth aimlessly. Second, because the molecule meanders, its net movement away from its starting location

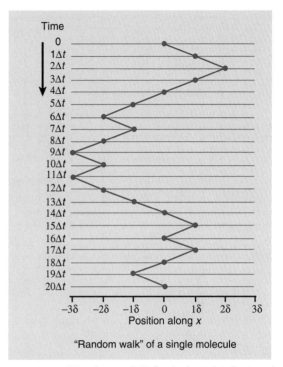

"Random walk" of a single molecule

FIGURE 2-3 ■ "Random walk" of a single molecule. A molecule is initially at position $x = 0$. During each increment of time, Δt, the molecule can take a step of size δ, either to the left or to the right. The position occupied by the molecule after each time increment is marked by a *dot*. A typical series of 20 steps is shown.

is not rapid. These two features manifest themselves in important ways when we consider the aggregate behavior of a large number of molecules. Figure 2-4 presents the results of a numerical simulation of diffusive spreading of 2000 molecules initially confined at $x = 0$ (Figure 2-4A). At each time point, each molecule takes a random step (forward and backward steps are equally probable). After each molecule has taken 10 random steps (Figure 2-4B), some molecules are seen to have moved away from the initial position, and the number of molecules remaining at precisely $x = 0$ has dropped to approximately 250. After 100 steps have been taken (Figure 2-4C), many molecules have moved farther afield, with a corresponding drop in the number remaining at $x = 0$ to approximately 100. The trend continues in Figure 2-4D (after 1000 steps). Note the change in magnitude of the vertical axis in each panel to rescale the spatial

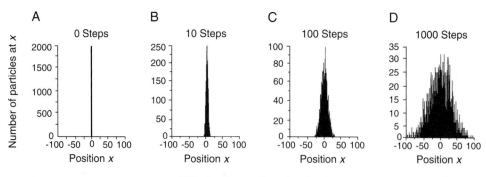

Diffusional spreading of molecules

FIGURE 2-4 ■ Spreading of molecules in space by random movements. The "experiment" is exactly the same as shown in Figure 2-3, except that 2000 molecules are being monitored. Initially 2000 molecules are located at $x = 0$. For each step in time, each of the molecules may move 1 step to the left or to the right. The number of molecules found at each position along the x-axis is shown at time $= 0$ (**A**) and after each molecule had taken 10 steps (**B**), 100 steps (**C**), and 1000 steps (**D**). The result of each molecule undergoing an independent random walk is to cause the entire ensemble of molecules to spread out in space.

distribution for visual clarity. Clearly, the spatial distribution of molecules is gradually broadened by diffusion.

One may ask what the average position of all the molecules is after diffusion has caused the spatial distribution to broaden. Figure 2-4 shows that as the molecules move randomly, they spread out progressively, but *symmetrically*, so that their average position is always centered on $x = 0$. This is reasonable: because moves to the right and left are equally probable, at any time, there should always be roughly equal numbers of molecules to the right and to the left of 0. The average position of such a distribution must be $x = 0$ at all times. This observation indicates that the average position is not an informative measure of the progress of diffusion.

The Root-Mean-Squared Displacement Is a Good Measure of the Progress of Diffusion

We seek a quantitative description of the fact that, with time, the molecules will cluster less and will progressively spread out in space. The desired measure is the **root-mean-squared (RMS) displacement**, d_{RMS} (see Appendix C). For diffusion in one dimension,

$$d_{RMS}^{1-D} = \sqrt{2Dt} \qquad [4]$$

where D is the diffusion coefficient (as in Fick's First Law) and t is time. For diffusion in two and three dimensions, the RMS displacements are given by, respectively,

$$d_{RMS}^{2-D} = \sqrt{4Dt} \qquad [5]$$

and

$$d_{RMS}^{3-D} = \sqrt{6Dt} \qquad [6]$$

An example of one-dimensional diffusion could be a repair enzyme randomly scanning DNA for single-strand breaks. A phospholipid molecule moving within a lipid bilayer undergoes two-dimensional diffusion. A glucose molecule moving in a volume of solution exemplifies three-dimensional diffusion.

Square-Root-of-Time Dependence Makes Diffusion Ineffective for Transporting Molecules over Large Distances

The most important aspect of the RMS displacement is that it does not increase linearly with time. Rather, random molecular movement involves meandering and thus causes spreading that increases only with the *square root* of time. Figure 2-5A shows the mathematical difference between displacement that varies directly with time and displacement that varies with the square root of time. The feature to notice is that over long distances the square root function seems to "flatten out." This means that to diffuse just a little

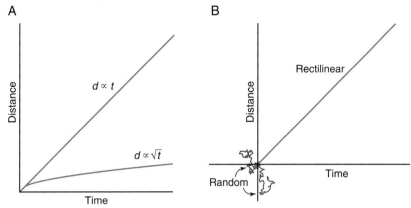

FIGURE 2-5 ■ Comparison of linear and square-root dependence of distance on time. **A,** With a linear time dependence, equal increments of time give equal increments of distance traveled. With a square-root time dependence, as the distance to be traveled becomes greater, the time required to cover the distance becomes disproportionately longer. **B,** A visually intuitive comparison of random and rectilinear motion. Starting from the origin, two molecules are allowed to take 50 steps of equal size, with each step taken in a random direction. A third molecule takes 50 steps of identical size but always in the same direction. Whereas the molecule undergoing rectilinear movement is far away from the origin after 50 steps, the randomly moving molecules meander and stay close to the origin.

farther takes a lot more time. In fact, because of the square-root dependence of the RMS displacement on time, to go 2 times farther takes 4 times as long, 10 times farther takes 100 times as long, and so on. A more intuitive illustration of the qualitative difference between random and rectilinear movement is shown in Figure 2-5B. The conclusion is that, over long distances, diffusion is an ineffective way to move molecules around.

Diffusion Constrains Cell Biology and Physiology

The practical significance of the fact that diffusion has a square-root dependence on time (Equations [4], [5], and [6]) can be shown by a simple calculation. Diffusion constants for biologically relevant small molecules (e.g., glucose, amino acids) in water are typically approximately 5×10^{-6} cm^2/second. For such molecules to diffuse a distance of 100 μm (0.01 cm) would take $(0.01)^2/6D = 3.3$ seconds (use Equation [6] and solve for t). For the same molecules to diffuse a distance of 1 cm (slightly less than the width of a fingernail), however, would take $1^2/6D = 33,000$ seconds $= 9.3$ hours! These results show that diffusion is sufficiently fast for transporting molecules over microscopic distances but

is extremely slow and ineffective over even moderate distances. Not surprisingly, therefore, most cells in the body are within 100 μm of a capillary and thus only seconds away from both a source of nutrient molecules and a sink for metabolic waste (Box 2-1). These calculations also demonstrate why even small insects (e.g., a mosquito) must have a circulatory system to transport nutrients into, and waste out of, the body.

FICK'S FIRST LAW CAN BE USED TO DESCRIBE DIFFUSION ACROSS A MEMBRANE BARRIER

A membrane typically separates two compartments in which the concentrations of some solutes can be different. We may designate the two compartments as i (inside) and o (outside), corresponding, for example, to the cytosol and extracellular fluid, respectively. The concentration difference between the two compartments, $\Delta C = C_i - C_o$, gives rise to a concentration gradient across the membrane, which has a certain thickness, say Δx. The concentration gradient, $\Delta C/\Delta x$, drives the diffusion of the solute across the membrane, thus leading to a flux of material, J, through the membrane. This description suggests that Fick's First Law

BOX 2-1
THE DENSITY OF CAPILLARIES IS A FUNCTION OF THE METABOLIC RATE OF A TISSUE

Oxygen (O_2) diffuses passively from tissue capillaries to cells in the tissue. To provide adequate O_2 to meet cellular metabolic needs, capillaries must be spaced closely enough in tissue to ensure that that O_2 concentration does not fall below the level required for mitochondrial function. We would expect capillary density in a particular tissue to depend on the metabolic rate of that tissue. Thus in slowly metabolizing tissue (e.g., subcutaneous), cells are typically separated by larger average distances from tissue capillaries. In contrast, in metabolically active tissues, cells are much closer to capillaries. In the cerebral cortex or the heart, for example, cells are typically only 10 to 20 μm from a capillary. In skeletal muscle the density of active capillaries depends strongly on the level of physical activity. At rest, skeletal muscle fibers are, on average, 40 μm from a functioning capillary. During strenuous exercise, many more capillaries are "recruited" and the average separation between muscle fibers and capillaries falls to less than 20 μm.

The necessity of capillaries in delivering O_2 to cells can be exploited clinically. Solid tumors require an adequate supply of O_2 for growth. Angiogenesis (growth of new blood vessels) is therefore essential for tumor growth. As a result of the pioneering research of Dr. Judah Folkman, new therapeutic regimens, involving drugs that inhibit angiogenesis, are being developed to promote the destruction of solid tumors.

in the form of Equation [2] would be well suited for analyzing such a situation:

$$J = -D\frac{\Delta C}{\Delta x} = -D\frac{C_i - C_o}{\Delta x} \qquad [7]$$

In this form the equation applies to a solute diffusing across a membrane of thickness Δx, provided that the solute dissolves as well in the membrane as it does in water (i.e., the concentration of the solute just inside the membrane matches the solute concentration in the adjacent aqueous solution; Figure 2-6A).

In Figure 2-6A, C_o^{mem} is the concentration of solute in the part of the membrane in immediate contact with the outside aqueous solution; C_i^{mem} is the concentration of solute in the part of the membrane in immediate contact with the inside aqueous solution. Realistically, because biological membranes are hydrophobic and nonpolar, whereas the aqueous solution is highly polar, solutes typically show different solubilities in the membrane relative to aqueous solution. To take such differential solubilities into account, we can define a quantity, β, the **partition coefficient**:

$$\beta = \frac{C^{mem}}{C^{aq}} \qquad [8]$$

where C^{aq} is the solute concentration in aqueous solution and C^{mem} is the solute concentration just inside the membrane. With the use of the partition coefficient, the solute concentrations just inside either face of the membrane can be written:

$$C_i^{mem} = \beta \times C_i \text{ and } C_o^{mem} = \beta \times C_o$$

The diffusion equation can now be cast in the following form:

$$J = -D\left[\frac{\Delta C}{\Delta x}\right]_{\text{Across membrane}} = -D\frac{C_i^{mem} - C_o^{mem}}{\Delta x}$$

$$= -D\frac{\beta C_i - \beta C_o}{\Delta x} = -D\beta\frac{C_i - C_o}{\Delta x} \qquad [9]$$

This form of the equation shows that the partition coefficient serves to modulate the solute concentration gradient within the membrane: when β is greater than 1 (solute dissolves better in the membrane than in aqueous solution), the concentration gradient in the membrane is enhanced and flux is proportionally increased (Figure 2-6B). Conversely, when β is less than 1 (solute dissolves better in aqueous solution than in the membrane), the concentration gradient in the membrane is diminished and flux is proportionally decreased (Figure 2-6C).

Equation [9] also predicts that when β equals 0, the flux, J, through the membrane would also be 0. In other words, if a substance is completely insoluble in the membrane, its flux through the membrane would be 0; that is, the membrane is completely *impermeable* to a substance that is not soluble in the membrane.

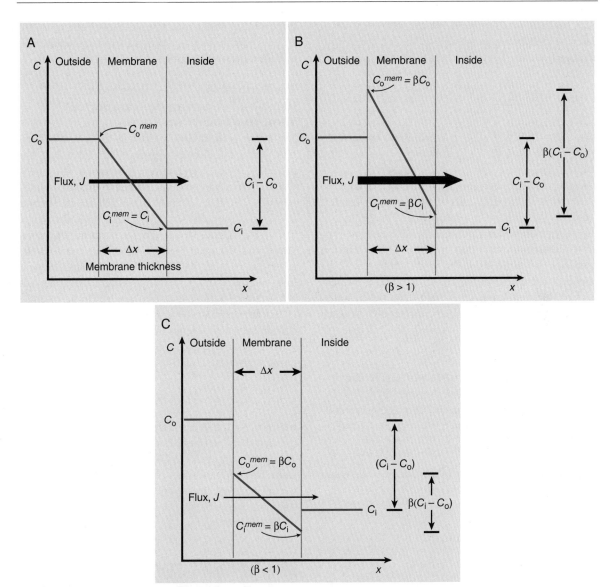

FIGURE 2-6 ■ Diffusion of a solute across a membrane is driven by the solute concentration gradient in the membrane. A solute is present in the outside solution at concentration C_o, and in the inside solution at concentration C_i. C_o^{mem} and C_i^{mem} are the solute concentrations in the part of the membrane immediately adjacent to the outside and inside solutions, respectively. The partition coefficient, β, is the ratio of the solute concentration in the membrane to the solute concentration in the aqueous solution in contact with the membrane ($\beta = C_o^{mem}/C_o = C_i^{mem}/C_i$). **A,** A solute that dissolves equally well in the membrane and in aqueous solution is characterized by $\beta = 1$. **B,** A solute that preferentially dissolves in the membrane has $\beta > 1$. **C,** A solute that dissolves better in aqueous solution than in the membrane has $\beta < 1$. In **B,** a larger β makes the solute concentration gradient steeper in the membrane and leads to a larger flux of the solute through the membrane. Conversely, in **C,** a smaller β makes the solute concentration gradient shallower in the membrane and leads to a smaller flux of the solute through the membrane.

This suggests that Equation [9] can also be used to describe **membrane permeability.** Indeed, we can rearrange Equation [9] to yield the following form:

$$J = -\left(\frac{D\beta}{\Delta x}\right)\left(C_i - C_o\right) = -P\left(C_i - C_o\right) \qquad [10]$$

where $P = (D\beta/\Delta x)$ is the **permeability** (or **permeability coefficient**) of the membrane for passage of a solute.* The dimensions of P are cm/second (i.e., a velocity), so that when P is multiplied by the concentration difference (with units of mol/cm^3), the result is (mol/cm^2)/second —the appropriate units for flux. In the mathematical description earlier, P is seen to contain microscopic properties such as D, the diffusion coefficient of solute inside the lipid membrane; β, the partition coefficient of the solute; and Δx, the thickness of the membrane. In actuality, the permeability coefficient can be determined empirically for each solute, without the need to measure the microscopic parameters described previously.

The Net Flux Through a Membrane Is the Result of Balancing Influx Against Efflux

An alternative way of looking at fluxes and permeabilities is suggested by Equation [10]:

$$J = -P(C_i - C_o) = PC_o - PC_i = J_{out \to in} - J_{in \to out} \qquad [11]$$

In other words, the **net flux** of a solute, J, is the result of balancing the inward flux (**influx**),

$$J_{out \to in} = P \times C_o \qquad [12]$$

against the outward flux (**efflux**),

$$J_{in \to out} = P \times C_i \qquad [13]$$

The two individual fluxes are *unidirectional* fluxes. Influx can thus be viewed as the inward flux being driven by the presence of solute on the outside at concentration, C_o, whereas efflux can be viewed as the outward flux being driven by the presence of solute on the inside at concentration, C_i. Mathematically, it is useful to note that multiplying a permeability and a concentration yields a unidirectional flux. It is also important to notice that, given the way Equations [10] and [11] are defined, a net flux that brings solute *into* a cell is a positive quantity.

The Permeability Determines How Rapidly a Solute Can Be Transported Through a Membrane

Membrane permeability coefficients for several biologically relevant molecules are shown in Table 2-1. The permeability coefficients give a fairly good idea of the relative permeability of a membrane to different solutes. In general, small neutral molecules, such as water, oxygen (O_2), and carbon dioxide (CO_2), with molecular weight (MW) 18, 32, and 44, respectively, permeate the membrane readily. Larger, highly hydrophilic organic molecules (e.g., glucose, MW 180) barely permeate. Inorganic ions, such as sodium (Na^+), potassium (K^+), chloride (Cl^-), and calcium (Ca^{2+}), are essentially impermeant.

An approximate description of how fast a solute concentration difference across a membrane is abolished as solute molecules diffuse through the membrane is given by Equation [B8] in Box 2-3:

$$\Delta C(t) = \Delta C^0 e^{-\frac{t}{\tau}}$$

where $\tau = 1/(P \times$ surface-to-volume ratio) is a *time constant* that describes the time scale on which concentration differences change. With the aid of Equation [B8] from Box 2-3 and the permeabilities in Table 2-1, we can calculate corresponding values of the time constant and immediately get a sense of how fast concentration differences for a given substance across the cell

TABLE 2-1		
Permeability of Plain Lipid Bilayer Membrane to Solutes		
SOLUTE	P (cm/sec)	τ*
Water	10^{-4}–10^{-3}†	0.5–5 sec
Urea	10^{-6}	~8 min
Glucose, amino acids	10^{-7}	~1.4 hr
Cl^-	10^{-11}	~1.6 yr
K^+, Na^+	10^{-13}	~160 yr

*Calculated for a spherical cell with a diameter of 30 μm (see Box 2-3). τ is the time constant that indicates how rapidly a solute concentration difference across the membrane can be dissipated by diffusion.
†Plain phospholipid bilayers are relatively permeable to water, but water permeability is reduced by the presence of cholesterol, which is found in all animal cell membranes.

---

*Equation [10] describes the diffusion of a substance across a membrane barrier. As such, it is applicable to many physiological situations, including gas exchange in the lung between the air space of an alveolus and the blood in a capillary (Box 2-2).

membrane would be evened out. For a spherical cell that is 30 μm in diameter, the surface area is $A_{cell} = 2.83 \times 10^{-5}$ cm^2, and the volume is $V_{cell} = 1.41 \times 10^{-8}$ cm^3, giving a surface-to-volume ratio of 2000 cm^{-1}. The τ values corresponding to the various P values are given in Table 2-1. Now the permeability properties of lipid bilayer membranes are easier to grasp. Small neutral molecules such as water, O_2, and CO_2 permeate readily and *fast*—on the order of seconds. Common nutrients such as glucose and amino acids take more than an hour to

permeate, which means that in any real biological context, they may as well be considered impermeant. Ions such as Cl^-, Na^+, K^+, and Ca^{2+} are so impermeant that years to centuries are required for them to permeate through a simple lipid bilayer membrane. Although measurements have never been made, we can infer that proteins, being very large molecules and often carrying multiple ionic charges, are also essentially impermeant.

The main point of this discussion is that, except simple, small, neutral molecules, essentially everything

---

BOX 2-2
FICK'S FIRST LAW OF DIFFUSION IS USED TO DESCRIBE GAS TRANSPORT IN THE LUNG

Ventilation delivers O_2 to, and removes CO_2 from, the lungs. Exchange of O_2 and CO_2 between the lung and pulmonary blood occurs through a thin (~0.3 μm) membranous barrier separating the alveolar air space from the blood inside capillaries apposed to the outer surface of the alveolus (see Figure B-1 in this box).

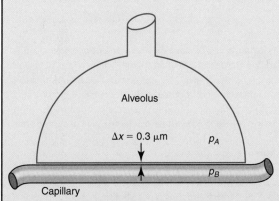

FIGURE B-1 ■ Schematic representation of a capillary apposed to an alveolus. The O_2 partial pressures in the alveolar air space and the capillary blood are symbolized by p_A and p_B, respectively. The diffusion barrier between the air space and the blood is typically ~0.3 μm.

The concentration (or partial pressure) of a physiologically important gas typically differs between the alveolar space and the blood. This concentration (or partial pressure) difference drives the diffusion of the gas between the two compartments. Pulmonologists

use a variant form of Fick's First Law to describe gas exchange across the alveolocapillary barrier:

$$\dot{V}_{gas} = A \times \left[\frac{-D \cdot \beta_m}{t} (p_B - p_A) \right] \quad [B1]$$

where $\dot{V}_{gas}$ is the volume of a gas transported per unit time across a membrane barrier of area, A, and thickness, t; D is the diffusion coefficient, and β_m the solubility, of the gas in the membrane barrier; and p_B and p_A are the partial pressures of the gas in the blood and in the alveolus, respectively. Comparison of Equation [B1] with Equation [10] in the text immediately shows their similarity of form (i.e., the amount of substance transported is driven by a concentration difference). All the proportionality factors in Equation [B1] can be grouped together as the *diffusing capacity of the lung* (D_L) for a particular gas. Equation [B1] then takes the very simple form

$$\dot{V}_{gas} = D_L(p_A - p_B) \quad [B2]$$

We note that this equation has the same form as the equation for flux in the text (Equation [10]). Inspection of Equation [B1] shows how various physiological or environmental changes could alter the amount of oxygen transported into the body. For example, if edema (accumulation of fluid) occurs to some extent in the alveoli, thus increasing the total thickness (t) of the alveolocapillary barrier, $\dot{V}O_2$ would decrease. Similarly, destruction of alveoli by disease would reduce the total surface area (A) across which gas exchange may take place and thus lower $\dot{V}O_2$. Finally, at high altitudes, where the partial pressure of O_2 is diminished, p_A of O_2 is correspondingly lower, leading also to decreased $\dot{V}O_2$.

BOX 2-3

HOW RAPIDLY CAN DIFFUSION ABOLISH CONCENTRATION DIFFERENCES ACROSS A MEMBRANE?

With a little bit of mathematics, we can improve our understanding of relative permeabilities and better appreciate what is meant when something is called permeant or impermeant. Recall that the permeability coefficient, P, has dimensions of *cm/second*. That is, the concept of *time* (and thus *rate*) is somehow embodied in the description of permeability presented in the text. We now make this connection explicit.

Text Equation [10] stated that the flux (J) of molecules across a cell membrane is driven by the concentration difference of that molecule between the inside and the outside:

$$J = -P(C_i - C_o) = -P\Delta C \qquad [B1]$$

When the membrane is permeable to a particular species, for that species, any concentration difference between inside and outside cannot persist. As molecules start to permeate the membrane from one side to the other, any concentration difference between inside and outside will gradually diminish. To determine how the concentration difference is gradually abolished, we need only to figure out how the concentration inside the cell is changed by the flux of molecules. Given that flux (J) is in units of *moles per cm² per second* (amount of molecules passing through a unit area of membrane in unit time), the number of moles of molecules entering the cell through its entire surface area (A_{cell}) in unit time must be:

$$J \times A_{cell} \quad \text{with units of mol/sec}$$

These moles of molecules are added to the total internal volume of the cell (V_{cell}). The resulting concentration change in the cell must be:

$$(J \times A_{cell})/V_{cell} \quad \text{with units of } (mol/cm^3)/sec$$

Merging this with the foregoing Equation [B1], we can write the rate of change of the concentration difference as:

$$\frac{\Delta[\Delta C]}{\Delta t} = J \times \frac{A_{cell}}{V_{cell}} = -P\frac{A_{cell}}{V_{cell}}\Delta C \qquad [B2]$$

The ratio A_{cell}/V_{cell} is the *surface-to-volume ratio* of the cell. If we redefine the product of the permeability

coefficient and the surface-to-volume ratio as k, Equation [B2] can be written more compactly as:

$$\frac{\Delta[\Delta C]}{\Delta t} = -k\Delta C \qquad [B3]$$

or, with derivative notation, as:

$$\frac{d[\Delta C]}{dt} = -k\Delta C \qquad [B4]$$

Equation [B4] can be rearranged and integrated:

$$\int_{t=0}^{t} \frac{1}{\Delta C}\frac{d[\Delta C]}{dt}dt = \int_{t=0}^{t} -k dt \qquad [B5]$$

The result of integration is:

$$\ln\frac{\Delta C(t)}{\Delta C^0} = -kt \qquad [B6]$$

In terms of exponentials, Equation [B6] is:

$$\frac{\Delta C(t)}{\Delta C^0} = e^{-kt} \text{ or } \Delta C(t) = \Delta C^0 e^{-kt} \qquad [B7]$$

Equation [B7] describes how the concentration difference across a cell membrane (initially at ΔC^0) will change with time if the membrane is permeable: the concentration difference will decrease exponentially with time. The time course of such a change is shown in Figure B-1. In the previous discussion, we assumed that the cell volume does not change significantly during the course of equilibration.

The constant k is called a *rate constant*, and its magnitude governs how fast the concentration difference is abolished (large k means rapid abolition of concentration difference between inside and outside). If we recall that:

$$k = P \times (\text{surface-to-volume ratio})$$

this makes sense: the higher the permeability of the membrane, the faster molecules will be able to move through the membrane, and the more rapidly the concentration difference across the membrane will be abolished. The reciprocal of the rate constant is called the *time constant* and is given the symbol τ (Greek letter "tau"):

$$\tau = 1/k$$

BOX 2-3

HOW RAPIDLY CAN DIFFUSION ABOLISH CONCENTRATION DIFFERENCES ACROSS A MEMBRANE?—cont'd

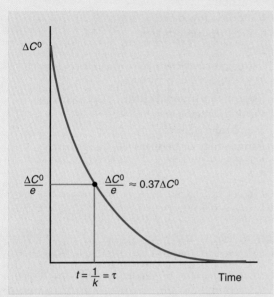

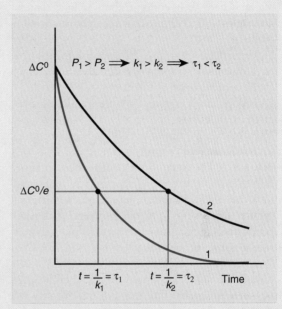

FIGURE B-1 ■ Exponential decay of a solute concentration difference across a cell membrane. Initially ($t = 0$), the concentration difference, ΔC, is at initial value, ΔC^0. With time, ΔC diminishes and asymptotically approaches 0. The time constant, τ (which is equal to the inverse of the rate constant, k), is the time at which ΔC has dropped to $1/e$ (or ~37%) of its initial value.

τ is the time it takes for the concentration difference to drop to $1/e$ (~37%) of its initial value. Put another way, when $t = \tau$, $\Delta C = 0.37 \Delta C^0$. A solute that has higher permeability has a shorter τ, whereas one with lower permeability has a longer τ. These relationships are illustrated in Figure B-2.

FIGURE B-2 ■ The higher the membrane permeability, the faster the disappearance of a solute concentration gradient across the cell membrane. For two solutes (1 and 2) with the same initial concentration difference (ΔC^0) across the membrane, if the membrane is more permeable to solute 1 than to solute 2 ($P_1 > P_2$), the concentration difference of solute 1 will decrease faster than that of solute 2. This is equivalent to saying that the rate constants and time constants for the two solutes have the following relationships: $k_1 > k_2$ and $\tau_1 < \tau_2$.

Because τ is just the reciprocal of k, Equation [B7] can also be written as:

$$\Delta C(t) = \Delta C^0 e^{-\frac{t}{\tau}} \qquad [B8]$$

that is biologically relevant and important cannot readily pass through a simple lipid bilayer membrane. For this reason a diversity of special mechanisms has evolved to transport a broad spectrum of biologically important species across cellular membranes. Ion pumps and channels permit influx and efflux of Na^+, K^+, Ca^{2+}, and Cl^-. A range of carrier proteins allows movement of sugars and amino acids across membranes.

Elaborate and highly regulated machinery governs endocytosis and exocytosis to bring large molecules like proteins into and out of cells. Endocytosis and exocytosis lie in the realm of cell biology and are not discussed in this text. Ion channels, pumps, and carriers are basic to cellular functions that underlie physiology and neuroscience and are discussed in later chapters.

SUMMARY

1. Diffusion causes the movement of molecules from a region where their concentration is high to a region where their concentration is low; that is, molecules tend to diffuse down their concentration gradient.

2. Fick's First Law describes diffusion in quantitative terms: the flux of molecules is directly proportional to the concentration gradient of those molecules.

3. Diffusion results entirely from the random movement of molecules.

4. The distance that molecules diffuse is proportional to the square root of time.

5. Because of the square-root dependence on time, diffusion is effective in transporting molecules and ions over short distances that are on the order of cellular dimensions (i.e., micrometers). Diffusion is extremely ineffective over macroscopic distances (i.e., a millimeter or greater).

6. The net flux of molecules diffusing across a cell membrane may be viewed as the net balance between an inward flux (influx) and an outward flux (efflux).

7. The ease with which a species may diffuse through a membrane barrier is characterized by the membrane permeability, P. Higher permeability permits a larger flux.

8. The product of a permeability and a concentration is a flux. For example, $P_{Na} \times [Na^+]_o$ represents a flux of Na^+ into the cell (an influx), whereas $P_{Cl} \times [Cl^-]_i$ represents a flux of Cl^- out of the cell (an efflux).

9. With the exception of small neutral molecules such as O_2, CO_2, water, and ethanol, essentially no biologically important molecules and ions can spontaneously diffuse across biological membranes.

KEY WORDS AND CONCEPTS

- Diffusion
- Random movement
- Concentration gradient
- Flux
- Diffusion coefficient (or diffusion constant)
- Fick's First Law of Diffusion
- Root-mean-squared (RMS) displacement
- Partition coefficient
- Membrane permeability
- Permeability or permeability coefficient
- Net flux
- Influx
- Efflux

STUDY PROBLEMS

1. If a collection of molecules diffuses 5 μm in 1 second, how long will it take for the molecules to diffuse 10 μm?

2. The permeability of the plasma membrane to K^+ is given the symbol P_K. If the intracellular and extracellular K^+ concentrations are $[K^+]_i$ and $[K^+]_o$, write the expressions representing influx (J_{inward}) and efflux ($J_{outward}$) of K^+. What is the expression for the net flux of K^+?

BIBLIOGRAPHY

Ferreira HG, Marshall MW: *The biophysical basis of excitability*, Cambridge, England, 1985, Cambridge University Press.

Feynman RP: *Feynman lectures on physics*, New York, NY, 1970, Addison-Wesley.

3

OSMOTIC PRESSURE AND WATER MOVEMENT

OBJECTIVES

1. Understand the nature of osmosis.

2. Define osmotic pressure in terms of solute concentration through van't Hoff's Law.

3. Define the driving forces that control water movement across membranes.

4. Understand that fluid movement across a capillary wall is determined by a balance of hydrostatic and osmotic pressures.

5. Know how cell volume changes in response to changing concentrations of permeant and impermeant solutes in the extracellular fluid.

OSMOSIS IS THE TRANSPORT OF *SOLVENT* DRIVEN BY A DIFFERENCE IN *SOLUTE* CONCENTRATION ACROSS A MEMBRANE THAT IS IMPERMEABLE TO SOLUTE

Because of diffusion, a net movement of molecules occurs *down* concentration gradients, and substances tend to move in a way that abolishes concentration differences in different regions of a solution. Alternatively, it could be said that diffusion results in *mixing*. We now examine the consequences when a solute is prevented from diffusing down its concentration gradient. Figure 3-1A, shows two aqueous compartments, 1 and 2, separated by a semipermeable membrane. Initially compartment 1 contains pure water and compartment 2 contains a solute dissolved in water. The membrane is permeable to water but impermeable to the solute. Because the solute concentration differs between the two compartments (higher in 2 than in 1), normally a net flux of solute molecules from 2 into 1

would occur. Because the membrane is impermeable to the solute, however, such a flux cannot take place; that is, the solute molecules cannot move from 2 into 1 to abolish the concentration difference between the two compartments.

The situation can be viewed from another perspective. When a solute is dissolved in water to form an aqueous solution, as the concentration of solute in the solution is increased, the concentration of *water* in the same solution must correspondingly decrease. In the two compartments shown in Figure 3-1A, whereas the solute concentration is higher in 2 than in 1, the water concentration is higher in 1 than in 2. Therefore, although the membrane does not permit the solute to move across from 2 into 1, it does allow water to move from 1 into 2. The presence of the water concentration difference allows us to predict, correctly, that a net flux of water will occur from 1 into 2. This movement of water through a semipermeable membrane, from a region of higher water concentration to a region of lower water concentration, is called **osmosis**.

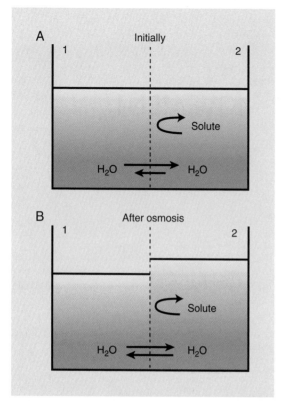

FIGURE 3-1 ■ Two aqueous compartments separated by a semipermeable membrane that allows passage of water but not solute. Compartment 1 contains pure water; compartment 2 contains a solute dissolved in water. **A,** Before any osmosis has taken place. **B,** After osmosis has occurred.

WATER TRANSPORT DURING OSMOSIS LEADS TO CHANGES IN VOLUME

In light of the foregoing, we can think about simple diffusion again. Because an increase in solute concentration dictates a corresponding decrease in water concentration, when a solute concentration gradient exists, a water concentration gradient running in the opposite direction must also be present. We can therefore view simple diffusion as a net flux of solute molecules down their concentration gradient occurring *simultaneously* with a net flux of water molecules down the water concentration gradient.

In osmosis, the membrane does not permit solute movement, so the only flux is that of water. Because the net flux of water in one direction is not balanced by a flux of solute in the opposite direction, the net movement of water would be expected to contribute to a change in volume: as water moves from compartment 1 to compartment 2, the volume of solution in 2 should increase, as shown in Figure 3-1B.

OSMOTIC PRESSURE DRIVES THE NET TRANSPORT OF WATER DURING OSMOSIS

In the two-compartment situation depicted in Figure 3-1, as osmosis proceeds, the solution volume in 2 will increase and a "head" of solution will build up (Figure 3-1B). The head of solution will exert **hydrostatic pressure,** which will tend to "push" the water across the membrane back to compartment 1, thus reducing the net flux of water from 1 into 2. As the solution volume in 2 increases, a point will be reached when the hydrostatic pressure from the solution head is sufficient to counteract exactly the net flow of water from 1 into 2.

This view suggests that we could also think of osmotic flow of water as being driven by some "pressure" that forces water to flow from 1 into 2. This pressure, arising from unequal solute concentrations across a semipermeable membrane, is termed **osmotic pressure**. From the earlier discussion, we expect that the larger the solute concentration difference between two solutions separated by a semipermeable membrane, the larger the osmotic pressure difference driving water transport. Therefore the osmotic pressure of a solution should be proportional to the concentration of solute. Indeed, for any solution, the osmotic pressure can be described fairly accurately by **van't Hoff's Law**[*]:

$$\pi = RTC_{solute} \qquad [1]$$

where π is the osmotic pressure, C_{solute} is the solute concentration, T is the absolute temperature (T = Celsius temperature + 273.15, in absolute temperature units, i.e., Kelvins), and R is the universal gas

---

[*]Derived by Jacobus Henricus van't Hoff, who received the Nobel Prize in Chemistry in 1901.

constant (0.08205 liter·atmosphere/mole/Kelvin). The various units of measurement used in the study of osmosis are described in Box 3-1.

It is important to stress that the osmotic pressure of a solution is a *colligative* property—that is, a property that depends only on the total concentration of dissolved particles and not on the detailed chemical nature of the particles. Therefore, in Equation [1], C_{solute} is the total concentration of all solute particles. For example, a nondissociable molecular solute such as glucose or mannitol at a concentration of 1 M gives $C_{solute} = 1\ M$. In contrast, a 1 M solution of an ionic solute such as NaCl, which can fully dissociate into equal numbers of Na^+ and Cl^- ions, actually has a solute particle concentration of $C_{solute} = 2\ M$. Similarly,

for a 1 M solution of $MgCl_2$, which can dissociate into constituent Mg^{2+} and Cl^- ions, $C_{solute} = 3\ M$.

In Figure 3-1 the osmotic pressure of the solution in compartment 1 is 0 ($C_{solute} = 0$), whereas the osmotic pressure of the solution in compartment 2 is equal to RTC_{solute}. In practice, the magnitude of the osmotic pressure is also operationally equal to the amount of hydrostatic pressure that must be applied to the compartment with the higher solute concentration to stop the net flux of water into that compartment. The magnitude of osmotic pressures in physiological solutions is typically very high (Box 3-2).

BOX 3-1
UNITS OF MEASUREMENT USED IN THE STUDY OF OSMOSIS

Osmotic pressure is proportional to the total concentration of dissolved particles. Each mole of osmotically active particles is referred to as an osmole. One mole of glucose is equivalent to 1 osmole, because each molecule of glucose stays as an intact molecular particle when in solution. One mole of NaCl is equivalent to 2 osmoles, because when in solution, each mole of NaCl dissociates into 1 mole of Na^+ ions and 1 mole of Cl^- ions, both of which are osmotically active.

One osmole in 1 liter of solution gives a 1 *osmolar* solution. Alternatively, one osmole in 1 kilogram of solution gives a 1 *osmolal* solution. Because *osmolarity* is defined in terms of solution volume, whereas *osmolality* is defined in terms of the weight of solution, osmolarity changes with temperature, whereas osmolality is independent of temperature. For simplicity, in this chapter all concentrations of osmotically active solutes are given in units of molar (M = mol/L) or millimolar (mM = 10^{-3} mol/L).

With commonly used factors, van't Hoff's Law (Equation [1] in text) gives the osmotic pressure in units of atmospheres. Physiological pressures are typically given in units of millimeters of mercury (mm Hg). The conversion factor is 1 atmosphere = 760 mm Hg.

BOX 3-2
PHYSIOLOGICALLY RELEVANT OSMOTIC PRESSURES ARE HIGH

All fluid compartments in the body contain dissolved solutes. Extracellular fluid is high in Na^+ and Cl^- and low in K^+, whereas intracellular fluid is high in K^+ and phosphate in various forms and low in Na^+ and Cl^-. For cells in the body to maintain constant volume, the osmotic pressure arising from solutes inside cells must be equal to the osmotic pressure arising from solutes in the extracellular fluid. The total solute concentration in the fluids is typically close to 300 mM. The osmotic pressure resulting from this solute concentration at 37°C (310.15 Kelvin) can be estimated with van't Hoff's Law:

$$\pi = RTC = 0.08205\frac{L\cdot atm}{mol\cdot K}\times 310.15K \times$$
$$300\times 10^{-3}\frac{mol}{L} = 7.63\ atm$$

At 7.63 atmospheres (atm), the osmotic pressure of intracellular and extracellular fluids is quite high, especially in light of the fact that air pressure at sea level is just 1 atm. Indeed, to experience 7.63 atm, one needs to dive to a depth of ~67 m (~220 ft) in the ocean. The high osmotic pressure of physiological solutions is the reason that red blood cells, which are much more permeable to water than to solutes, will swell very rapidly and burst (lyse) when they are placed into water or dilute solutions.

A note about units of measurement: 1 atm is equivalent to 760 mm of mercury (Hg).

When impermeant solute is present on both sides of a semipermeable membrane, water flux across the membrane will depend on the osmotic pressure difference between the two compartments. In turn, the osmotic pressure difference depends on the net imbalance of solute concentrations across the membrane:

$$\Delta\pi = RT\Delta C_{solute} \qquad [2]$$

where $\Delta\pi$ is the osmotic pressure difference across the membrane, ΔC_{solute} is the difference in solute concentrations across the membrane, and R and T are as defined for Equation [1].

In real life, membranes are never completely impermeable to solute. What happens when the membrane is only partially impermeable to solute? We can deduce the answer by considering the two extreme cases we already know: (1) if the membrane allows the solute molecules to pass through as freely as water molecules can, the flux of water is counterbalanced by the flux of solute in the opposite direction and the situation is no different from free diffusion; there would be no osmotic pressure difference; and (2) if the membrane is completely impermeable to solute molecules but completely permeable to water molecules, the maximum osmotic pressure that can be achieved is given by Equation [2]. The behavior of real-life membranes must lie somewhere between these two extremes. Because the relative permeability of the membrane to solute and water determines the actual behavior, we can define a parameter, called the **reflection coefficient**, represented by the symbol σ:

$$\sigma = 1 - \frac{P_{solute}}{P_{water}} \qquad [3]$$

When the membrane permeability to solute, P_{solute}, has the value of 0, the reflection coefficient takes on the value of 1. In this case the membrane "reflects" all solute molecules and does not allow them to pass through, and full osmotic pressure should be achieved. When the solute molecules permeate the membrane as readily as do water molecules ($P_{solute}/P_{water} = 1$), the reflection coefficient takes on the value of 0. In this case the membrane allows free passage of both water and solute, and the osmotic pressure should be 0. By incorporating the reflection coefficient into the osmotic pressure equation, we can more accurately describe real-life behavior:

$$\Delta\pi = \sigma RT\Delta C_{solute} \qquad [4]$$

BOX 3-3
TONICITY AND OSMOLARITY

Each mole of dissolved solute particles contributes 1 osmole to the solution; therefore any dissolved solute contributes to the osmolarity of a solution. With regard to the ability to drive water flow by osmosis, not all solutes are equal. Solutes with low membrane permeability (σ close to 1) have far greater osmotic effect than those with high membrane permeability (σ close to 0). Therefore two solutions of equal osmolarity can have different osmotic effects on cells. As an example, take a cell initially equilibrated with extracellular fluid (ECF). When a solute is added to the ECF, the ECF becomes *hyperosmolar* with respect to the intracellular fluid (ICF). If the added solute has $\sigma = 1$ (is impermeant), the cell will lose water and shrink; the ECF is then said to be *hypertonic* with respect to the cell. If the added solute has $\sigma = 0$ (is completely membrane permeant), the osmotic pressure of the ECF will not change and neither will the cell volume. In this case the ECF is said to be *isotonic* with respect to the cell. Similarly, a solution that causes the cell to gain water and swell is said to be *hypotonic*.

---

where $\Delta\pi$ is the effective osmotic pressure difference across a membrane.

The fact that, even at the same concentration, solutes with different reflection coefficients can generate different osmotic pressures gives rise to a distinction between **osmolarity** and **tonicity**, which is explained in Box 3-3.

OSMOTIC PRESSURE AND HYDROSTATIC PRESSURE ARE FUNCTIONALLY EQUIVALENT IN THEIR ABILITY TO DRIVE WATER MOVEMENT THROUGH A MEMBRANE

In the previous section we noted that hydrostatic pressure can counteract water movement driven by osmotic pressure. This suggests that as far as water movement through a membrane is concerned, hydrostatic and osmotic pressures act equivalently—both are capable of driving water movement through a membrane. When water movement is described, it is customary to consider the *volume* of water that passes through a unit area of membrane per unit time. **Volume flow** through a membrane, given the symbol J_v, is quantitatively described by the following equation (sometimes called the **Starling equation***):

$$J_v = L_p(\Delta\pi - \Delta P) = L_p(\sigma RT\Delta C_{solute} - \Delta P) \qquad [5]$$

wherein ΔP is the hydrostatic pressure difference across the membrane and L_p is a proportionality constant called the **hydraulic conductivity** (a measure of the ease with which a membrane allows water to pass through). Equation [5] emphasizes the equivalence of hydrostatic and osmotic pressures in driving water volume flow through membranes. Furthermore, it indicates that the direction of water volume flow across a membrane is determined by the *balance* of hydrostatic and osmotic pressures across the membrane.**

Once again, it is instructive to check the dimensions of the various quantities in Equation [5]. The pressure terms (hydrostatic and osmotic) naturally have units of pressure and can be expressed in units such as atmospheres (atm), millimeters of mercury (mm Hg), or dynes per square centimeter (dyne·cm^{-2}). L_p reflects the ability of a certain volume of water to pass through a certain area of membrane, driven by a certain amount of pressure in a certain amount of time, and thus could have units of cm^3·cm^{-2}·atm^{-1}·sec^{-1} (equivalent to cm·atm^{-1}·sec^{-1}) or cm^3·cm^{-2}·(dyne·cm^{-2})$^{-1}$·sec^{-1} (equivalent to cm^3·dyne^{-1}·sec^{-1}). We can figure out that J_v must have units of cm/sec—seemingly bizarre units

---

*Named after Ernest Starling, a 19th-century physiologist who first investigated fluid movement driven by osmotic and hydrostatic forces and who is also known for the Frank-Starling Law of heart function.

**That the balance of hydrostatic and osmotic forces determines the direction of water movement across a membrane is the basis for a water purification process called reverse osmosis. By application of high pressure to salty water on one side of a semipermeable membrane, water can be made to flow by "reverse osmosis" from the side of high salt concentration to become water with low salinity on the other side. Reverse osmosis is used on an industrial scale to generate fresh water from seawater in some parts of the world where fresh water is in short supply.

for *volume* flow! However, if we recognize that cm/sec is exactly equivalent to cm^3 per cm^2 per second, we see that J_v does indeed have the right physical meaning of *volume* of water flowing through unit *area* of membrane per unit *time*.

To use Equation [5] to calculate a volume flow, one must know L_p as well as the surface area of the membrane through which the fluid flows. In an actual situation, these two parameters may be very difficult to measure. Therefore an alternative form of the Starling equation is often used:

$$J_v = K_f(\Delta\pi - \Delta P) = K_f(\sigma RT\Delta C_{solute} - \Delta P) \qquad [6]$$

where K_f is the **filtration constant** and is equal to the product of L_p and the membrane area through which fluid flows. K_f may be regarded as an empirical proportionality factor that relates the volume of fluid through a membrane barrier and the driving force for fluid flow ($\Delta\pi - \Delta P$) across that barrier. We note that in Equation [6], K_f has units of volume per unit time per unit pressure (e.g., cm^3·sec^{-1}·(mm Hg)$^{-1}$), whereas J_v has units of volume per unit time (e.g., cm^3·sec^{-1}).

The Direction of Fluid Flow Through the Capillary Wall Is Determined by the Balance of Hydrostatic and Osmotic Pressures, as Described by the Starling Equation

In analyzing fluid movement across capillary walls, Starling recognized the importance of hydrostatic and osmotic forces. The Starling equation (Equations [5] and [6]) succinctly summarizes how net fluid movement is determined by hydrostatic and osmotic pressures. Figure 3-2 shows the four pressures that are in play; the direction of water movement driven by each pressure component is indicated by an arrow associated with that pressure. The capillary hydrostatic pressure (blood pressure) is P_c; the hydrostatic pressure in the interstitium is P_i. The principal contributors to osmotic pressure are dissolved proteins, which are too large to pass through the capillary walls. The osmotic pressure of the capillary is termed the capillary **colloid osmotic pressure**, or the capillary **oncotic pressure**, and is symbolized as π_c. The osmotic pressure resulting from dissolved proteins in the interstitial fluid is called the interstitial colloid osmotic pressure, or the interstitial oncotic pressure, and is symbolized as π_i.

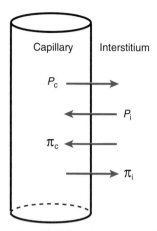

FIGURE 3-2 ■ Pressures that determine fluid movement between the interstitium and the lumen of a capillary. Capillary hydrostatic pressure, P_c, and interstitial oncotic (colloid osmotic) pressure, π_i, drive fluid movement from the capillary into the interstitium (filtration). Capillary oncotic pressure, π_c, and interstitial hydrostatic pressure, P_i, drive fluid movement from the interstitium into the capillary (absorption).

The capillary hydrostatic pressure tends to push water out of the capillary, whereas any interstitial hydrostatic pressure tends to push water into the capillary. The osmotic components operate in just the opposite way. The capillary osmotic pressure results from impermeant solutes inside the capillary and would tend to retain water inside the capillary (or "pull" water from the interstitium into the capillary). Similarly, the interstitial osmotic pressure tends to retain water in the interstitium (or "pull" water out of the capillary into the interstitium). As blood passes through a capillary, whether net movement of water occurs into or out of the capillary is determined by the balance of the four pressure components. The net movement of fluid out of the capillary is called filtration, and the net movement of fluid into the capillary is called absorption.

If we consider the net movement of fluid out of the capillary (filtration) as a positive quantity, the meaning of the Starling equation can be summarized as follows:

$$\text{Fluid movement} \propto (\text{Pressures that drive fluid out})$$
$$- (\text{Pressures that drive fluid in})$$

but

$$(\text{Pressures that drive fluid out}) = P_c + \pi_i$$

and

$$(\text{Pressures that drive fluid in}) = P_i + \pi_c$$

The Starling equation (Equation [6]) can now be written as follows:

$$\text{Fluid movement} = J_v = K_f \left[(P_c + \pi_i) - (P_i + \pi_c)\right] \quad [7]$$

Analysis of the situation in Figure 3-3A, illustrates the principles involved in applying the Starling equation.

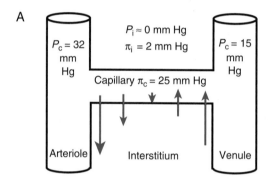

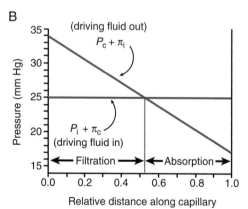

FIGURE 3-3 ■ A, Capillary connecting an arteriole and a venule. The magnitudes of all hydrostatic and osmotic pressures are marked. Arrows indicate the direction (in or out) and magnitude of the fluid movement at various distances along the capillary. B, Plot showing how the driving force ($P_c + \pi_i$) for outward fluid flow, as well as the driving force ($P_i + \pi_c$) for inward fluid flow, vary along the length of the capillary. The portions along the capillary length where filtration (net outward fluid flow) and absorption (net inward fluid flow) occur are indicated.

Typical pressure values are shown in Figure 3-3. The arteriolar and venular hydrostatic pressures are 32 and 15 mm Hg, respectively. The interstitial hydrostatic pressure is typically approximately 0. The capillary and interstitial oncotic pressures are $\pi_c = 25$ mm Hg and $\pi_i = 2$ mm Hg, respectively (Box 3-4).

We wish to determine whether fluid moves into or out of the capillary at two key points along the capillary, the arteriolar and venular ends of the capillary. Two observations are useful: (1) at the point where the capillary is connected to the arteriole, the capillary hydrostatic pressure should be essentially identical to the blood pressure in the arteriole, that is, $P_c = 35$ mm Hg; and (2) at the point where the

capillary is connected to the venule, the capillary hydrostatic pressure should be essentially the same as the blood pressure in the venule, that is, $P_c = 15$ mm Hg. All other pressure components (P_i, π_c, and π_i) do not vary with location. Determination of the net effect of the various pressure components is now a straightforward calculation. At the arteriolar end,

$$(P_c + \pi_i) - (P_i + \pi_c) = (32 + 2) - (0 + 25) = +9 \text{ mm Hg}$$

This means that a net positive pressure is driving fluid out of the capillary near the arteriolar end. At the venular end,

$$(P_c + \pi_i) - (P_i + \pi_c) = (15 + 2) - (0 + 25) = -8 \text{ mm Hg}$$

BOX 3-4

DISSOLVED PROTEINS GIVE RISE TO THE COLLOID OSMOTIC (ONCOTIC) PRESSURE IN THE PLASMA AND THE INTERSTITIUM

By far the most abundant solutes in blood plasma are low-molecular-weight ions and molecules (e.g., Na^+, Cl^-). The total concentration of such small solutes is close to 300 mM. If a membrane barrier were completely impermeable to these solutes, an osmotic pressure close to 6000 mm Hg (i.e., ~7.6 atm; see Box 3-1) would develop. In reality, the wall of a typical capillary is quite permeable to small solutes (reflection coefficient, $\sigma \approx 0$). Thus with respect to small solutes, the interstitial fluid has approximately the same composition as plasma, with the consequence that small solutes exert very little net osmotic effect across the capillary wall. Incidentally, it is reasonable that the walls of most capillaries are quite permeable to low-molecular-weight solutes, because the circulating blood brings small nutrient molecules (e.g., glucose, amino acids) that must leave the capillaries to nourish the cells in tissue. These cells, in turn, generate metabolic waste (e.g., carbon dioxide) that must enter the capillaries and be borne away by the blood.

Two types of proteins are found in significant amounts in the plasma. Albumin, with a molecular weight of ~69,000, is present at ~4.5 g/dL (grams/deciliter, same as grams/100 milliliters), or ~0.65 mM. Globulins, with a molecular weight of ~150,000, are present at ~2.5 g/dL, or ~0.17 mM. These proteins, being macromolecules, do not readily pass through the

capillary wall ($\sigma \gtrsim 0.9$); therefore the plasma is protein rich, whereas the interstitial fluid is low in proteins. Because the total protein concentration in plasma is close to 0.82 mM, we expect the maximum osmotic pressure contributed by the proteins to be:

$$\pi_c = \sigma RTC = 0.9 \times 0.08205 \frac{L \cdot atm}{mol \cdot K} \times 310.15K \times$$

$$0.82 \times 10^{-3} \frac{mol}{L} = 0.0216 \text{ atm}$$

or 15.8 mm Hg. This number is still quite a bit less than the $\pi_c = 25$ mm Hg that is typically measured. This suggests that each protein molecule exists in solution not as a single particle, but rather as an ionic macromolecule with some associated small ions. Because the proteins cannot leave the capillary, the population of small "counter-ions" that are associated with them must also remain within the capillary (see Box 4-4), thereby increasing the total amount of effectively impermeant solute. In other words, if we ignore the fact that proteins can dissociate into more than one ionic particle, we underestimate the osmotic contribution of proteins.

The most important point is that proteins, because they do not readily pass through the capillary wall, are the predominant contributor to osmotic forces across the capillary wall.

This means that a net negative pressure is driving fluid out, which is the same as saying that the net pressure effect is to drive fluid back into the capillary near the venular end. In this capillary, filtration occurs toward the arteriolar end and absorption takes place toward the venular end. It is worth emphasizing that because P_i, π_c, and π_i tend to be relatively constant, the capillary hydrostatic pressure, P_c, being the only variable, becomes the primary determinant of whether filtration or absorption occurs.

If the capillary hydrostatic pressure decreases linearly from the arteriolar to the venular end, we can estimate the magnitude of the Starling forces along the entire length of the capillary. The result is shown in Figure 3-3B. At the arteriolar end, where the capillary hydrostatic pressure is high, the forces driving fluid out override the forces driving fluid in, and the result is filtration of fluid out of the capillary. Advancing along the capillary, capillary hydrostatic pressure wanes and fluid filtration correspondingly decreases. Eventually the capillary hydrostatic pressure declines to the point where the forces driving fluid in overtake the forces driving fluid out, and the result is absorption of fluid back into the capillary. In the example shown in Figure 3-3, filtration occurs over a little more than the first half of the length of the capillary and absorption takes place over the remainder. The end result is that over the length of the capillary, there is a slight excess of filtration over absorption, with a net flow of a small amount of fluid into the interstitium. The fluid that "leaks" into the interstitium is gathered by the lymphatic system and eventually returned into circulation. During inactivity, lymph production in humans amounts to 3 to 4 liters over a 24-hour period. Because blood circulates at the rate of approximately 5 liters per minute, over a 24-hour period more than 7000 liters of blood will pass through the capillaries. This shows that capillary filtration and fluid reabsorption typically are finely balanced. Under certain circumstances the relative balance of filtration and absorption is disrupted, leading to **edema,** or excess accumulation of fluid in tissue (Box 3-5).

BOX 3-5

EDEMA: EXCESS ACCUMULATION OF FLUID

The Starling equation (Equation [6] in main text) shows that whether fluid leaves or enters the capillary is determined by the balance of capillary and interstitial hydrostatic and osmotic pressures. Altered pressure balance leads to altered fluid flow. When venous blood pressure is raised as a result of venous blood clots or congestive heart failure, the result is elevated hydrostatic pressure (P_c) in the capillary, which leads to more fluid filtration. Liver disease and severe starvation can both lead to greatly reduced albumin production by the liver. Because albumin is the predominant contributor to plasma colloid osmotic pressure (π_c), loss of albumin drastically lowers the retention of fluid in the capillaries. Finally, some factors in insect venom (e.g., mellitin from bees), as well as endogenous biochemical agents secreted during the allergic response (e.g., histamine released from mast cells), markedly increase the permeability of capillary walls. Plasma proteins that are normally impermeant can then leave the capillary and enter the interstitium, which lowers π_c and raises π_i. All the processes described here lead to increased movement of fluid from the capillaries into the interstitium and give rise to edema.

Accumulation of fluid in the brain (cerebral edema) is dangerous. The brain is encased by the cranium; therefore enlargement of the brain resulting from edema can give rise to excessive intracranial pressure, which can lead to abnormal neurological symptoms. In contrast to capillaries in the peripheral circulation, cerebral capillaries have exceedingly low permeability to most solutes, including even small molecules such as glucose (MW 180). This blood-brain barrier makes it possible to reduce brain edema by introducing a small solute such as mannitol (MW 182) into the circulation. Because mannitol cannot cross the blood-brain barrier, it increases the osmotic pressure of blood plasma relative to the cerebrospinal fluid. Water is thus drawn out of the brain into the circulatory system, with a corresponding reduction in brain volume. Neurosurgeons routinely use this technique to treat brain edema or to reduce brain volume during neurosurgery.

BOX 3-6

FILTRATION IN THE GLOMERULI OF THE KIDNEY

Capillaries in the glomeruli of the kidney have the highest hydraulic conductivity of any vessels in the body. Blood passing through the glomerulus is filtered through the capillaries; the resulting "ultrafiltrate" contains no blood cells and is largely free of proteins. Each glomerulus measures ~90 μm in diameter but encompasses capillaries with a combined length of ~20,000 μm. These glomerular capillaries present a total of ~350,000 μm², or ~0.0035 cm², of surface through which filtration can occur. A pair of healthy adult kidneys contains ~2.2 million glomeruli, which afford a total filtration area of ~7,700 cm². The hydraulic conductivity of glomerular capillaries has been estimated at 3.3×10^{-5} cm·sec^{-1}·(mm Hg)$^{-1}$, whereas the average driving force for fluid filtration is $\Delta\pi - \Delta P \approx$ 7.5 mm Hg. Combining these quantitative estimates yields a volume flow of 1.9 cm³ (or mL) of ultrafiltrate per second. Thus, over the course of 24 hours, ~165 liters of ultrafiltrate are formed. Because urine production is only ~1.5 liters per day, we can immediately deduce that greater than 99% of the volume of the initially formed ultrafiltrate is ultimately reabsorbed back into the circulation.

In different tissues in the body, capillaries can vary widely in their ability to allow fluid filtration (the hydraulic conductivity can vary by more than 100-fold). The highest filtration rates are in the glomerular capillaries of the kidney (Box 3-6).

ONLY IMPERMEANT SOLUTES CAN HAVE PERMANENT OSMOTIC EFFECTS

Transient Changes in Cell Volume Occur in Response to Changes in the Extracellular Concentration of *Permeant* Solutes

The discussion on solute permeability in Chapter 2 shows that equilibration of a **permeant solute** across a membrane is only a matter of time. As long as the membrane has some finite permeability for a solute, that is, $P_{solute} \neq 0$, or equivalently, $\sigma \neq 1$, the solute will ultimately penetrate, and become equilibrated across, the membrane. In the long term, then, there can be no permanent concentration difference for a permeant solute across the membrane (i.e., with time, $\Delta C_{solute} \rightarrow 0$). Because Equation [4] indicates that the net osmotic pressure exerted by a solute depends on ΔC_{solute}, and, ultimately, $\Delta C_{solute} \rightarrow 0$ for a permeant solute, we understand that permeant solutes can give rise to temporary, but not permanent, changes in osmotic pressure. Moreover, because osmotic pressure drives the movement of water and consequent changes in volume (i.e., volume flow, Equation [5]), we can also say that permeant solutes can give rise only to temporary, not permanent, changes in volume.

Suppose a permeant solute is added to the extracellular fluid (ECF) bathing a cell. If the cell membrane is equally permeable to the added solute and to water, the reflection coefficient for the solute is $\sigma = 0$ and the solute should have no osmotic effect. If the cell membrane is less permeable to the added solute than to water, that is, $0 < \sigma < 1$, the solute will increase the osmotic pressure of the ECF relative to the intracellular fluid (ICF). In response, water will move out of the cell, with consequent cell shrinkage, which leads to a corresponding increase in the concentration of the impermeant solutes already trapped in the cell. Because the added extracellular solute is permeant, however, even as water is leaving the cell, the solute penetrates the cell to increase the solute concentration of the ICF. As a result, water will begin to follow the permeant solute back into the cell. The process continues until the cell has expanded back to its original volume, at which point the intracellular and extracellular concentrations of permeant, *as well as* impermeant, solutes are equalized. The time course of the entire process is shown qualitatively in Figure 3-4, which also shows recovery of the cell volume to its original value when normal ECF composition is restored. That permeant solutes can have no permanent osmotic effect and cannot cause permanent volume changes is demonstrated quantitatively in Box 3-7.

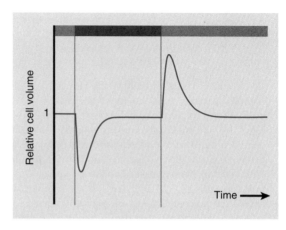

FIGURE 3-4 ■ Schematic representation of the time course of cell volume changes in response to changes in the concentration of permeant solute in the extracellular fluid (ECF). Solution composition is indicated by the *long bar: light gray* is normal ECF; *dark gray* is ECF with increased permeant solute concentration. The graph shows the response to a step increase in solute concentration and subsequent restoration of normal ECF.

BOX 3-7

DETERMINING VOLUME CHANGES IN RESPONSE TO OSMOTIC CHANGES: EFFECT OF *PERMEANT* SOLUTES

For a cell initially in osmotic equilibrium with a bath containing impermeant solute, what happens when the bath osmolarity is increased by adding a *permeant* solute, P? Qualitatively, we know that permeant solutes eventually equilibrate across the cell membrane until the inside and outside concentrations are equal. Thus the addition of permeant solutes to the bath should initially cause some water to leave the cell. Ultimately, however, the permeant solute concentration inside and outside should become equal. This means that, in the end, the permeant solute should have no permanent effect on the cell volume. We will demonstrate this quantitatively.

The cell is initially equilibrated with a bath containing impermeant (NP) solute. The total volume is:

$$V_{\text{Total}} = V_{\text{Cell,Initial}} + V_{\text{Bath,Initial}} \qquad [\text{B1}]$$

Because the osmotic pressures (and hence osmolarities) must be equal at equilibrium, we can say that:

$$C_{\text{NP,Cell,Initial}} = C_{\text{NP,Bath,Initial}}$$

which also means that:

$$\frac{n_{\text{NP,Cell}}}{V_{\text{Cell,Initial}}} = \frac{n_{\text{NP,Bath}}}{V_{\text{Bath,Initial}}} \qquad [\text{B2}]$$

where n_{NP} is the number of moles of NP solute. Note that because NP solute cannot move from one compartment into another, the total number of moles in each compartment must remain the same. Thus $n_{\text{NP,Cell}}$ and

$n_{\text{NP,Bath}}$ remain unchanged regardless of osmotic conditions. If a permeant solute, P, is added to the outside and a new osmotic equilibrium is established, the intracellular and bath concentrations of P should be equal (and could be symbolized as C_P). The new osmotic balance between inside and outside can be written as:

$$C_{\text{NP,Cell,Final}} + C_P = C_{\text{NP,Bath,Final}} + C_P$$

which, after canceling the C_P terms, is the same as:

$$C_{\text{NP,Cell,Final}} = C_{\text{NP,Bath,Final}} \qquad [\text{B3}]$$

If we assume that some permanent volume change ΔV *had* taken place as a result of equilibration, then, because any volume increase or decrease in either the cell or the bath is accompanied by the opposite change in the other compartment, Equation [B3] becomes:

$$\frac{n_{\text{NP,Cell}}}{V_{\text{Cell,Initial}} + \Delta V} = \frac{n_{\text{NP,Bath}}}{V_{\text{Bath,Initial}} - \Delta V} \qquad [\text{B4}]$$

Combining Equations [B2] and [B4] yields:

$$(V_{\text{Cell,Initial}} + V_{\text{Bath,Initial}}) \cdot \Delta V = 0 \qquad [\text{B5}]$$

The term in parentheses is the total volume, V_{Total}, which is constant and nonzero; therefore ΔV must be 0—no volume change could have taken place. We are thus forced to conclude that permeant solutes have no permanent osmotic effect and cannot cause permanent volume changes.

Persistent Changes in Cell Volume Occur in Response to Changes in the Extracellular Concentration of *Impermeant* Solutes

When the concentration of impermeant solutes in the ECF is increased, the extracellular osmotic pressure increases. Typically the volume of the ECF bathing a cell is much larger than the volume of the cell (this is known as the **infinite bath** condition). Because the cell membrane is essentially impermeable to most solutes present inside or outside the cell (see Chapter 2), only water will move out of the cell in response to the increase in ECF osmotic pressure, with a consequent drop in cell volume. This is illustrated in Figure 3-5A, which also shows recovery of the cell volume to its original value when normal ECF composition is restored.

Conversely, if the impermeant solute concentration in the ECF is decreased, water would enter the cell and cause a lasting increase in cell volume, which would recover when normal ECF composition is restored. The corresponding time course is shown in Figure 3-5B. That changes in impermeant solute concentration can cause persistent volume changes is demonstrated quantitatively in Box 3-8.

The Amount of *Impermeant* Solute Inside the Cell Determines the Cell Volume

Because impermeant solutes remain trapped inside the cell, when a cell is challenged by changing external osmotic conditions, the only changes that it can undergo rapidly are volume changes brought about by gain or loss of water. In other words, the cell gains or loses water to dilute or concentrate its impermeant solutes appropriately to match the osmolarity outside. For this reason it is the amount of impermeant intracellular solutes that ultimately determines cell volume.

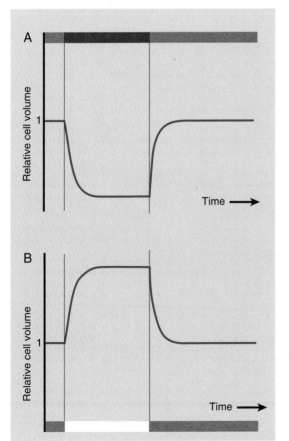

FIGURE 3-5 ■ Schematic representation of the time course of cell volume changes in response to changes in the concentration of impermeant solute in the extracellular fluid (ECF). Solution composition at any given time is indicated by the *long bar: light gray* is normal ECF; *dark gray* is ECF with increased impermeant solute concentration; *white* is ECF with decreased impermeant solute concentration. **A,** Response to a step increase in solute concentration and subsequent restoration of normal ECF. **B,** Response to a step decrease in solute concentration and subsequent restoration of normal ECF.

BOX 3-8

DETERMINING VOLUME CHANGES IN RESPONSE TO OSMOTIC CHANGES: EFFECT OF *IMPERMEANT* SOLUTES

"INFINITE BATH" CONDITION

A cell containing 200 mM impermeant (NP) solute is initially in osmotic equilibrium with a bath also containing 200 mM of NP solute. The bath volume is so large compared with the volume inside the cell that the bath can be considered "infinite." The osmolarity of the bath is then raised to 400 mM NP. What volume change will the cell undergo? We can deduce what should happen qualitatively. An impermeant solute cannot move into or out of the cell. Therefore the difference in NP concentration inside and outside of the cell means that an osmotic driving force should move water out of the cell until a new osmotic equilibrium is established. That is, the cell should shrink. To determine the magnitude of the volume change, we have to do some calculation.

When osmotic equilibrium is established, no *net* movement of water should occur into or out of the cell, that is, the osmotic pressures (and osmolarities) outside and inside are equal. Therefore the initial condition of equilibrium (before the bath osmolarity was raised) and the final condition of equilibrium (after the cell had equilibrated with the new solution) are:

$$C_{NP,Cell,Initial} = C_{NP,Bath,Initial} \quad \text{and}$$
$$C_{NP,Cell,Final} = C_{NP,Bath,Final} \quad \text{[B1]}$$

We also know that (1) $C_{NP,Bath,Initial}$ = 200 mM, and (2) because the bath is "infinite" ($V_{Bath} >> V_{Cell}$), the bath volume and concentration would essentially remain unchanged when a small amount of water moves out of the cell into the bath, that is, $C_{NP,Bath,Final}$ = 400 mM. Substituting these two pieces of information into Equation [B1] gives:

$$C_{NP,Cell,Initial} = 200 \quad \text{and} \quad C_{NP,Cell,Final} = 400 \quad \text{[B2]}$$

The osmolarity, C, however, is just the number of millimoles (mmol), symbolized by n, divided by the volume, V. Moreover, because impermeant solute cannot enter or leave the cell, $n_{NP,Cell}$ cannot change. Therefore Equation [B2] can be written as:

$$\frac{n_{NP,Cell}}{V_{Cell,Initial}} = 200 \quad \text{and} \quad \frac{n_{NP,Cell}}{V_{Cell,Final}} = 400 \quad \text{[B3]}$$

Dividing the first equation in Equation [B3] by the second gives:

$$\frac{V_{Cell,Final}}{V_{Cell,Initial}} = \frac{1}{2} \quad \text{[B4]}$$

When the osmolarity of an infinite bath is doubled by the addition of impermeant solute, the cell shrinks to one half of its initial volume.

FINITE (NONINFINITE) BATH CONDITION

Assume the solution conditions are the same as in the infinite bath case, except that the bath is no longer infinite. Now any water movement into or out of the cell will cause a corresponding volume change in the bath as well as the cell. If the bath osmolarity is suddenly doubled from 200 to 400 mM, we expect water movement out of the cell so that the bath osmolarity will drop as the cell osmolarity increases, until the osmolarity inside and outside is equalized.

The equilibrium conditions are always the same:

$$C_{NP,Cell,Initial} = C_{NP,Bath,Initial} \quad \text{and}$$
$$C_{NP,Cell,Final} = C_{NP,Bath,Final} \quad \text{[B5]}$$

Because the bath is of finite size, any water flow into or out of the cell will change the bath volume. This means that after the new osmotic equilibrium is attained, the bath osmolarity will have changed from 400 mM. The way to take this change into consideration is as follows. If the cell undergoes volume change ΔV, the bath must undergo a corresponding volume change of $-\Delta V$. Thus the final cell volume will be ($V_{Cell,Initial} + \Delta V$). Correspondingly, the final bath volume is ($V_{Bath,Initial} - \Delta V$).

Because of this volume change, the final bath osmolarity will be modified by the ratio $V_{Bath,Initial}/(V_{Bath,Initial} - \Delta V)$:

$$C_{NP,Bath,Final} = 400 \frac{V_{Bath,Initial}}{V_{Bath,Initial} - \Delta V} \quad \text{[B6]}$$

Moreover, because $V_{Cell,Final} = V_{Cell,Initial} + \Delta V$ and $n_{NP,Cell}$ remains unchanged, the conditions of equilibrium (Equation [B5]) become:

BOX 3-8

DETERMINING VOLUME CHANGES IN RESPONSE TO OSMOTIC CHANGES: EFFECT OF IMPERMEANT SOLUTES—cont'd

$$\frac{n_{NP,Cell}}{V_{Cell,Initial}} = 200 \quad \text{and}$$

$$\frac{n_{NP,Cell}}{V_{Cell,Initial} - \Delta V} = 400 \frac{V_{Bath,Initial}}{V_{Bath,Initial} - \Delta V} \qquad [B7]$$

Solving the two equations in Equation [B7] for ΔV gives the result:

$$\Delta V = \frac{-V_{Cell,Initial} V_{Bath,Initial}}{2V_{Bath,Initial} + V_{Cell,Initial}} \qquad [B8]$$

For example, assume that initially the bath and cell volumes are equal: $V_{Cell,Initial} = V_{Bath,Initial} = V$. Substitution into Equation [B8] gives $\Delta V = -V/3$. This means that the cell will shrink by $V/3$ and, correspondingly, the bath will expand by $V/3$. The bath osmolarity will be diluted from 400 by a factor of $V/(V + V/3) = 3/4$, to 300 mM. The cell osmolarity will be concentrated from 200 by a factor of $V/(V - V/3) = 3/2$, to 300 mM.

In arriving at Equation [B8], we did not actually specify how large the bath was in relation to the cell volume. Equation [B8] should therefore apply to any volume. As a check, we apply Equation [B8] to the infinite bath case. Infinite bath means that $V_{Bath,Initial} \gg V_{Cell,Initial}$, which allows the denominator in Equation [B8] to be simplified: $(2V_{Bath,Initial} + V_{Cell,Initial}) \approx 2V_{Bath,Initial}$. This leads to the result $\Delta V \approx -V_{Cell,Initial}/2$. That is, doubling the osmolarity of an infinite bath causes the cell to shrink by one half of its initial volume—exactly what we determined earlier.

GENERAL EXPRESSION FOR IMPERMEANT SOLUTES

When the concentration of impermeant solutes in the bathing solution is changed from $C_{NP,Initial}$ to $C_{NP,New}$, the volume change that occurs in a cell is given by:

$$\Delta V = \frac{(C_{NP,Initial} - C_{NP,New}) V_{Cell,Initial} V_{Bath,Initial}}{C_{NP,Initial} V_{Cell,Initial} + C_{NP,New} V_{Bath,Initial}} \qquad [B9]$$

Equation [B9] is general; it applies to arbitrary bath and cell volumes and applies regardless of whether the bath osmolarity is increased or decreased. When the appropriate values are used in this general expression, we can easily verify that we get the same results as were obtained for the two specific examples discussed earlier in this box.

SUMMARY

1. If a membrane is permeable to water but not to solute, and the solute concentration differs on the two sides of the membrane, water will move across the membrane from the side where the solute concentration is lower to the side where it is higher. This movement of water across a semipermeable membrane is called osmosis.

2. Osmotic movement of water leads to changes in fluid volume—the volume increases on the side of the membrane with higher solute concentration, and the volume decreases on the side with lower solute concentration.

3. Osmotic movement of water can be thought of as being driven by a difference in osmotic pressure on the two sides of a membrane. Osmotic pressure is proportional to solute concentration. Water moves from the side with low osmotic pressure (i.e., the side with higher water concentration) to the side with high osmotic pressure.

4. A difference in hydrostatic pressure on the two sides of a membrane can also drive water movement across the membrane barrier. Water moves from the side with high hydrostatic pressure to the side with low hydrostatic pressure.

5. The direction of net fluid flow across a capillary wall is controlled by the balance of hydrostatic and osmotic pressures, as described by the Starling equation.

6. A permeant solute can cross a membrane barrier and eventually abolish its own concentration gradient. Therefore a change in the extracellular concentration of a permeant solute can cause only a transient change in cell volume.

7. An impermeant solute cannot cross a membrane barrier to abolish its own concentration gradient. Therefore a change in the concentration of an impermeant solute can cause a persistent change in cell volume.

KEY WORDS AND CONCEPTS

- Osmosis
- Osmotic pressure
- Osmolarity
- Tonicity
- van't Hoff's Law
- Reflection coefficient
- Hydrostatic pressure
- Volume flow
- Starling equation
- Colloid osmotic pressure (oncotic pressure)
- Hydraulic conductivity (filtration constant)
- Edema
- Permeant solute
- Impermeant solute
- Infinite bath

STUDY PROBLEMS

1. A cell is initially equilibrated with a very large volume of plasma that contains 300 mM *permeant* solute and 10 mM *impermeant* solute. The initial volume of the cell is V_0. Knowing that membrane permeability to water is higher than to solute, consider the three separate situations described here.

   a. The concentration of *impermeant* solute in the plasma is increased to 20 mM. Will water move into or out of the cell? When osmotic equilibrium is reestablished, what will be the final volume of the cell?

   b. The concentration of *permeant* solute in the plasma is increased to 400 mM. Describe the movement of water that is expected to take place. When equilibrium is reestablished, what will be the final volume of the cell?

   c. The concentration of *impermeant* solute in the plasma is increased to 20 mM *and* the concentration of *permeant* solute is decreased to 200 mM. Describe how the cell volume will change with time.

BIBLIOGRAPHY

Atkins PW: *Physical chemistry*, ed 5, New York, NY, 1994, WH Freeman.

Gennis RB: *Biomembranes: molecular structure and function*, New York, NY, 1989, Springer-Verlag.

Weiss TF: *Cellular biophysics*, Cambridge, MA, 1996, MIT Press.

ELECTRICAL CONSEQUENCES OF IONIC GRADIENTS

OBJECTIVES

1. Understand how the movement of ions can generate an electrical potential difference across a membrane.

2. Learn the concept of the equilibrium potential and how to use the Nernst equation to calculate it.

3. Understand how the resting membrane potential is generated in a cell and how to use the Goldman-Hodgkin-Katz (GHK) equation to calculate membrane potential.

4. Understand the relationship between the GHK equation and the Nernst equation.

5. Know how changes in membrane permeability to permeant ions can change the membrane potential.

6. Understand the Donnan effect and its consequences for living cells.

IONS ARE TYPICALLY PRESENT AT DIFFERENT CONCENTRATIONS ON OPPOSITE SIDES OF A BIOMEMBRANE

For any membrane in a living cell, biologically important ions are distributed asymmetrically on opposite sides of the membrane. In this chapter we focus on the plasma membrane (PM) of a cell, which separates the intracellular and extracellular environments. Across the PM, concentrations of the three common monovalent ions, Na^+, K^+, and Cl^-, are different. Ionic distributions for a "typical" mammalian cell are shown in Table 4-1. It is clear that the asymmetrical distributions of Na^+, K^+, and Cl^- ions give rise to concentration gradients of these ions across the PM. Such **ion concentration gradients** can drive the diffusional movement of the ions across the PM, which is selectively permeable to these ions. However, because ions carry electrical charge, their diffusional movement across the PM

gives rise to electrical effects, which we now examine.

SELECTIVE IONIC PERMEABILITY THROUGH MEMBRANES HAS ELECTRICAL CONSEQUENCES: THE NERNST EQUATION

Consider a cell with the ion distributions shown in Table 4-1. Because the K^+ concentration is higher inside the cell than outside, if the PM is selectively permeable *only* to K^+, we expect that K^+ ions will move down their concentration gradient, out of the cell (Figure 4-1).

When the positive K^+ ions leave the cell, however, they introduce positive charges to the exterior of the PM while leaving behind an equal number of negative charges on the intracellular side. This means that an electrical potential difference develops as a result of the diffusional movement of K^+ ions out of the cell. As K^+ ions exit the cell, the interior of the cell becomes

TABLE 4-1

Intracellular and Extracellular Concentrations of the Common Monovalent Ions for a Typical Mammalian Cell

ION	INTRACELLULAR (mM)	EXTRACELLULAR (mM)
K^+	140	5
Na^+	10	145
Cl^-	6	106

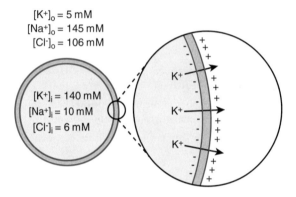

$[K^+]_o = 5\,mM$
$[Na^+]_o = 145\,mM$
$[Cl^-]_o = 106\,mM$

$[K^+]_i = 140\,mM$
$[Na^+]_i = 10\,mM$
$[Cl^-]_i = 6\,mM$

Plasma membrane permeable to K^+

FIGURE 4-1 ■ When the plasma membrane is selectively permeable only to K^+ ions, the K^+ concentration gradient (high concentration inside and low concentration outside) drives net movement of K^+ ions out of the cell.

progressively more negative while the exterior becomes correspondingly more positive. The effect of the developing electric field is to *oppose* further movement of K^+ ions (i.e., the negative interior tends to attract positive K^+, whereas the positive exterior tends to repel K^+). This analysis suggests that the "leakage" of K^+ ions cannot continue indefinitely because eventually a strong enough electric field will build up to balance exactly the tendency of K^+ to move out of the cell, down its concentration gradient. When the electrical forces exactly balance the driving force of the concentration gradient, we say that **electrochemical equilibrium** is reached. At electrochemical equilibrium, there can be no further *net* movement of K^+ ions into or out of the cell. The preceding discussion tells us that when K^+ ions are in electrochemical equilibrium

across the PM, an electrical potential difference exists across the PM, with the inside being more negative than the outside. This electrical potential difference at which no net movement of K^+ occurs is the **equilibrium potential** for K^+ and is given the symbol E_K. For any ion whose extracellular and intracellular concentrations are C_{out} and C_{in}, respectively, the equilibrium potential can be calculated using the **Nernst equation*** (Box 4-1):

$$E_{eq} = \frac{RT}{zF} \ln \frac{C_{out}}{C_{in}} \qquad [1]$$

where R is the universal gas constant, T is the absolute temperature in Kelvins (Celsius temperature plus 273.15), z is the electrical charge on the ion (+1 for K^+, –1 for Cl^-), and F is Faraday's constant (96,485 coulombs/mol). Some alternative forms of the Nernst equation that may be more convenient for use in computation are shown in Box 4-2. The **membrane potential (V_m)** for a cell is defined as the electrical potential *inside* the cell measured relative to the electrical potential *outside*. Because the extracellular electrical potential is a reference level against which the intracellular potential is measured, we can define the extracellular electrical potential to be zero (0).

Using the concentrations given in Table 4-1, we calculate the equilibrium potential for K^+ to be:

$$E_K = \frac{61.5}{+1} \log \frac{5}{140} = 61.5\,(-1.45) = -89.1\,mV$$

This is the potential inside the cell relative to the outside, and it is negative, as we deduced earlier.

The Nernst equation can be used to calculate the equilibrium potential for any permeant ion, as long as the inside and outside concentrations for that ion are known. For example, for the ionic distributions shown in Table 4-1, if the PM were permeable only to Na^+ ions, the sodium equilibrium potential, E_{Na}, for our cell would be:

$$E_{Na} = \frac{61.5}{+1} \log \frac{145}{10} = 61.5\,(+1.16) = +71.5\,mV$$

---

*Named after Hermann Walther Nernst, who first derived the equation in 1889. Nernst received the Nobel Prize in Chemistry in 1920. Because the equilibrium potential for an ion is defined by the Nernst equation, the equilibrium potential is also known as the **Nernst potential.**

BOX 4-1

MOVEMENT OF IONS DRIVEN BY AN ELECTRICAL POTENTIAL GRADIENT AND THE ORIGIN OF THE NERNST EQUATION

In Chapter 2, diffusion, or the movement of molecules resulting from the presence of concentration gradients, was discussed. In an analogous way, we now examine the movement of electrically charged molecules (ions). Because an ion is charged, when placed in an electric field, it will experience a force, thus causing it to move. It is reasonable that the speed at which an ion moves in solution should depend on the strength of the electric field (the stronger the electric field, the faster the ion will move) and the charge on the ion (the higher the electrical charge on an ion, the faster it will move in an electric field). In an electric field, a change in electrical potential, E, occurs with distance, x, that is, $\Delta E/\Delta x$. Analogous to the case of diffusion, where $\Delta C/\Delta x$ was a concentration gradient, the change in electrical potential with distance, $\Delta E/\Delta x$, is an *electrical potential gradient*. If the speed of an ion is s, the relationship between ion speed, the electrical potential gradient, and the ionic charge is:

$$s = uz\frac{\Delta E}{\Delta x} \qquad [B1]$$

where z is the ionic charge (e.g., $+1$ for Na^+, $+2$ for Ca^{2+}, -2 for SO_4^{2-}) and u is a proportionality constant known as the "ionic mobility." Because $\Delta E/\Delta x$ has dimensions of volts per centimeter (V/cm) and s must have units of centimeters per second (cm/sec) and z is just the number of charges on an ion and therefore is dimensionless, u must have units of $(cm^2/sec)/V$ for all the units to work out in Equation [B1].

Knowing the speed of ion movement, we can easily figure out the flux of ions moving under the influence of an electric field. Imagine a cylindrical volume of solution containing the ions of interest (Figure B-1). In the figure the electrical potential is more negative at the right, so positive ions (cations) would naturally move toward the right. Because flux is the quantity of ions passing through unit area per unit time, to derive an expression for the flux, J, we need only to find out the quantity of ions flowing through area, A, in a given period of time, Δt. Because the ions are moving at speed s toward the right, within the period Δt, any ion within a distance of $s \times \Delta t$ to the left of the area A would pass through A. The cylindrical volume containing these ions that would

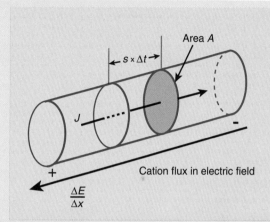

FIGURE B-1 ■ Flux of positive ions (cations) being driven by an electric field, $\Delta E/\Delta x$. The speed of movement of the ions is symbolized as s.

pass through A is $s \times \Delta t \times A$. The number of moles of ions in this volume is $C \times s \times \Delta t \times A$, where C is the concentration of the ion of interest. Taking the number of moles of ions that would pass through A and dividing by the area, A, and by the time interval, Δt, gives the flux of ions driven by the electrical field:

$$J_{electr} = -[(C \times s \times \Delta t \times A)/A]/\Delta t = -C \times s \quad [B2]$$

Substituting for s (Equation [B1]) gives:

$$J_{electr} = -zuC\frac{\Delta E}{\Delta x} \qquad [B3]$$

The minus sign takes into account the fact that for cations (z being a positive number), ion drift is toward the negative direction of the electric field, whereas for anions (z being a negative number), ion drift is toward the positive direction of the electric field.

If the electric field is not linear, $\Delta E/\Delta x$ can be replaced with dE/dx (a derivative):

$$J_{electr} = -zuC\frac{dE}{dx} \qquad [B4]$$

Continued

BOX 4-1

MOVEMENT OF IONS DRIVEN BY AN ELECTRICAL POTENTIAL GRADIENT AND THE ORIGIN OF THE NERNST EQUATION—cont'd

Equation [B4] bears a strong resemblance to Equation [3] from Chapter 2 that describes diffusion flux. Whereas a concentration gradient drives the diffusive flux of molecules or ions, an electrical potential gradient (an electric field) drives the electrical flux of ions.

In view of the foregoing, the total (net) flux of an ion is the sum of the flux caused by diffusion and the flux driven by an electric field:

$$J_{total} = J_{diffusion} + J_{electr} = -D\frac{dC}{dx} - zuC\frac{dE}{dx} \quad \text{[B5]}$$

This is the Nernst-Planck equation quantifying the ionic flux driven by a concentration gradient and an electrical potential gradient.

At electrochemical equilibrium the flux driven by the concentration gradient is exactly balanced by the flux driven in the opposite direction by the electrical potential gradient, so the net flux must be zero. Therefore:

$$J_{total} = J_{diffusion} + J_{electr} = -D\frac{dC}{dx} - zuC\frac{dE}{dx} = 0 \quad \text{[B6]}$$

which means that:

$$zuC\frac{dE}{dx} = -D\frac{dC}{dx} \quad \text{[B7]}$$

Einstein derived the relationship between the diffusion coefficient (D) and the mobility (u) of an ion:

$$D = \frac{uRT}{F} \quad \text{[B8]}$$

where R is the universal gas constant, T is the absolute temperature in Kelvins (Celsius temperature plus 273.15), and F is Faraday's constant (96,485 coulombs/mol). Using Equation [B8], we can rewrite Equation [B7]:

$$zuC\frac{dE}{dx} = -\frac{uRT}{F}\frac{dC}{dx} \quad \text{[B9]}$$

which rearranges to:

$$\frac{dE}{dx} = \frac{-RT}{zF}\frac{1}{C}\frac{dC}{dx} \quad \text{[B10]}$$

Equation [B10] can be integrated across the thickness of the membrane:

$$\int_{x_1}^{x_2}\frac{dE}{dx}\,dx = \frac{-RT}{zF}\int_{x_1}^{x_2}\frac{1}{C}\frac{dC}{dx}\,dx \quad \text{[B11]}$$

The result of integration is:

$$E_2 - E_1 = \frac{-RT}{F}\left(\ln C_2 - \ln C_1\right)$$
$$= \frac{-RT}{zF}\ln\frac{C_2}{C_1} = \frac{RT}{zF}\ln\frac{C_1}{C_2} \quad \text{[B12]}$$

In other words, if a membrane is selectively permeable to a particular ion, and the ion is in electrochemical equilibrium across the membrane, we can calculate the membrane potential, ($E_2 - E_1$), that would be established just by knowing the concentration of the ion on the two sides of the membrane (C_1 and C_2).

The membrane potential of a cell is defined to be the potential of the inside relative to the outside (i.e., $E_{in} - E_{out}$; subscripts 2 and 1 taken to be *in* and *out*, respectively). The membrane potential that is established when an ion is in electrochemical equilibrium across the membrane is referred to as the equilibrium potential for that ion and is given the symbol E_i, where the subscript *i* designates the particular ion under discussion (e.g., E_K, E_{Cl}, and E_{Na} are the equilibrium potentials for K^+, Cl^-, and Na^+, respectively). By adopting these conventions, we can rewrite Equation [B12] in a form that is one of the most important equations in cellular physiology—the *Nernst equation*:

$$E_{eq} = \frac{RT}{zF}\ln\frac{C_{out}}{C_{in}} \quad \text{[B13]}$$

at 37°C. The sign for E_{Na} is positive because, as positively charged Na^+ ions leak into the cell, down their concentration gradient, they make the inside of the cell more positive while leaving behind a corresponding excess of negative charges on the outside. That is, the inside of the cell becomes more positive relative to the outside, hence E_{Na} is positive. It is equally straightforward to verify that for chloride ions, E_{Cl} equals −76.8 mV at 37°C for our cell.

BOX 4-2

ALTERNATIVE FORMS OF THE NERNST EQUATION THAT MAY BE MORE CONVENIENT FOR CALCULATIONS

If desired, the natural logarithm, ln, in the Nernst equation can be converted to base-10 logarithm: $\ln(C_{out}/C_{in}) = 2.303 \cdot \log(C_{out}/C_{in})$, to give the equivalent expression:

$$E_{eq} = \frac{2.303\,RT}{zF} \log \frac{C_{out}}{C_{in}} \qquad [B1]$$

At 37°C, the group of constants $RT/F = 26.7$ mV. For computation at 37°C, either of the following two forms of the Nernst equation could be used (to give E in units of mV):

$$E_{eq} = \frac{26.7}{z} \ln \frac{C_{out}}{C_{in}} \quad \text{or} \qquad [B2]$$

$$E_{eq} = \frac{61.5}{z} \log \frac{C_{out}}{C_{in}}$$

Some ion movement is required to establish physiological membrane potentials. Therefore we are justified in asking whether such movements significantly alter the ion concentrations inside the cell. After all, if ions enter or leave the cell, the intracellular ion concentration *must* change. In turn, we may ask whether the concentrations used in the Nernst equation should be corrected for the effect of such ion movements. The calculation in Box 4-3 shows that the number of ions that move into or out of the cell to establish a V_m is so small that the cellular ion concentrations are essentially undisturbed.

At first sight the magnitudes of the electrical potentials calculated earlier seem somewhat small. However, it must be remembered that these potentials exist across the PM, which is only approximately 50 angstroms thick (1 angstrom = 1×10^{-10} meter). Box 4-4 gives some observations about the V_m and the strength of electrostatic forces acting on oppositely charged ions separated by the PM.

The most important observation from the preceding discussion is that selective permeability of the PM to ions can have profound electrical consequences. For a typical cell (with ionic concentrations similar to those in Table 4-1), if the PM is selectively permeable to K+, the resulting K+ efflux will tend to drive the cell's V_m to more negative values. Alternatively, if the PM is selectively permeable to Na+, the resulting Na+ influx will tend to drive the cell's V_m to more positive values. The linkage between the V_m and the selective ionic permeabilities of a cell underlies the mechanism by which living cells regulate

their electrical properties. This linkage is more fully explored in Chapter 7.

THE STABLE RESTING MEMBRANE POTENTIAL IN A LIVING CELL IS ESTABLISHED BY BALANCING MULTIPLE IONIC FLUXES

Cell Membranes Are Permeable to Multiple Ions

The concept of the equilibrium, or Nernst, potential for a particular ion (e.g., Na+, K+, or Cl−) was developed by examining the fluxes of that ion in an idealized cell whose membrane is permeable *only* to that ion. The PM of a real cell, however, has moderate to significant permeability to all three common monovalent ions. During the earlier discussion on the Nernst potential, we noted that it is the selective permeability of the PM to various ions that allows the cell to regulate its V_m. Just how does this regulation take place? How do the permeabilities of K+, Na+, and Cl− contribute to the cell's **resting membrane potential**?

If the cell is permeable to all three ionic species, fluxes of all three ions will occur across the PM. As we have seen, **ion fluxes** into and out of the cell have electrical consequences; namely, they change the V_m of the cell. For example, efflux of K+ tends to drive V_m toward more negative values, whereas influx of Na+ tends to drive V_m toward more positive values. Similarly, influx of Cl− brings negative charges into the cell and would drive V_m toward more negative values. With all these fluxes occurring simultaneously

BOX 4-3

THE ION MOVEMENT NEEDED TO ESTABLISH A PHYSIOLOGICAL MEMBRANE POTENTIAL DOES NOT SIGNIFICANTLY CHANGE ION CONCENTRATIONS INSIDE THE CELL

Realizing that ions must move across the plasma membrane to establish a membrane potential (V_m), we may ask whether such ion movements (e.g., leakage of K^+) will significantly alter the intracellular concentration of the ion of interest. To answer this question, we need to know one important property of biological membranes: the membrane capacitance, C. The capacitance is a measure of the amount of charge, q, that is separated by the membrane at a given V_m:

$$C = \frac{q}{V_m} \qquad [B1]$$

The amount of charge is then just $q = C \cdot V_m$. The relevant units are the coulomb for electrical charge, the volt for electrical potential, and the farad (symbol F) for capacitance; 1 farad equals 1 coulomb per volt. The capacitance of biological membranes is typically 1 $\mu F/cm^2$ of membrane area ($1 \times 10^{-6} F/cm^2$). A spherical cell with a radius of 10 μm has a membrane surface area of:

$$A_{cell} = 4\pi r^2 = 1257 \ \mu m^2 = 1.257 \times 10^{-5} \ cm^2$$

The capacitance for such a cell is:

$$C = 1 \times 10^{-6} \ F/cm^2 \times (1.257 \times 10^{-5} \ cm^2)$$
$$= 1.257 \times 10^{-11} \ F$$

If the V_m of this cell is equal to the potassium equilibrium potential, $E_K = -89.1$ mV (i.e., -0.0891 V; see main text), the amount of charge separated by the cell membrane is:

$$q = C \cdot E_K = (1.257 \times 10^{-11} \ F) \times (0.0891 \ V)$$
$$= 1.120 \times 10^{-12} \ coulombs$$

To convert the amount of electrical charge into the quantity of K^+ ions that had to move to establish E_K, we make use of Faraday's constant ($F = 96,485$ coulombs/mol; note the distinction between Faraday's constant, F, and the farad, F):

$$\text{Amount of } K^+ \text{ moved} = q/F = (1.120 \times 10^{-12} \ coulombs)/$$
$$(96,485 \ coulombs/mol) = 1.161 \times 10^{-17} \ mol$$

To assess whether the leakage of 1.161×10^{-17} mol of K^+ ions out of the cell significantly lowers the K^+ content of the cell, we need to calculate the moles of K^+ originally present in the cell. The number of moles of K^+ present inside the cell is just $[K^+]_{in} \times Vol_{cell}$. The volume of the cell is:

$$Vol_{cell} = 4\pi r^3/3 = 4189 \ \mu m^3 = 4.189 \times 10^{-12} \ L$$

(1 Liter = $10^{15} \ \mu m^3$). The total amount of K^+ initially present in the cell must have been:

$$K^+ \text{ content} = (0.140 \ mol/L) \times (4.189 \times 10^{-12} \ L)$$
$$= 5.864 \times 10^{-13} \ mol$$

The fraction of the K^+ content that had to move out of the cell to establish E_K is simply:

$$\frac{\text{Amount of } K^+ \text{ moved}}{K^+ \text{ content in cell}} = \frac{1.160 \times 10^{-17} \ mol}{5.864 \times 10^{-13} \ mol}$$

$$= 1.979 \times 10^{-5} \approx \frac{20}{1,000,000} = 0.002\%$$

Thus, for every 1 million K^+ ions in the cell, roughly 20 must leak out to establish $E_K = -89.1$ mV. This K^+ leakage would diminish the intracellular K^+ content by only ~0.002%—a negligibly small fraction. Therefore it is clear that the ion movement needed to establish a V_m, although electrically significant, does not change ion concentrations much.

(and "fighting" with each other), eventually a steady state will be established—a steady state in which the cell will have a stable, nonvarying V_m. What is the implication of a stable, nonvarying V_m? We know that whenever *net* movement of electrically charged ions into or out of the cell occurs, V_m will change. We can therefore conclude that a stable, nonvarying V_m implies that no *net* charge movement occurs into or out of the cell in the steady state. In other words, all fluxes tending to make the cell more negative are exactly balanced by fluxes that tend to make the cell more positive.

BOX 4-4

THE STRENGTH OF ELECTRICAL FORCES
AND THE PRINCIPLE OF ELECTRONEUTRALITY

Typical physiological membrane potentials (V_m) are on the order of many tens of millivolts across a membrane that is approximately 50 angstroms in thickness ($1 \text{ Å} = 10^{-8} \text{ cm} = 10^{-10} \text{ m}$). Because the *electric field*, $\mathscr{E}$, is defined as the electrical potential per unit distance, the electric field across the PM when $V_m = 80$ mV is:

$$\mathscr{E} = \frac{80 \times 10^{-3} \text{ V}}{50 \times 10^{-10} \text{m}} = 16,000,000 \, \frac{\text{V}}{\text{m}}$$

This is about a thousand times stronger than typical field strengths used in electrophoresis.

One can appreciate the strength of the electrostatic (or "coulombic") interaction by calculating the attractive force between oppositely charged ions separated on the two sides of the PM. The magnitude of the electrostatic force, $F_{electrostatic}$, depends on q, the amount of charge separated by the membrane, and $\mathscr{E}$, the electric field across the membrane:

$$F_{electrostatic} = \frac{q \times \mathscr{E}}{2} \qquad [B1]$$

In Box 4-3, it was shown that to establish $E_K = -89.1$ mV (or -0.0891 V) across the PM of a 20-μm

spherical cell, 1.12×10^{-12} C of charges are separated on the two sides of the membrane. Using Equation [B1], we can estimate the magnitude of the attractive force (in newtons) exerted by the separated charges on each other across the 50 Å thickness of the PM:

$$F_{electrostatic} = \frac{1}{2} \cdot \left(1.12 \times 10^{-12} \text{ C}\right) \times \frac{0.0891 \text{ V}}{50 \times 10^{-10} \text{ m}}$$

$$= 9.98 \times 10^{-6} \text{ N}$$

We recall from Box 4-3 that the membrane area of the 20-μm spherical cell is 1.257×10^{-9} m^2. Therefore the force per unit area of membrane is 7940 N·m^{-2}. Because each newton (N) is equal to 0.225 pounds of force, this means that for a square meter of membrane area, the attractive force between the charges would be ~1800 pounds, or nine-tenths of a ton! Incidentally, this remarkable strength of the electrical forces underlies the *principle of electroneutrality*, which states that, in a solution, the number of positive ions is balanced by an equal number of negative ions, so that overall, the solution carries no net electrical charge.

The Resting Membrane Potential Can Be Quantitatively Estimated by Using the Goldman-Hodgkin-Katz Equation

The stable, resting V_m of a cell that is permeable to all three of the common monovalent ions is given quantitatively by the **Goldman-Hodgkin-Katz (GHK) equation**[*]:

$$V_m = \frac{RT}{F} \ln \frac{P_K[\text{K}^+]_o + P_{Na}[\text{Na}^+]_o + P_{Cl}[\text{Cl}^-]_i}{P_K[\text{K}^+]_i + P_{Na}[\text{Na}^+]_i + P_{Cl}[\text{Cl}^-]_o} \qquad [2]$$

---

[*]The GHK equation was first derived by the biophysicist David E. Goldman for his PhD dissertation research. Later, it was rederived and cast into its present, more physiologically useful form by the physiologists Alan L. Hodgkin and Bernard Katz, recipients of the Nobel Prize in Physiology or Medicine in 1963 and 1970, respectively.

where P_K, P_{Na}, and P_{Cl} are the cell membrane **ionic permeabilities** for K$^+$, Na$^+$, and Cl$^-$, respectively.

We can examine the GHK equation to understand its meaning in terms of a physical picture. Recall that the product of a membrane permeability coefficient and a concentration is a unidirectional flux (see Chapter 2); for example:

$$\text{K}^+ \text{ efflux} = J_K^{in \to out} = P_K[\text{K}^+]_i \text{ and}$$

$$\text{K}^+ \text{ influx} = J_K^{out \to in} = P_K[\text{K}^+]_o$$

Keeping this in mind, we see that the three terms summed in the numerator of the GHK equation correspond to K$^+$ influx, Na$^+$ influx, and Cl$^-$ *efflux*. All three fluxes represent ion movements that tend to drive V_m to more *positive* values (positive K$^+$ and Na$^+$ entering and negative Cl$^-$ leaving the cell). The three terms summed in the denominator, however, correspond to K^+ efflux, Na^+ efflux, and Cl^- *influx*, all of which represent ion movements that tend to drive V_m to more *negative* values (positive K^+ and Na^+ leaving and negative Cl^- entering). The GHK equation therefore describes the behavior of V_m when all the ion fluxes that tend to drive V_m in the positive direction are balanced against all the ion fluxes that tend to drive

V_m in the negative direction. These observations can also be stated in electrical terms. Because a flux of ions is equivalent to a flow of electrical charges, an ionic flux is also an ionic current. Therefore we can say that the GHK equation gives the value of V_m when no *net* current is flowing through the membrane. The precise **relationship between ionic flux and ionic current** is defined in Box 4-5.

BOX 4-5

THE RELATIONSHIP BETWEEN IONIC FLUXES AND IONIC CURRENTS

Up to this point, the movement of ions through a membrane has been discussed in terms of flux, J, which is the number of moles of ion moving through a unit area of membrane per unit time. In future discussions that deal with the electrical behavior of cells, it is more convenient to use the concept of electrical *current* flowing through the membrane (symbolized as I), rather than ion flux. The two concepts are equivalent and are related in a simple way. Electrical current is the movement of *charges* per unit time. Converting flux into current involves figuring out the relationship between moles of ions and the amount of charge they carry. Each mole of ions represents z moles of electrical charge if each ion has charge z. Moreover, to convert molar units to electrical units, we need to use the Faraday constant, $F = 96,485$ coulombs/mol of charge.

The relationship between ionic flux and current through the membrane is then:

$$I = zF \times J \times A_{mem} \qquad [B1]$$

where A_{mem} is the membrane area across which the flux/current is occurring. The two quantities, flux and current, are seen to be directly related through conversion factors and can be thought of as the same quantity in different units.

Different sign conventions are used in describing fluxes and currents in cellular physiology. Whereas in flux theory, flow of any kind of "particle" (i.e., positive ions, negative ions, or neutral molecules) *into* the cell is defined to be a positive flux, flow of *positive* charges *out of* the cell is defined to be positive current in electrical theory. The four scenarios that are physically possible are summarized in Table B-1.

TABLE B-1		
Sign Conventions for Fluxes and Currents		
FLOW OF POSITIVE OR NEGATIVE IONS RELATIVE TO CELL	**DIRECTION AND SIGN OF FLUX, J**	**DIRECTION AND SIGN OF CURRENT, I**
⊖→⊕ (positive ion moving out)	Outward, negative ($J < 0$)	Outward, positive ($I > 0$)
⊖←⊕ (positive ion moving in)	Inward, positive ($J > 0$)	Inward, negative ($I < 0$)
⊖←⊖ (negative ion moving in)	Inward, positive ($J > 0$)	Outward, positive ($I > 0$)
⊖→⊖ (negative ion moving out)	Outward, negative ($J < 0$)	Inward, negative ($I < 0$)

If we know the intracellular and extracellular concentrations, as well as the permeabilities, of the permeant ions, it is straightforward to use the GHK equation to calculate the resting V_m of the cell. It is customary and convenient to take the permeability coefficient for K^+ as a reference and normalize the other ionic permeabilities relative to that of K^+. The membrane of a resting cell has relatively high permeability to K^+ and Cl^- ions and relatively low permeability to Na^+ ions. Thus typical relative permeabilities for a resting cell may have the following values: $P_K = 1$, $P_{Na} = 0.02$, and $P_{Cl} = 0.5$. Knowing these permeability coefficients and the fact that $RT/F = 26.7$ mV at 37°C, we can use the GHK equation and the concentrations given in Table 4-1 to calculate V_m for our typical cell:

$$V_m = 26.7 \ln \frac{1(5) + 0.02(145) + 0.5(6)}{1(140) + 0.02(10) + 0.5(106)} = -76.8 \text{ mV}$$

The value of the resting V_m calculated from the GHK equation can be compared with the equilibrium potential for K^+ calculated previously through the Nernst equation: $E_K = -89.1$ mV. The resting V_m is approximately 12 mV more positive than E_K. This situation is typical; most cells have a resting V_m that is more positive than E_K by 5 to 20 mV.

A Permeant Ion Already in Electrochemical Equilibrium Does Not Need to Be Included in the Goldman-Hodgkin-Katz Equation

Earlier in the chapter the equilibrium potential for Cl^- was calculated for a cell with the ionic distributions shown in Table 4-1, and the result was $E_{Cl} = -76.8$ mV. Comparing E_{Cl}, calculated with the Nernst equation, with V_m calculated with the foregoing GHK equation, we see that E_{Cl} happens to be the same as V_m—both are equal to -76.8 mV. Thus, at the resting V_m, Cl^- ions are in electrochemical equilibrium in this cell. This situation is characteristic of many cells in which Cl^- is not actively transported and yet its membrane permeability is high: Cl^- simply distributes *passively* in accordance with the resting V_m until it is in electrochemical equilibrium.

We can use the fact that Cl^- ions are in electrochemical equilibrium in our cell to illustrate another aspect of the GHK equation. Recall that a stable resting V_m is achieved by balancing the fluxes of various permeant ions. When an ion is already in electrochemical equilibrium, however, no net flux occurs for that ion. Because the GHK equation incorporates the balance of permeant ion fluxes to arrive at a resting V_m, any ion whose net flux is already zero need not be included in the calculation. In our cell, because Cl^- is already in electrochemical equilibrium, the Cl^- terms can be left out of the GHK equation when calculating the resting V_m. This is easy to verify:

$$V_m = \frac{RT}{F} \ln \frac{P_K[K^+]_o + P_{Na}[Na^+]_o}{P_K[K^+]_i + P_{Na}[Na^+]_i}$$
$$= 26.7 \ln \frac{1(5) + 0.02(145)}{1(140) + 0.02(10)} = -76.8 \text{ mV}$$

Indeed, in this particular case, the GHK equation gives the same resting V_m even when the Cl^- terms are left out of the numerator and denominator.

We can conclude that if a permeant ion is already in electrochemical equilibrium, that ion need not be included in the GHK equation for calculating the V_m. Said in another way, if V_m is equal to the equilibrium potential for a particular permeant ion, terms involving that ion can be dropped from the GHK equation without any effect.

The Nernst Equation May Be Viewed as a Special Case of the Goldman-Hodgkin-Katz Equation

The GHK equation predicts the V_m when the cell membrane is permeable to all three common monovalent ions. The Nernst equation, conversely, predicts the V_m when the membrane is permeable to only *one* ion. Therefore the Nernst equation should be a limiting case of the GHK equation. In other words, if the permeability coefficients of all but one ion are made zero in the GHK equation (corresponding to the membrane being permeable only to a single type of ion), the GHK equation should give the equilibrium potential for that permeant ion. This expectation is easy to verify. For example, if $P_{Na} = P_{Cl} = 0$ (so the membrane is permeable only to K^+), the GHK equation can be simplified:

$$V_m = \frac{RT}{F} \ln \frac{P_K[K^+]_o + 0 \times [Na^+]_o + 0 \times [Cl^-]_i}{P_K[K^+]_i + 0 \times [Na^+]_i + 0 \times [Cl^-]_o}$$

$$V_m = \frac{RT}{F} \ln \frac{P_K[K^+]_o}{P_K[K^+]_i} = \frac{RT}{F} \ln \frac{[K^+]_o}{[K^+]_i} = E_K$$

Indeed, when the membrane is permeable only to K^+, the GHK equation simplifies to the Nernst equation for K^+. Similarly, if $P_K = P_{Cl} = 0$ (membrane permeable only to Na^+), the GHK equation simplifies to the Nernst equation for Na^+.

E_K is the "Floor" and E_{Na} is the "Ceiling" of Membrane Potential

From the foregoing discussions, we can draw several inferences about the V_m of mammalian cells. E_K, the K^+ equilibrium potential, is very negative (typically approximately -90 mV), and E_{Na}, the Na^+ equilibrium potential, is very positive (typically at least $+60$ mV). These two equilibrium potentials define the range of voltages that is practically accessible to a living mammalian cell: E_K is the "floor" and E_{Na} is the "ceiling" of accessible potentials. E_{Cl}, the Cl^- equilibrium potential, is always negative, and its value usually hovers around the resting V_m. Depending on detailed conditions in a particular cell, E_{Cl} can be slightly more positive than, equal to, or slightly more negative than the resting V_m. These conclusions are summarized graphically in Figure 4-2.*

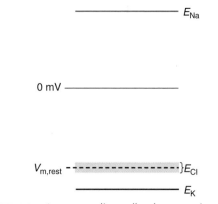

FIGURE 4-2 ■ In mammalian cells, the potassium equilibrium potential (E_K) is the lower bound and the sodium equilibrium potential (E_{Na}) is the upper bound of the range of accessible membrane potentials (V_m). The resting V_m ($V_{m,rest}$) is always more positive than E_K, and the chloride equilibrium potential (E_{Cl}) usually has a value close to $V_{m,rest}$.

The Difference Between the Membrane Potential and the Equilibrium Potential of an Ion Determines the Direction of Ion Flow

An ion will always flow in such a way as to drive V_m toward its own equilibrium potential. This knowledge allows us to deduce the direction of ion flux. For example, our typical cell has resting $V_m = -76.8$ mV and $E_{Na} = +71.5$ mV. Thus Na^+ ions will move and "attempt to" shift V_m from -76.8 mV toward $+71.5$ mV, that is, to make the cell interior more positive. Because Na^+ can make the cell more positive only by entering the cell, we deduce that Na^+ influx will occur. Analogously, the cell has $E_K = -89.1$ mV. Therefore any K^+ flux would tend to shift V_m from -76.8 mV toward -89.1 mV, that is, to make the cell more negative. Because K^+ can make the cell more negative only by leaving the cell, we deduce that K^+ efflux will occur.

THE CELL CAN CHANGE ITS MEMBRANE POTENTIAL BY SELECTIVELY CHANGING MEMBRANE PERMEABILITY TO CERTAIN IONS

The relationship between the GHK and Nernst equations immediately suggests that if Na^+ permeability predominates, V_m should approach E_{Na}. If K^+ permeability

---

*Among the physiologically important ions, Ca^{2+} has the most asymmetrical distribution: $[Ca^{2+}]_o \approx 1$ mM, $[Ca^{2+}]_i \approx 0.1$ μM. These values give a Ca^{2+} equilibrium potential of $E_{Ca} = +123$ mV, which is far more positive than E_{Na}. Thus it is fair to ask why E_{Na} is the voltage ceiling and E_{Ca} is not. The reason can be understood by comparing Na^+ and Ca^{2+} fluxes. When cells are stimulated, the ion fluxes that would drive V_m to more positive values are Na^+ influx and Ca^{2+} influx, quantitatively represented by $P_{Na}[Na^+]_o$ and $P_{Ca}[Ca^{2+}]_o$, respectively. Typically, $P_{Na} \gg P_{Ca}$ and $[Na^+]_o \gg [Ca^{2+}]_o$; therefore Ca^{2+} influx is much, much smaller than Na^+ influx. The electrical effect of Ca^{2+} influx is thus comparatively insignificant.

predominates, then V_m should approach E_K. This inference holds true for any permeant ion. As a demonstration, assume that the relative permeability coefficients are $P_K = 1$, $P_{Na} = 20$, and $P_{Cl} = 0.5$, so that now the membrane permeability to Na^+ is 20 times that of K^+, whereas in the earlier case the Na^+ permeability was only 1/20 that of K^+. Using the same ionic distributions as before, the GHK equation now yields:

$$V_m = 26.7 \ln \frac{1(5) + 20(145) + 0.5(6)}{1(140) + 20(10) + 0.5(106)} = +53.5 \text{ mV}$$

Recalling that in this cell $E_{Na} = +71.5$ mV, we see that with the vastly increased Na^+ permeability, V_m is now quite positive and rather close to E_{Na}, as anticipated.

The preceding observations suggest that the cell can change its V_m just by manipulating the relative permeabilities of the common permeant ions through the cell membrane—an important mechanism that underlies the ability of nerve cells to generate and transmit electrical signals (see Chapter 7).

THE DONNAN EFFECT IS AN OSMOTIC THREAT TO LIVING CELLS

If a cell were merely a bag of simple ions, such as K^+, Na^+, and Cl^-, immersed in an extracellular solution containing similar ions, nothing complex or interesting could ever happen; real biology requires something else. To carry out any real biological process, a cell must have in it, in addition to simple ions, more complex machinery. Biological machinery takes the form of proteins and nucleic acids, all of which are macromolecules that (1) carry multiple electrical charges and (2) are membrane-impermeant and therefore trapped inside the cell. It is important to know the ionic consequences of the presence of impermeant, multiply charged macromolecules for the cell.

The situation to be considered is schematically represented in Figure 4-3, where M^+ represents a singly charged cation (e.g., Na^+, K^+), A^- represents a singly charged anion (e.g., Cl^-), and P^{n-} represents a macromolecule bearing n negative charges. The PM of a living cell always has finite permeability to the common small ions; therefore M^+ and A^- can permeate the PM,

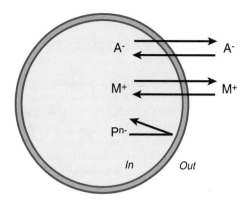

FIGURE 4-3 ■ A cell containing permeant cations and anions (M^+ and A^-) and impermeant polyanions (P^{n-}) bathed in extracellular solution containing only permeant cations and anions.

but the large polyanion P^{n-} cannot. Because P^{n-} cannot leave the cell and yet each negative charge on P^{n-} must be balanced with a positive charge, each molecule of P^{n-} will retain n M^+ ions inside the cell, just to maintain electroneutrality. The remaining "free" M^+ and A^- will equilibrate across the PM. Therefore a cell contains impermeant solute inside (all the P^{n-} macromolecules with their entourage of M^+ ions) while being bathed in a solution of permeant ions (M^+ and A^-). Recalling the discussion of osmosis from Chapter 3, we recognize that the imbalance of impermeant solute will tend to drive water into the cell continuously. Thus the presence of impermeant solutes inside the cell (but not outside) puts the cell in danger of uncontrolled swelling and rupture. This and other consequences of having multiply charged macromolecules trapped inside a compartment enclosed by a semipermeable membrane are collectively referred to as the **Donnan effect** (for a more quantitative treatment, see Box 4-6).

Because the Donnan effect gives rise to osmotic imbalance, water will enter the cell by osmosis and cause swelling and rupture. To forestall such osmotic catastrophe, the cell can potentially adopt one of two survival strategies. The first is to pump water out as quickly as it enters, but no evidence indicates that this occurs in the cells of higher organisms. The second strategy is, in effect, to make an extracellular solute impermeant, so that the impermeant solute inside the cell is balanced by impermeant solute outside. How

BOX 4-6

A CALCULATION ILLUSTRATING THE DONNAN EFFECT

We use the Nernst equation and the concept of the equilibrium potential for a permeant ion to examine the Donnan effect. With reference to Figure 4-3, we adopt the following symbols: $[M^+]_i$ and $[M^+]_o$ are the concentrations of M^+ inside and outside the cell, respectively, whereas $[A^-]_i$ and $[A^-]_o$ are the concentrations of A^- inside and outside the cell, respectively. $[P^{n-}]$ is the concentration of impermeant macromolecules inside the cell. We know that if a permeant ion is allowed to move and redistribute across a cell membrane, a membrane potential will be established that counteracts any further movement of the ion driven by its concentration gradient. At this balance point there is no net flux of the permeant ion across the membrane, and the membrane potential attained is the Nernst, or equilibrium, potential for that ion. In the present situation both M^+ and A^- can permeate the cell membrane; therefore the equilibrium potential for each ion will be established eventually:

$$E_M = \frac{RT}{(+1)F} \ln \frac{[M^+]_o}{[M^+]_i} \quad \text{and} \qquad [B1]$$

$$E_A = \frac{RT}{(-1)F} \ln \frac{[A^-]_o}{[A^-]_i}$$

Because both ionic species (M^+ and A^-) are simultaneously in equilibrium across the *same* cell membrane, the equilibrium potentials achieved must be *identical*. That is, $E_M = E_A$:

$$\frac{RT}{(+1)F} \ln \frac{[M^+]_o}{[M^+]_i} = \frac{RT}{(-1)F} \ln \frac{[A^-]_o}{[A^-]_i} \qquad [B2]$$

Algebraic simplification gives:

$$\frac{[M^+]_i}{[M^+]_o} = \frac{[A^-]_o}{[A^-]_i} \qquad [B3a]$$

or, the equivalent expression:

$$[M^+]_i\,[A^-]_i = [M^-]_o\,[A^-]_o \qquad [B3b]$$

Equation [B3] is known as the Donnan condition. It can be seen from the Donnan condition (Equation [B3a]) that the distribution of permeant cations across the membrane is the *inverse* of the distribution of permeant anions across the same membrane. Thus the presence of impermeant

macromolecules that carry multiple negative charges results in a higher concentration of permeant cations inside the cell relative to the outside, whereas the concentration of permeant anions is correspondingly lower inside relative to the outside of the cell. Any system in which the permeant cations and anions obey the Donnan condition is said to be in Donnan equilibrium.

We now analyze the situation with respect to the principle of electroneutrality (see Box 4-4): in any solution the total number of positive and negative charges must be equal, so that the solution remains electrically neutral overall. Electroneutrality thus dictates that, inside the cell (see Figure 4-3):

$$[M^+]_i = [A^-]_i + n[P^{n-}] \qquad [B4]$$

because each A^- needs only one M^+, whereas each P^{n-} must have n M^+ for charge balance. Charge balance must also hold for the extracellular fluid; therefore:

$$[M^+]_o = [A^-]_o \qquad [B5]$$

If the ratio in Equation [B3a], known as the Donnan ratio, is given the symbol R_D, Equation [B4] can be rewritten as:

$$R_D\,[M^+]_o = \frac{[M^+]_o}{R_D} + n[P^{n-}] \qquad [B6]$$

Multiplying through by R_D and rearranging the terms gives the quadratic equation:

$$[M^+]_o\,R_D^2 - n[P^{n-}]R_D + [M^+]_o = 0 \qquad [B7]$$

which has the solution:

$$R_D = \frac{n[P^{n-}]}{2[M^+]_o} + \sqrt{1 + \left(\frac{n[P^{n-}]}{2[M^+]_o}\right)^2} \qquad [B8]$$

A reasonable assumption is that $[M^+]_o = [A^-]_o = 150$ mM. In addition, we assume that $n = 50$ and $[P^{n-}] = 5$ mM (i.e., the cell contains approximately 5 mM of macromolecules bearing 50 negative charges each, on average). For these estimates:

$$R_D = \frac{50(5)}{300} + \sqrt{1 + \left(\frac{50(5)}{300}\right)^2} = 2.14 \qquad [B9]$$

BOX 4-6

A CALCULATION ILLUSTRATING THE DONNAN EFFECT—cont'd

This means that:

$$R_D = \frac{[M^+]_i}{[M^+]_o} = \frac{[A^-]_o}{[A^-]_i} = 2.14 \qquad \text{[B10]}$$

This calculation verifies that the presence of the negatively charged macromolecules inside the cell causes excess accumulation of permeant cations in the cell relative to the extracellular fluid, whereas there is a corresponding deficit of permeant anions in the cell relative to the extracellular fluid. This reciprocal asymmetrical distribution of the permeant cations and anions resulting from the presence of impermeant charged macromolecules is one aspect of the *Donnan effect*.

The second, more important, consequence of having impermeant charged macromolecules in the cell can be seen by examining the total concentration of solutes inside and outside the cell. Because $[M^+]_o = [A^-]_o = 150$ mM, the calculated Donnan ratio (Equation [B10]) requires that:

$$[M^+]_i = 2.14[M^+]_o = 2.14(150) = 321 \text{ mM}$$

$$[A^-]_i = [A^-]_o/2.14 = 150/2.14 = 70 \text{ mM}$$

In addition, we still have $[P^{n-}] = 5$ mM. Therefore total solute concentration inside the cell is:

$$[\text{Solute}]_i = [M^+]_i + [A^-]_i + [P^{n-}]_i$$
$$= 321 + 70 + 5 = 396 \text{ mM}$$

whereas the total solute concentration outside the cell is:

$$[\text{Solute}]_o = [M^+]_o + [A^-]_o = 150 + 150 = 300 \text{ mM}$$

It is clear that the total intracellular solute concentration is significantly higher than the total extracellular solute concentration. We conclude that, because of the Donnan effect, the presence of multiply charged macromolecules inside the cell will always cause the intracellular osmolarity to *exceed* the extracellular osmolarity. This osmotic imbalance will always drive water movement into the cell, which leads to increased cell volume and eventual rupture. Therefore, to survive, a living cell must have evolved mechanisms to counteract this osmotic imbalance. The principal mechanism for osmoregulation and cell volume control is the plasma membrane sodium pump (Na^+/K^+-ATPase; see Chapter 11).

can the cell "transform" a permeant solute (e.g., Na^+ ions) into an impermeant solute? The cell can render a solute *effectively* impermeant by pumping the solute out as soon as it enters. In this way no net gain or loss of the solute occurs, which means that the solute behaves *as if* it is impermeant. Indeed, this is the major mechanism for regulating cell volume—all living cells pump out permeant cations as quickly as they enter the cell. For living cells at steady state, Na^+

ions, the major permeant cations outside the cell, are extruded from the cell by active transport as rapidly as they leak into it. This is functionally equivalent to making the cell membrane impermeable to Na^+ ions. Thus the PM sodium pump (Na^+/K^+-ATPase; see Chapter 11), by constantly removing a small ionic solute from the interior of the cell with the expenditure of ATP energy, maintains cellular osmotic balance.

SUMMARY

1. Biologically important ions (e.g., Na^+, Ca^{2+}, Cl^-) typically are asymmetrically distributed across a biological membrane. That is, an ionic species is typically present at different concentrations on opposite sides of a biomembrane. This implies that a concentration gradient exists for each type of ion across the membrane.
2. Movement of ions (which carry electrical charge) across a membrane changes the electrical potential across the membrane.

3. If a cell membrane is selectively permeable to only a *single* type of ion, the concentration gradient of the ion will drive diffusion of that ion across the membrane. Within a very short time such ion movement will generate an electrical potential across the membrane that will be strong enough to oppose any further net movement of ions across the membrane. The V_m that is reached is known as the equilibrium potential of the ion. At the equilibrium potential of an ion, no *net* flux of that ion can occur across the membrane.

4. The equilibrium potential of an ion can be calculated by using the Nernst equation, as long as the concentrations of the ion on the two sides of the membrane are known. The equilibrium potential is also known as the Nernst potential.

5. In reality, a cell membrane is permeable to multiple ionic species, each of which will have a flux across the membrane. The steady-state *resting* V_m of a cell is achieved when all the ionic fluxes balance each other so that no *net* movement of ionic charges across the membrane occurs.

6. The resting V_m of a cell can be quantitatively estimated by using the *GHK equation*, as long as the concentrations of the relevant ions, as well as the relative membrane permeabilities for the ions, are known.

7. A cell can change its V_m by controlling the relative permeabilities of the cell membrane to certain ions (principally Na^+, K^+, and Cl^-).

8. The presence of impermeant, multiply charged macromolecules (e.g., nucleic acids, proteins) inside the cell gives rise to the Donnan effect, one aspect of which is that intracellular osmolarity will tend to be higher relative to the extracellular environment. This would cause water to move into the cell, which would swell and rupture. The sodium pump (Na^+,K^+-ATPase; see Chapter 11), by continually pumping out Na^+ ions, reduces the intracellular osmolarity to match the extracellular osmolarity and thus counteracts the osmotic consequences of the Donnan effect.

KEY WORDS AND CONCEPTS

- Ion concentration gradient
- Electrochemical equilibrium
- Equilibrium (Nernst) potential
- Nernst equation
- Membrane potential
- Ionic permeability
- Resting membrane potential
- Ion flux
- Goldman-Hodgkin-Katz (GHK) equation
- Donnan effect
- Relationship between ionic flux and ionic current

STUDY PROBLEMS

1. In a particular cell, $E_{Na} = +20$ mV. If the intracellular concentration of Na^+ is found to be $[Na^+]_i = 5$ mM, what would the extracellular Na^+ concentration be?

2. For a particular cell, the intracellular and extracellular concentrations of the common monovalent ions are shown in the following table:

Ion	Intracellular (mM)	Extracellular (mM)
K^+	140	3
Na^+	15	145
Cl^-	5	105

   a. If the relative permeabilities of the cell membrane to the three ions are 0.8 : 1.0 : 0.5 (P_K : P_{Na} : P_{Cl}), what is the V_m of this cell?

   b. What is the equilibrium potential for Cl^-?

   c. What is the expected direction of the net *flux* of Cl^-?

   d. What are the direction and sign of the Cl^- *current*?

3. If Na^+ extrusion by the Na^+ pump in the plasma membrane of a cell is inhibited, how is the volume of the cell expected to change? Explain.

BIBLIOGRAPHY

Atkins PW: *Physical chemistry*, ed 5, New York, NY, 1994, WH Freeman.
Byrne JH, Schultz SG: *An introduction to membrane transport and bioelectricity*, ed 2, New York, NY, 1994, Raven Press.
Ferreira HG, Marshall MW: *The biophysical basis of excitability*, Cambridge, England, 1985, Cambridge University Press.
Gennis RB: *Biomembranes: molecular structure and function*, New York, NY, 1989, Springer-Verlag.
Weiss TF: *Cellular biophysics*, Cambridge, MA, 1996, MIT Press.

Ion Channels and Excitable Membranes

ION CHANNELS

OBJECTIVES

1. Understand that ion channels are gated, water-filled pores that increase the permeability of the membrane to selective ions.

2. Describe the function of the selectivity filter in an ion channel.

3. Understand that ion channels can be grouped into gene families on the basis of structural homology.

4. Describe the structural features of the voltage-gated channel superfamily.

ION CHANNELS ARE CRITICAL DETERMINANTS OF THE ELECTRICAL BEHAVIOR OF MEMBRANES

This chapter and the following three chapters focus on the properties of the cell membrane that determine the overall electrical behavior of the cell. To put this material in its proper context, consider a typical neuron, such as the α motor neuron illustrated in Figure 5-1. The cell body (soma) contains the nucleus, mitochondria, and the endoplasmic reticulum, which is the site of protein synthesis. Two types of processes usually extend from the cell body. *Dendrites* are relatively short, small-diameter processes that branch extensively and receive signals from other neurons. The *axon* is a long cylindrical process that can be more than 3 m in length and is responsible for transmitting signals to other neurons or effector cells. The axon begins at a region of the soma called the *axon hillock* and terminates in small branches that make contact with as many as 1000 other neurons. Specialized junctions *(synapses)* are formed at the points of contact between neurons and are sites of communication between the cells (see Chapters 12 and 13). The main electrical functions of this type of cell are as follows: (1) to sum, or integrate, electrical inputs from a large number of other neurons; (2) to generate an *action potential* (AP), which is a rapid, transient membrane depolarization, if the inputs reach a critical level; and then (3) to propagate the AP signal to the nerve terminals. All these processes depend on the activity of several types of **ion channels**. The channels are integral membrane proteins that form water-filled (aqueous) pores through which ions can permeate.

The primary role of the neuronal cell body and dendrite membranes (Figure 5-1) is to integrate, over both space and time, the activity of all synaptic inputs impinging on the cell. The characteristics of this integrative process are determined largely by the passive electrical properties of the membrane, which are described in Chapter 6. When the V_m at the axon hillock (Figure 5-1) reaches threshold, an AP is generated. Threshold behavior and the generation of the AP are

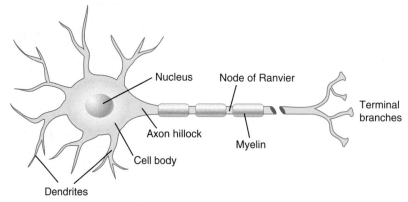

FIGURE 5-1 ■ Structure of a myelinated neuron. Two types of processes extend from the cell body: dendrites and axons. The dendrites and the cell body (or soma) form the receptive surfaces of the neuron—they receive inputs from other neurons. The axon begins at the axon hillock and can be up to a few meters in length in some animals. The neuron sends output signals through the axon. Many axons, like the one shown here, are myelinated: they are wrapped by a fatty sheath of myelin that insulates the axon from the surrounding solution. The myelin is interrupted by the nodes of Ranvier. The terminal branches of the axon make synaptic contacts with other neurons, or in the case of the α motor neuron, with skeletal muscle cells.

caused primarily by two types of ion channels in nerves, voltage-gated Na^+ and K^+ channels. The properties of these channels and their roles in the generation of the AP are presented in Chapter 7. Once the AP has been generated, it is conducted, or *propagated,* at full amplitude (i.e., it is *"all-or none"*; see Chapter 7) along the axon to the nerve terminals. AP propagation depends on both the passive properties of the membrane and the dynamic activity of the voltage-gated Na^+ and K^+ channels.

DISTINCT TYPES OF ION CHANNELS HAVE SEVERAL COMMON PROPERTIES

Ion Channels Increase the Permeability of the Membrane to Ions

The permeability of a pure phospholipid bilayer membrane to ions (e.g., Na^+, K^+, Cl^-, and Ca^{2+}) is extremely small: the permeability coefficients for these ions are in the range of 10^{-11} to 10^{-13} cm/sec (see Chapter 2). Because of this low intrinsic permeability, ions can cross membranes only by two special mechanisms: by reversibly binding to a carrier protein (see Chapters 10 and 11) or by diffusion through an aqueous pore (Figure 5-2). The maximum transport rate

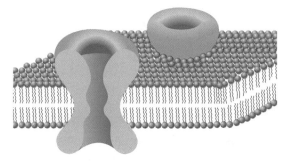

FIGURE 5-2 ■ An open ion channel embedded in a lipid bilayer membrane. The ion channel is a protein macromolecule that extends across the membrane and is in contact with the aqueous environment on both sides of the membrane. The channel provides an aqueous pathway for selected ions to move through the membrane. *(Redrawn from Hille B: Ionic channels of excitable membranes, ed 3, Sunderland, Mass, 2001, Sinauer.)*

for carriers is on the order of 5000 ions per second. This is much too slow to generate the rapid changes in V_m that are required for neuronal signaling. Excitation of nerve and muscle (i.e., the generation and propagation of the AP; see Chapter 7) and neuronal signaling require much faster ion movements. The rate of ion movement by diffusion through a small pore in the membrane is usually several orders of magnitude faster than the transport rate of carriers (Box 5-1).

Ion Channels Are Integral Membrane Proteins That Form Gated Pores

Ion channels are large macromolecular proteins that often consist of several peptide subunits. These channel proteins extend across the lipid bilayer and are in contact with the aqueous environment on both sides of the membrane. The channel protein forms a water-filled pore that allows ions to cross the membrane by diffusion. A single channel can conduct ions at the rate of 1 to 100 million ions per second.

The pore in an ion channel is not open all the time. Channels can open and close spontaneously and in response to various stimuli. The channel functions as if it had a gate that could close and prevent ions from moving through the pore or could open to allow ion movement. **Voltage-gated** ion channels have an *open probability* (the fraction of time the channel is open) that depends on V_m (see Chapter 7). **Ligand-gated** channels are activated after the binding of a specific type of molecule (the ligand) to a receptor site located on the channel (see Chapter 12). In the latter case, open probability is related to ligand binding, which in turn is related to ligand concentration.

Ion Channels Exhibit Ionic Selectivity

One of the more remarkable properties of ion channels is their ability to conduct ions *selectively* at very high rates. For example, K^+-selective ion channels that conduct approximately 30 million K^+ ions per second are 100 times more permeable to K^+ than to Na^+. This, at first, seems remarkable, in view of the fact that the crystal radius of Na^+ (0.095 nm) is actually less than that of K^+ (0.133 nm). If the channel is simply an aqueous pathway for ion movement, how is ion selectivity achieved?

The explanation of selectivity is that the channel has a narrow region within the pore that acts as a **selectivity filter**. The selectivity filter has a certain size and shape and acts as a molecular sieve to prevent larger ions from passing through. However, selectivity also requires *specific interaction* between the ion and the selectivity filter. Ions bind water molecules tightly (i.e., they are *hydrated*), and they must shed some waters of hydration to fit through the narrow selectivity filter. A specific type of ion will move through the channel only if the ion's binding interaction with the selectivity filter compensates for the loss of waters of hydration (Box 5-2).

BOX 5-1

CALCULATION OF THE RATE OF ION MOVEMENT THROUGH AN AQUEOUS PORE

The diffusion equation (see Chapter 2) can be used to calculate the rate of ion movement across a membrane through an aqueous pore. The equation is

$$J = -D\frac{\Delta C}{\Delta x}$$

where J is the flux per unit area (in mol/cm²/sec), D is the diffusion coefficient, ΔC is the concentration difference across the membrane, and Δx is the pore length. We will assume that the pore is a cylinder with radius r_p. To calculate a flux in units of mol/sec, we multiply both sides of the flux equation by the cross-sectional area of the pore ($\pi \times r_p^2$):

$$J(\text{mol/sec}) = -\pi \times r_p^2 \times D\frac{\Delta C}{\Delta x} \qquad [\text{B1}]$$

We assume that (1) D for the ion in the pore is the same as in bulk solution (2×10^{-5} cm²/sec), (2) r_p is 3×10^{-8} cm, (3) ΔC is 100 mM, and (4) Δx is 5×10^{-7} cm. Plugging these values into Equation [B1] gives

$$J = \pi \times (3 \times 10^{-8}\,\text{cm})^2 \times 2 \times 10^{-5}\frac{\text{cm}^2}{\text{sec}}$$

$$\times \frac{0.1\dfrac{\text{mol}}{10^3\,\text{cm}^3}}{5 \times 10^{-7}\,\text{cm}} \approx 1 \times 10^{-17}\frac{\text{mol}}{\text{sec}}$$

Multiplying by Avogadro's number (6.022×10^{23} ions/mol) gives the flux as ~6 million ions per second, which is ~1000 times faster than the maximum rate of carrier-mediated transport (see Chapter 11).

SELECTIVITY INVOLVES INTERACTION OF AN ION WITH THE SELECTIVITY FILTER

Water is a dipolar molecule because electrical charge is asymmetrically distributed within the molecule—the oxygen atom is slightly negative and the hydrogen atoms are slightly positive. As a result, an ion in aqueous solution is hydrated; that is, the ion has an entourage of water molecules that are electrostatically attracted to it. This is a stable configuration: the hydration energy is of the same magnitude as a covalent bond. Thus energy must be expended to remove waters of hydration from an ion. The selectivity filter in an ion channel is a narrow region containing carboxyl or carbonyl groups that can interact with the ion in place of water molecules. An ion will move through the narrow region only if the energy of interaction with the selectivity filter compensates for the loss of waters of hydration. For example, a K^+ ion in a rigid 0.3-nm diameter pore lined with carbonyl oxygen atoms may have the same energy as a K^+ ion in water. However, a Na^+ ion, being smaller, would have suboptimal interactions with the selectivity filter and would thus have a higher energy in the pore than in water. Therefore it would not shed its water molecules and thus could not enter the pore.

ION CHANNELS SHARE STRUCTURAL SIMILARITIES AND CAN BE GROUPED INTO GENE FAMILIES

Channel Structure Is Studied with Biochemical and Molecular Biological Techniques

To understand fully how an ion channel works, we must know the **ion channel structure** in detail. In the early 1970s, channels were identified as polypeptides. Various biochemical and molecular biological techniques were then developed to isolate and characterize the structure of channel proteins. By making use of their ability to bind specific ligands, channel proteins were purified by affinity chromatography. Channels isolated in this way were found to be large, glycosylated proteins often composed of more than one protein subunit. For example, the voltage-gated Na^+ channel (see Chapter 7) was first purified on the basis of its ability to bind tetrodotoxin* (TTX) with high affinity. The principal (α) subunit of the Na^+ channel has a molecular weight of approximately 250 kDa and consists of a linear sequence of approximately 1800 amino acids. The α-subunit contains all

the functional characteristics of Na^+ channels, including the pore-forming region and the TTX-binding site. The Na^+ channel also contains two smaller, auxiliary peptide subunits. The function of the auxiliary subunits is unknown.

In 1984, the α-subunit of the voltage-gated Na^+ channel was cloned, revealing its amino acid sequence. On the basis of this sequence a model of the secondary structure (i.e., the protein folding pattern) of the channel was developed, which was later confirmed by various biochemical and functional studies. The Na^+ channel α-subunit (Figure 5-3) contains four homologous domains (designated I, II, III, and IV), each with six α-helical membrane-spanning segments (S1 to S6). Segment S4 has a positively charged amino acid at every third residue and is the voltage sensor (see Chapter 7). A sequence of residues connecting S5 to S6 on the extracellular side of the channel, called the P region or P loop, dips partway into the membrane to form part of the channel pore.

Voltage-gated Ca^{2+} and K^+ channels are structurally similar to Na^+ channels: they are composed of four repeats of the basic motif containing S1 to S6 and the P loop. Voltage-gated Ca^{2+} channels, like Na^+ channels, have the four repeats linked in a single α-subunit. In contrast, **voltage-gated K^+ channels** are composed of four peptide subunits, each of which is a single instance of the basic motif (Figure 5-4A).

---

*Tetrodotoxin is a puffer fish toxin that selectively blocks voltage-gated Na^+ channels with high affinity.

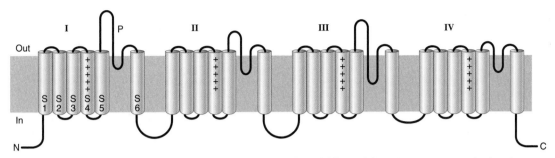

FIGURE 5-3 ■ The voltage-gated Na$^+$ channel: representation of the folding of the primary structure in the plasma membrane. The Na$^+$ channel α-subunit is a single polypeptide chain that contains four homologous domains (I, II, III, and IV). Each domain contains six α-helical segments (S1 to S6, shown as cylinders) that span the membrane. Segment S4 contains a positive amino acid at every third position. Each domain also contains a P region that connects S5 and S6 on the extracellular side, and that dips into the membrane to form part of the pore.

The remarkably similar amino acid sequences of the voltage-gated ion channels indicate that they are members of a **gene superfamily** and probably evolved from a common ancestral gene.

The voltage-gated K$^+$ channels are also related to two other families of K$^+$-selective ion channels (Figure 5-4). The inward-rectifier K$^+$ channel* subunit has only two α-helical transmembrane segments connected by a pore-forming P loop; four of these subunits assemble into a functional channel. The two-pore K$^+$ channel subunit has four transmembrane segments and two P loops, and the functional channel comprises two of these subunits.

Structural Details of a K$^+$ Channel Are Revealed by X-Ray Crystallography

The most direct method for determining protein structure is analysis of the x-ray diffraction pattern obtained from a protein crystal. Ion channel proteins have been difficult to crystallize, in part because of the large amount of protein required. MacKinnon and his colleagues crystallized a member of the inward-rectifier K$^+$ channel family, KcsA, from the bacterium *Streptomyces lividans.*† This channel protein

has homology to mammalian K$^+$ channels, a finding that indicates that the bacterial and mammalian channels were derived from a common ancestor. The KcsA channel protein was relatively easy to crystallize because it is not glycosylated, has a small size, and could be produced in large quantities by overexpression in *Escherichia coli.*

The crystal structure of KcsA provides a detailed picture of the channel structure and critical clues to the mechanism of K$^+$-selective permeation. The KcsA channel is a tetramer of four identical subunits that form a central aqueous pore (Figure 5-5A). Each subunit has two transmembrane segments: an inner helix that lines the pore near the intracellular surface of the membrane and an outer helix that faces the lipid bilayer (Figure 5-5B). The P (pore) region connects the two helices on the extracellular side and consists of three components: (1) the turret region, which is a chain of residues that surrounds the extracellular mouth of the channel; (2) the pore helix, which is inserted into the membrane between inner helices and provides contacts that hold the four subunits together; and (3) a loop of amino acids that forms the selectivity filter of the channel (Figure 5-5B and C). The selectivity filter is formed by the carbonyl oxygen atoms of three consecutive amino acids: glycine-tyrosine-glycine. The side chains of these amino acids point away from the pore and interact with other residues to stabilize the pore at the optimum diameter for K$^+$ permeation.

---

*In electronics, a rectifier is a device that allows current to flow in one direction, but not in the opposite direction. Inward-rectifier K$^+$ channels allow K$^+$ ions to flow freely into, but not out of, the cell.
†The physiologist Roderick MacKinnon shared the 2003 Nobel Prize in Chemistry for this work.

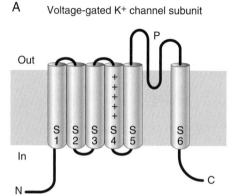

A   Voltage-gated K+ channel subunit

FIGURE 5-4 ■ Potassium channels are formed by separate subunits. **A,** A voltage-gated K⁺ channel subunit is homologous to the domains of the voltage-gated Na⁺ channel. It contains six membrane-spanning segments and a P region connecting S5 and S6. Four of these subunits assemble to form a functional voltage-gated K⁺ channel. **B,** An inward rectifier K⁺ channel subunit has only two membrane-spanning segments linked by a P region; four of these subunits assemble into a functional channel. **C,** A two-pore domain K⁺ channel subunit has four transmembrane segments and two P regions: two of these subunits form a functional K⁺ channel.

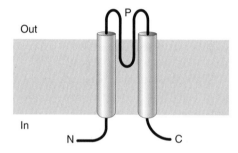

B   Inward rectifier K+ channel subunit

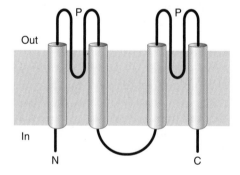

C   Two-pore domain K+ channel subunit

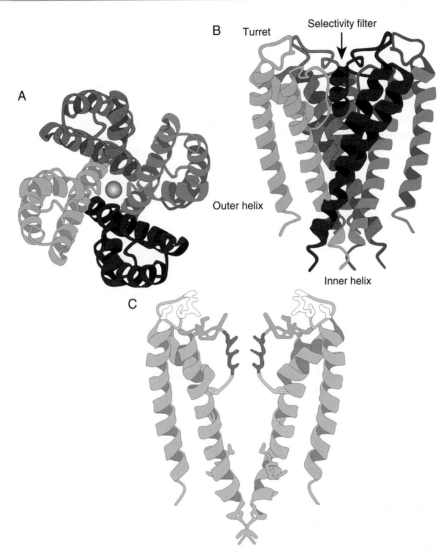

FIGURE 5-5 ■ Structure of an inward-rectifying K+ channel. **A,** View of the channel looking down at the pore from the extracellular side. Each of the four subunits is shown in a different shade of blue or gray, and each contributes a P region to the lining of the pore. A K+ ion is shown as a blue sphere in the middle of the pore. **B,** View of the channel parallel to the plane of the membrane. An inner helix from each subunit forms the inner part of the pore, and they are arranged as an inverted teepee. **C,** The same view as in **B** with two of the subunits removed. The gray region is the selectivity filter formed by three carbonyl oxygen atoms from three consecutive amino acids, glycine-tyrosine-glycine. *(Modified from Doyle DA, Cabral JM, Pfuetzner RA, et al:* Science 280:69, 1998.)

SUMMARY

1. An ion channel is a large protein that extends across the lipid bilayer and forms a water-filled pore, which allows ions to cross the membrane by diffusion. By controlling the membrane permeability to ions, channels play a primary role in the electrical behavior of the cell.

2. All ion channels function to increase membrane permeability to specific ions by allowing those ions to pass through *selectively* at very high rates.

3. All ion channels function as if they have a gate that can open or close to regulate ion movement. Voltage-gated channels are opened by changes in V_m. Ligand-gated channels are opened after the binding of a specific type of molecule to a receptor site on the channel.

4. Voltage-gated Na^+, K^+, and Ca^{2+} channels have similar amino acid sequences, indicating that they are members of a gene superfamily. These channels contain four homologous domains, each with six α-helical membrane-spanning segments (S1 to S6). Segment S4 is the voltage sensor. The P loop linking S5 to S6 forms part of the channel pore.

5. The detailed structure of a bacterial K^+ channel (KcsA) has been determined by x-ray crystallography. The KcsA channel is a tetramer of four identical subunits that form a central aqueous pore. The P region of each KcsA subunit contains a loop of amino acids that forms the selectivity filter.

KEY WORDS AND CONCEPTS

- Ion channel
- Voltage-gated channel
- Ligand-gated channel
- Selectivity filter
- Ion channel structure
- Voltage-gated K^+ channel
- Gene superfamily

STUDY PROBLEMS

1. The resting K^+ permeability of the pancreatic β-cell membrane is determined mainly by a specific population of K^+ channels. Describe at least three ways that the β-cell could change the properties of these K^+ channels and thereby change the K^+ permeability of the membrane.

2. A point mutation in the gene encoding a voltage-gated Na^+ channel results in a single amino acid substitution in the channel protein and causes the channel to change from being Na^+-selective to being Ca^{2+}-selective. What is the most likely location of the substituted amino acid in the channel structure? Describe a possible mechanism that could explain the change in selectivity resulting from a single amino acid substitution.

BIBLIOGRAPHY

Catterall WA. Structure and function of voltage-gated ion channels, *Annu Rev Biochem* 64:493, 1995.

Doyle DA, Cabral JM, Pfuetzner RA, et al. The structure of the potassium channel: molecular basis of K^+ conduction and selectivity, *Science* 280:69, 1998.

Hille B: *Ionic channels of excitable membranes*, ed 3, Sunderland, MA, 2001, Sinauer.

PASSIVE ELECTRICAL PROPERTIES OF MEMBRANES

OBJECTIVES

1. Understand that passive membrane electrical properties refer to properties that are constant near the resting potential of the cell.

2. Understand that membranes behave, electrically, like a resistor in parallel with a capacitor.

3. Understand that open ion channels are electrically equivalent to conductors (or resistors).

4. Describe Ohm's Law as it relates to current flow through ion channels.

5. Understand that membranes have capacitive properties because the lipid bilayer is an insulator that allows ions to accumulate at the surface of the membrane.

6. Define membrane time constant and length constant, and describe the passive properties that influence them.

THE TIME COURSE AND SPREAD OF MEMBRANE POTENTIAL CHANGES ARE PREDICTED BY THE PASSIVE ELECTRICAL PROPERTIES OF THE MEMBRANE

The steady-state V_m of a membrane permeable to more than one ion can be estimated with the GHK equation (see Chapter 4). One of the main limitations of this approach, however, is that it cannot be used to predict how the V_m *changes* as a function of time or distance. In neurons, skeletal muscle cells, and other electrically excitable cells, certain changes in V_m have well-defined time courses. For example, the nerve AP, which is described in detail in Chapter 7, is a "spike" of depolarization that lasts 1 to 2 milliseconds. In addition, postsynaptic potentials (V_m changes that result from neurotransmitter release at chemical synapses; see Chapters 12 and 13) have relatively fast rising phases and exponential decays. The membrane properties that help to determine the time course of these signals are described in this chapter. The passive spread of a change in V_m with distance along a membrane surface is also discussed.

Passive electrical properties refer to properties that are fixed, or constant, near the resting potential of the cell. Three such properties play important roles in determining the time course and spread of electrical activity: the membrane resistance, the membrane capacitance, and the internal or "axial" resistance of long thin processes or cells such as nerve axons and dendrites and skeletal muscle cells. By examining the membrane as an electrical circuit, we can deduce how these parameters can be used to describe changes in V_m. Equivalent circuit models of membranes are used to analyze potentials that vary with time and distance in a manner that depends only on the passive membrane properties. These potentials are called **electrotonic potentials**.

THE EQUIVALENT CIRCUIT OF A MEMBRANE HAS A RESISTOR IN PARALLEL WITH A CAPACITOR

Membrane Conductance Is Established by Open Ion Channels

Many ion channels behave, in electrical terms, like conductors (or resistors, because **conductance** = 1/resistance). Each channel (Figure 6-1A) can be modeled as a resistor, or conductor, with a *single-channel* conductance, γ, when the channel is open (Figure 6-1B). Naturally, when the channel is closed, the conductance is zero.

Most permeant ions are distributed asymmetrically across the plasma membrane (see Chapter 4). This results in a chemical driving force that tends to push the ion through the open channel. This chemical force functions as a battery (with voltage equal to the equilibrium potential of the ion, E_K in the case of Figure 6-1). The battery is in series with the resistor (γ_K) representing the open channel, as shown for K^+ in Figure 6-1B. The current flow through the open channel obeys **Ohm's Law** (see Appendix D), which describes current flow through a resistor with a resistance of R in ohms (Ω), or with a conductance $g = 1/R$ in siemens(S):

$$V = I \times R \quad \text{or} \quad I = \frac{V}{R} = g \times V \quad [1]$$

where I is the current in amperes (A) for a potential difference of V in volts (V). For ion channels, Ohm's Law must be modified because the net ionic flux (and therefore the current) will be zero when V_m is equal to the equilibrium potential of the ion. Because the equilibrium potential is almost never zero mV, Ohm's Law for a single K^+ channel is

$$i_K = \gamma_K (V_m - E_K) \quad [2]$$

where i_K is the current through a single channel and $V_m - E_K$ is called the **driving force**. Ohm's Law predicts that the K^+ current is directly proportional to the driving force (Figure 6-1C).

Membranes usually contain several different types of ion channels that are each present in large numbers. In electrical terms, single channels in the membrane represent conductors arranged in parallel; in this case the individual conductances are additive. In other words,

$$g_{Na} = N_o \times \gamma_{Na} \quad [3]$$

where g_{Na} is the total conductance of the open **sodium** (Na^+) **channels** present in a unit area of membrane, γ_{Na} is the conductance of a single Na^+ channel, and N_o is the number of open Na^+ channels per unit area. In an equivalent circuit, we can then model a group of Na^+ channels as a resistor with conductance equal to g_{Na}, in series with a battery of voltage E_{Na} (Figure 6-2). A similar resistor-battery pair can be used to model a population of **potassium** (K^+) **channels,** or any other ion channels, in the membrane (Figure 6-2).

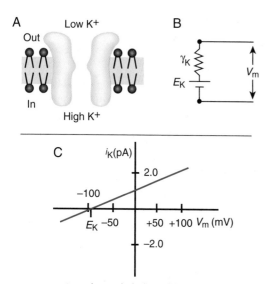

FIGURE 6-1 ■ Ion channels behave like electrical conductors. **A,** Schematic of a single, open K^+ channel, with a conductance of γ_K, embedded in a lipid bilayer that has an outwardly directed K^+ gradient. **B,** The equivalent electrical circuit of the channel is a resistor, or conductor, in series with a battery (E_K) that represents the concentration gradient for K^+ ions. **C,** Current through the channel (i_K) varies linearly with V_m.

Capacitance Reflects the Ability of the Membrane to Separate Charge

To complete the equivalent circuit, we need to account for the ability of the lipid bilayer to act as an electrical insulator that allows charges (ions such as K^+, Na^+, and Cl^-) to accumulate at the surface of the membrane. In electrical circuits a **capacitor** is an element

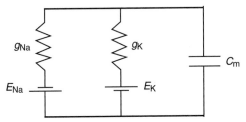

FIGURE 6-2 ■ Equivalent circuit of a membrane containing many open Na^+ and K^+ channels. The resistors labeled g_{Na} and g_K are the conductances of the membrane to Na^+ and K^+, respectively.

that stores, or separates, charges across an insulator. Thus the equivalent circuit of the membrane has a capacitor (C_m) connected in parallel with the elements representing the ion channels (Figure 6-2).

The amount of charge, q, in coulombs (C), that can be separated across the membrane is directly proportional to V_m:

$$q = C_m \times V_m \qquad [4]$$

where V_m is the potential difference in volts and C_m is the capacitance in farads (F). A 1-F capacitor can store 1 coulomb of charge per volt of potential difference. A farad is a very large quantity; all biological membranes have capacitances of approximately 1×10^{-6} F (1 μF)/cm² of membrane surface area (Box 6-1).

PASSIVE MEMBRANE PROPERTIES PRODUCE LINEAR CURRENT-VOLTAGE RELATIONSHIPS

The passive properties of cell membranes can be studied by the injection of current into the cell through a microelectrode (Box 6-2 and Figure 6-3A). When an inward pulse of current is passed across the membrane (Figure 6-3B), V_m approaches a more negative, or *hyperpolarized*, value following an exponential time course (Figure 6-3B) and eventually reaches a constant, steady-state level. The larger the applied current, the greater the hyperpolarization. A graph of the current versus the steady-state V_m is a straight line (Figure 6-3C). Therefore, in the steady state, the resting membrane behaves electrically like a resistor. The slope of the **linear current-voltage relationship** (i.e., $\Delta I/\Delta V_m$) is a measure of

the resting conductance of the membrane. The V_m at which the current-voltage line crosses the voltage axis is called the **reversal potential** (E_{rev}). At potentials negative to E_{rev} the current is negative (inward) and at potentials positive to E_{rev} the current is positive (outward); thus the current reverses direction at E_{rev}.

MEMBRANE CAPACITANCE AFFECTS THE TIME COURSE OF VOLTAGE CHANGES

Ionic and Capacitive Currents Flow When a Channel Opens

When current starts to flow across the membrane, V_m does not instantaneously reach a new steady-state level. Instead, the presence of **membrane capacitance** causes V_m to approach the steady-state level gradually (Figure 6-3B). To understand the effect of the membrane capacitance, we will first examine the currents that flow across the membrane when a channel opens. Consider a cell that contains only a single closed K^+ channel (Figure 6-4A). If the initial V_m is 0 mV when the K^+ channel opens, and the K^+ equilibrium

BOX 6-2

MEASURING AND MANIPULATING
THE MEMBRANE POTENTIAL

The most common technique for recording V_m involves the use of microelectrodes. The microelectrode is a glass capillary tube tapered to a fine, sharp tip with a diameter smaller than 1 μm. The electrode is filled with a concentrated salt solution, and a wire is placed in the back of the electrode to allow connection to electronic devices. The sharp tip of the microelectrode allows it to be pushed through the cell membrane without damaging the cell, thereby allowing the measurement of the intracellular potential. A pair of these electrodes (one intracellular and one extracellular) is connected to an amplifier to record V_m, as shown in Figure 6-3A.

A second pair of electrodes can be used to inject current into the cell. Figure 6-3A shows an inward current being passed into the cell: positive charges flow out of the extracellular electrode and are deposited on the outside surface of the cell, whereas positive charges move away from the inside surface of the membrane and enter the intracellular electrode. Thus this inward current makes V_m become more negative.

potential is −90 mV, K^+ ions will flow out of the cell down their electrochemical gradient. The removal of positive charge from the cell makes V_m move in the negative direction. Because the cell is permeable only to K^+, V_m will eventually reach E_K and stop changing. In terms of current flow in the equivalent circuit (Figure 6-4B), we say that an *outward* ionic K^+ current, I_K (K^+ ions moving out of the cell), produces an *inward* **capacitive current**, I_c. This capacitive current consists of positive charges moving away from the inside surface of the membrane and an equal number of positive charges moving up to the outside surface of the membrane. Because it takes time for ions to move through the channel and accumulate at the membrane surface, V_m can change only gradually. The general term **ionic current** (I_i) refers to current generated by ions crossing the membrane, such as through channels.

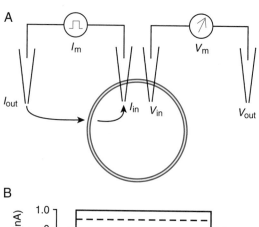

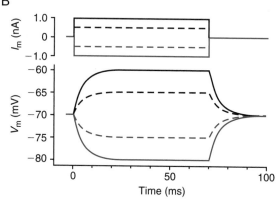

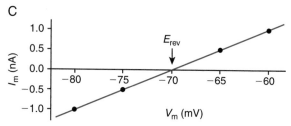

FIGURE 6-3 ■ Experimental arrangement used to study the passive properties of membranes. **A,** One intracellular microelectrode (I_{in}) is used to pass current across the membrane from a constant current source, I_m. The *arrows* illustrate an inward current flowing from the extracellular current electrode (I_{out}) through the membrane to I_{in}. A second intracellular electrode (V_{in}) is used to monitor the membrane potential (V_m). **B,** When a constant inward current is passed across the membrane, V_m approaches a new hyperpolarized steady-state level along an exponential time course, and the size of the hyperpolarization depends on the magnitude of the inward current. With small outward currents, V_m depolarizes along an exponential time course. **C,** The steady-state V_m is plotted as a function of the applied current *(filled circles),* and the *line* is a straight line fit to these data points. The reversal potential (E_{rev}) is indicated with an *arrow.*

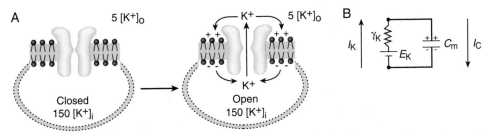

FIGURE 6-4 ■ Current flow through a single K^+ channel alters the charge distribution across the membrane. **A,** When the K^+ channel opens (*right*), K^+ ions flow out of the cell, thus making the outside more positive and leaving the inside of the cell more negative. **B,** In terms of the equivalent circuit, the outward K^+ current (I_K) produces an inward capacitive current (I_c).

The Exponential Time Course of the Membrane Potential Can Be Understood in Terms of the Passive Properties of the Membrane

The resting (passive) membrane can be represented by an equivalent circuit. This circuit (Figure 6-5A) can help to explain the role of the membrane capacitance in the exponential time course of the V_m change (see Figure 6-3B). All open ion channels have been combined into a single conducting pathway, R_m, in series with battery E_{RP}, which represents the resting potential of the cell. The membrane is connected to a constant current generator and to a device to monitor V_m. With no current flowing from the external source, V_m will be at E_{RP} with an excess of negative charges at the inside surface of the capacitor. A constant inward

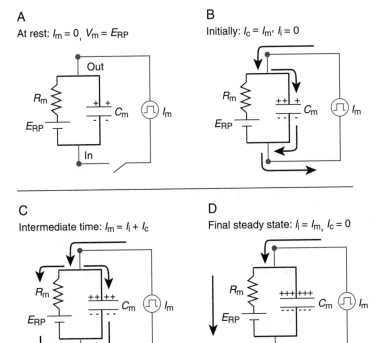

FIGURE 6-5 ■ An equivalent circuit of passive, resting membrane used to analyze current flow across the membrane. **A,** The membrane is modeled as a single resistor, R_m, which represents all open channels, in series with a battery that has a voltage equal to the resting potential of the cell (E_{RP}). C_m, membrane capacitance; I_m, total membrane current. **B,** At the instant a constant inward current, I_m, is turned on, all the current goes through the capacitor, and V_m begins to hyperpolarize. **C,** As soon as V_m begins to change, some of the current starts to flow through open channels (ionic current, I_i). **D,** Eventually, V_m reaches a new steady-state level, and all current flows through open ion channels.

current (I_m) is then passed into this circuit. At the instant the current is turned on, all of the current flows to the capacitor (Figure 6-5B). Perhaps the simplest way to understand this is to recognize that at the instant the current is turned on, there is no driving force for current flow through the resistor (i.e., $V_m - E_{RP} = 0$). The inward capacitive current further increases the charge separation across the capacitor because positive charges build up at the external surface and move away from the internal surface. As a result, V_m moves in the negative direction, which in turn produces a driving force for current flow through the resistor. Because the total current being controlled by the constant current generator must remain constant, the capacitive current (I_c) decreases in magnitude as the resistive (ionic) current (I_i) increases (Figure 6-5C). Finally, a new steady-state V_m is reached, and all the applied current flows through the resistor. Analysis of this circuit (see Appendix D) produces the following relationship between V_m and time:

$$\Delta V_m(t) = \Delta V_{m,\infty}\left[1 - e^{\frac{-t}{\tau_m}}\right]$$ [5]

where $\Delta V_m(t)$ is the change in V_m at time t, $\Delta V_{m,\infty}$ is the change in V_m in the steady state ($t = \infty$), and $\tau_m = R_m \times C_m$ is called the **membrane time constant**. The units for τ are seconds if R_m and C_m are in ohms and farads, respectively. Note that $\Delta V_{m,\infty} = I_m \times R_m$. This is a formal statement of the fact that in the steady state all of the current flows through the resistor and none flows through the capacitor. In addition to V_m, both the capacitive and ionic currents follow an exponential time course (Figure 6-6; see Appendix D).

Before the external current is turned off, all the applied current flows through channels (i.e., $I_i = I_m$; see Figure 6-5D). At this time the return pathway for current flow is through the external current generator. When the external current is turned off, the return pathway for the current is through the membrane capacitor; thus $I_i = -I_c$ (Figure 6-6). The resulting outward I_c causes V_m to become less negative; this reduces the driving force and thus decreases I_i. When $V_m = E_{RP}$, there is no driving force for current flow; thus I_i and I_c will both be zero and V_m will stop changing.

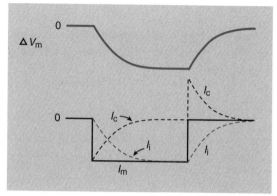

FIGURE 6-6 ■ In response to a pulse of constant current (I_m), ΔV_m (the change in V_m) follows an exponential time course to a new level. Initially, the capacitive current (I_c) is equal to I_m. I_c decays along an exponential time course as the ionic current (I_i) increases from zero to a steady-state level equal to I_m. When the current pulse is turned off ($I_m = 0$), the I_i and I_c are equal in magnitude, but opposite in direction. Both I_c and I_i then decline to zero as ΔV_m returns to its initial level.

The membrane time constant is a measure of the time scale over which changes in V_m occur. Consider the value of ΔV_m at a time equal to τ_m. If $t = \tau_m$,

$$\Delta V_m = \Delta V_{m,\infty}[1 - e^{-1}]$$
$$= \Delta V_{m,\infty}(1 - 0.37) = 0.63\Delta V_{m,\infty}$$ [6]

In other words, the time constant is the length of time required for 63% of the total change in V_m to occur.

MEMBRANE AND AXOPLASMIC RESISTANCES AFFECT THE PASSIVE SPREAD OF SUBTHRESHOLD ELECTRICAL SIGNALS

As small-amplitude changes in V_m spread along the surface of a cell, they decrease in amplitude. This effect can be conveniently studied in an elongated structure, such as an axon, with the experimental arrangement shown in Figure 6-7. One microelectrode is inserted into the axon to pass a constant subthreshold current across the membrane. The term "subthreshold" refers to the fact that V_m stays below the level required to initiate an AP. A second

A

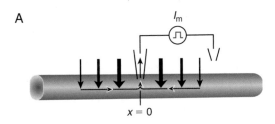

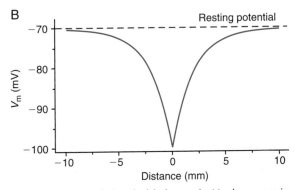

FIGURE 6-7 ■ A subthreshold change in V_m decreases in amplitude as an exponential function of distance along an axon. **A,** An external current source (I_m) passes a constant inward current pulse across the membrane. The current follows the path of least resistance: the largest membrane current density is closest to the current passing electrode. **B,** As a result, the change in V_m is largest at the site of current injection ($x = 0$), and it decreases as an exponential function of distance (x) away from this site.

electrode is used to measure the change in V_m at various distances away from the current-passing electrode. If an inward (hyperpolarizing) current pulse of long duration is used, V_m will reach a new, hyperpolarized steady-state level. A plot of the *steady-state* V_m as a function of *distance* reveals that the hyperpolarization is maximal at the site of current injection. The V_m becomes progressively less hyperpolarized as we move along the axon away from this point (Figure 6-7B). Far away from the site of current injection the membrane is at the resting potential and is unaffected by the current. Such passive spread of small-amplitude signals along the surface of the cell is called **electrotonic conduction**.

The Decay of Subthreshold Potentials with Distance Can Be Understood in Terms of the Passive Properties of the Membrane

Because the axon can be idealized as a cylindrical structure, it is convenient to normalize the units for resistance and capacitance to a 1-cm *length* of axon (we use lower case r and c to represent these parameters). The relevant parameters, and their units, are shown in Table 6-1, and Box 6-3 describes the origin of these units.

The axon can be modeled, in electrical terms, by the circuit depicted in Figure 6-8B. This circuit consists of several parallel r_m-c_m circuits connected together. Each r_m-c_m "patch" of membrane is connected through an internal resistance, r_i, and an external resistance, r_o, to the adjacent patch.

An intuitive explanation for the decay of amplitude with distance is provided by consideration of the circuit in Figure 6-8. As the applied current moves along the axon through r_i, some of the current "leaks" out through r_m, thus leaving less current available to affect more distant patches of membrane. Analysis of this circuit in the *steady state* predicts that the change in V_m, ΔV_m, should decrease exponentially with distance away from the current injection site, according to the relation,

$$\Delta V_m(x) = \Delta V_0 e^{\left(\frac{-x}{\lambda}\right)} \tag{7}$$

where $\lambda = \sqrt{\dfrac{r_m}{r_i + r_o}} \approx \sqrt{\dfrac{r_m}{r_i}}$

ΔV_0 is the change in V_m at the site of current injection (i.e., at $x = 0$). Nerve fibers are usually bathed in a

TABLE 6-1		
Parameters Used in the Equivalent Circuit of an Axon		
SYMBOL	CIRCUIT COMPONENT	UNITS
r_o	Extracellular resistance	ohm/cm
r_i	Intracellular resistance	ohm/cm
r_m	Membrane resistance	ohm × cm
c_m	Membrane capacitance	farad/cm

BOX 6-3

ORIGIN OF RESISTANCE UNITS USED IN THE CABLE EQUATION

The resistance of a conductor, such as a length of wire or a length of axoplasm, depends on the geometry of the conductor and on a parameter called the resistivity of the material (ρ). The resistivity depends on the physical properties of the material that determine its ability to conduct current. The resistance of a wire increases with the length of the wire and decreases with the diameter, or cross-sectional area. The following equation shows this relationship:

$$R = \rho \frac{L}{A_c} \qquad [B1]$$

In this equation R is the resistance of the conductor in ohms, ρ is the resistivity in ohm $\times$ cm, L is the length of the conductor in cm, and A_c is the cross-sectional area of the conductor in cm^2.

To study the passive properties of an axon, we can conveniently consider the axon in similar terms and express resistances for a 1-cm length of axon. For the internal, or axial, resistance of the axoplasm (which is equivalent to the resistance of a wire), we combine R and L (from Equation [B1]) into a single parameter r_i:

$$r_i = \frac{R}{L} = \frac{\rho}{A_c} \qquad \text{units:} \quad \frac{\text{ohm} \times \text{cm}}{\text{cm}^2} = \frac{\text{ohm}}{\text{cm}}$$

Thus r_i has units of ohm/cm. Identical units would apply to the longitudinal resistance of the extracellular solution (r_o). To convert the normalized resistance r_i or r_o into ohms, we

must multiply r_i or r_o by the length of the axon. In other words, the longitudinal internal (or external) resistance, in units of ohms, increases with the length of the axon.

The resistance equation (Equation [B1]) cannot be applied as easily to *membrane* resistance because the length of the conductor is now the membrane thickness, which is usually not known precisely. Therefore a new parameter is defined, $R_m = \rho \times L$, which combines the resistivity and length into a single parameter with units of ohm $\times$ cm^2. Thus, for membrane resistance, Equation [B1] can be rewritten:

$$R = \frac{\rho \times L}{A_c} = \frac{R_m}{A_s} = \frac{R_m}{\text{circumference} \times L}$$

where R_m has units of ohm $\times$ cm^2 and the cross-sectional area of the "resistor" (the membrane) is the membrane surface area in cm^2 (A_s). For a cylindrical axon the surface area is circumference $\times$ length (both parameters in cm). Finally, we can combine R and L into a single parameter (r_m) that represents the membrane resistance per cm length of axon:

$$r_m = R \times L = \frac{R_m}{\text{circumference}}$$

$$\text{units:} \quad \frac{\text{ohm} \times \text{cm}^2}{\text{cm}} = \text{ohm} \times \text{cm}$$

and clearly r_m has units of ohm $\times$ cm.

---

very large volume of conducting (low-resistance) extracellular fluid. Thus the total resistance to current flow in the external part of the loop (r_o) is negligible compared with the resistance along the corresponding stretch of axoplasm (r_i) inside the axon. Consequently $r_o \ll r_i$ and r_o can be dropped from the expression for λ. Equation [7] is often called a **cable equation** because a similar equation was derived to explain the decrease in the amplitude of signals transmitted through underwater transatlantic cables. We note that the membrane capacitance, c_m, has no effect on λ. This is because we are considering steady-state

V_m, and in the steady state no current flows through the capacitor.

The parameter λ is the **length constant** with units of cm. We can think of λ in physically descriptive terms. If the membrane resistance, r_m, is raised, the current has more difficulty escaping through the membrane. Hence a larger fraction of the total current flows farther from the stimulus site before escaping through the membrane; in quantitative terms, λ is increased. Similarly, if r_i is lowered, the current flows more easily along the axoplasmic path and λ is also increased.

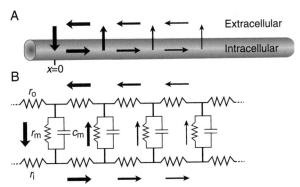

FIGURE 6-8 ■ The equivalent electrical circuit of an axon. **A,** The *arrows* indicate the steady-state current flowing in different regions of the axon during the injection of a constant current at the point $x = 0$. The relative thickness of the *arrow* represents the relative magnitude of the current. **B,** The equivalent circuit of the axon consists of patches of membrane, represented by parallel r_m-c_m circuits, connected together by resistors r_o and r_i. All parameters are normalized to a 1-cm length of axon: r_o is the resistance of extracellular fluid, r_i is the resistance of axoplasm, r_m is the membrane resistance, and c_m is the membrane capacitance.

The Length Constant Is a Measure of How Far Away from a Stimulus Site a Membrane Potential Change Will Be Detectable

At a distance λ cm from the point of current injection, ΔV_m would be reduced to approximately 37% of its value at the stimulus site:

$$\Delta V(\lambda) = \Delta V_0 e^{\left(\frac{-\lambda}{\lambda}\right)} = \Delta V_0 e^{-1} = 0.37 \Delta V_0 \qquad [8]$$

Thus, when λ is increased, the change in V_m requires a longer distance to decay by a given amount. In other words, a larger length constant means that the change in V_m will be detectable farther away from the stimulation site. In the next chapter we show that changes in λ affect the conduction velocity of the AP.

SUMMARY

1. Passive electrical properties refer to membrane properties that are constant near the resting potential of the cell. Three such properties, the membrane resistance, the membrane capacitance, and the internal resistance of long thin processes or cells, help to determine the time course and spread of electrical activity.

2. The cell membrane is electrically equivalent to a capacitor connected in parallel with a resistor and a battery connected in series.

3. Ion channels behave, in electrical terms, like conductors (or resistors). The current flow through an open channel obeys Ohm's Law.

4. The membrane is like a capacitor because the lipid bilayer acts as an electrical insulator, allowing charges to accumulate at the surface of the membrane. Capacitance is a measure of the amount of charge that can be separated across the membrane per unit of V_m. All biological membranes have a capacitance of approximately 1×10^{-6} F (1 μF) per cm^2 of membrane surface area.

5. When ionic current flows through channels, the membrane capacitance causes V_m to come to a new steady-state level gradually. In response to a pulse of constant current, V_m follows an exponential time course to a new level. The time constant of

this exponential curve is called the membrane time constant, τ_m, and is equal to the membrane capacitance times the membrane resistance.

6. The passive spread of small-amplitude, subthreshold signals along the surface of an axon is affected by membrane and axoplasmic resistances. The amplitude of these signals decreases as an exponential function of distance along the axon. The length constant, λ, provides a measure of how far subthreshold signals propagate. An increase in membrane resistance increases λ, and an increase in axoplasmic resistance decreases λ.

KEY WORDS AND CONCEPTS

- Passive electrical properties
- Electrotonic potentials
- Conductance
- Ohm's Law
- Driving force
- Sodium channels and potassium channels
- Capacitor
- Linear current-voltage relationship
- Reversal potential (E_{rev})
- Membrane capacitance
- Ionic and capacitive currents
- Membrane time constant (τ_m)
- Cable equation
- Length constant (λ)
- Electrotonic conduction

STUDY PROBLEMS

1. A small spherical cell has a total membrane surface area of 10 μm^2 (1 $\mu m = 10^{-4}$ cm) and a capacitance of 1×10^{-6} F/cm^2. Initially the cell has a V_m of zero mV, and there are no open channels in the membrane. Assume that $[Cl^-]_i = 10$ mM, $[Cl^-]_o = 140$ mM, $RT/F = 26.7$ mV, and $\gamma_{Cl} = 10^{-11}$ S.
   a. What happens to V_m when a single Cl^- channel opens?
   b. Draw a graph of V_m as a function of time following the opening of the channel. Indicate

the values of the initial and final V_m. What is the value of the membrane time constant in the presence of the open Cl^- channel?

2. A 100-μm diameter nerve axon has the following properties: $r_m = 2.5 \times 10^4$ ohm $\times$ cm, $r_i = 1 \times 10^5$ ohm/cm, $r_o = 0$, and $c_m = 3 \times 10^{-8}$ F/cm. For this axon the permeability to Cl^- (P_{Cl}) is very high and Cl^- ions are at equilibrium at the resting potential of -70 mV. A steady inward current is injected into the axon, which results in a V_m of -100 mV at the site of current injection.
   a. What would V_m be at a distance of 4 mm from the site of current injection?
   b. If P_{Cl} now became zero (by blocking all of the open Cl^- channels), what would happen to V_m at 4 mm from the site of current injection? (Assume that V_m remains at -100 mV at the site of current injection.)

3. Two different-diameter cylindrical dendrites on a cerebellar Purkinje neuron have the following passive membrane properties. Assume that the external resistance is zero and that the dendrites are cylindrical structures with homogeneous membrane properties along their entire length.

	r_m, ohm $\times$ cm	r_i, ohm/cm	c_m, F/cm
Dendrite 1	2×10^6	1×10^{10}	2×10^{-10}
Dendrite 2	2×10^4	1×10^6	2×10^{-8}

   a. An excitatory synaptic input impinges on each of these dendrites at a distance of 2 cm from the cell body. Activation of these synapses causes a depolarization of the postsynaptic cell, which is called an excitatory postsynaptic potential (EPSP). When a subthreshold EPSP occurs at this synapse, the amplitude of the EPSP decays along an exponential time course back to the resting potential. Would the decay rate in dendrite 1 be faster, slower, or the same as in dendrite 2? Explain your answer.
   b. The amplitude of the EPSP decreases as a function of distance away from the synapse. In which dendrite would the amplitude decrease more with distance? Explain your answer. Would a subthreshold EPSP in either dendrite produce a detectable depolarization of the cell body? Why?

BIBLIOGRAPHY

Aidley DJ. *The physiology of excitable cells*, ed 2, Cambridge, England, 1978, Cambridge University Press.

Hodgkin AL, Rushton WAH. The electrical constants of a crustacean nerve fibre, *Proc R Soc Lond Ser B* 133:444, 1946.

Jack JJB, Noble D, Tsien RW. *Electric current flow in excitable cells*, Oxford, England, 1975, Clarendon.

Katz B. *Nerve, muscle and synapse*, New York, NY, 1966, McGraw-Hill.

Rall W: Core conductor theory and cable properties of neurons. In Kandel ER, editor: *Handbook of physiology: a critical, comprehensive presentation of physiological knowledge and concepts. Sect 1. The nervous system. Vol 1. Cellular biology of neurons*, Part 1, pp 39-97, Bethesda, MD, 1977, American Physiological Society.

GENERATION AND PROPAGATION OF THE ACTION POTENTIAL

OBJECTIVES

1. Describe the properties of the voltage clamp, and explain why it is useful for the study of ion channels.

2. Describe the properties of voltage-gated Na^+ and K^+ channels.

3. Understand the terms "conductance," "ionic current," and "driving force," and use Ohm's Law to calculate these quantities.

4. Define inactivation and describe some functional properties of neurons that result from Na^+ channel inactivation.

5. Explain how the activity of voltage-gated Na^+ and K^+ channels generates the action potential.

6. Explain how local circuit currents produce action potential propagation in nonmyelinated axons.

7. Describe how propagation in myelinated axons differs from that in nonmyelinated axons, and explain why the conduction velocity is much faster as a result of myelination.

THE ACTION POTENTIAL IS A RAPID AND TRANSIENT DEPOLARIZATION OF THE MEMBRANE POTENTIAL IN ELECTRICALLY EXCITABLE CELLS

Action potentials (APs) are observed in "excitable cells" (neurons, muscle cells, and some endocrine cells). An AP is caused by a sudden selective alteration in the permeability of the membrane to small ions. In neurons or skeletal muscle cells the membrane rapidly increases its permeability to Na^+ ions, thereby allowing Na^+ to flow into the cell down its electrochemical gradient and making the inside potential more positive. The Na^+ permeability then decreases and the K^+ permeability rises. This allows K^+ to flow out of the cell and drive V_m back toward its resting level. The membrane permeability to Na^+ and K^+ ions is controlled, at the molecular level, by voltage-gated Na^+ and K^+ channels, respectively.

Properties of Action Potentials Can Be Studied with Intracellular Microelectrodes

Many properties of the AP in a nerve axon can be illustrated by use of the experimental arrangement shown in Figure 7-1A. One intracellular electrode is used to pass a current pulse across the membrane. A second electrode (the "recording electrode") is used to monitor the resulting changes in V_m. When a hyperpolarizing, or small depolarizing, current step is passed across the membrane, V_m exponentially approaches a new steady-state level (see Chapter 6). If a depolarizing stimulus exceeds a critical level (termed **threshold**), an AP is generated (Figure 7-1B). During an axonal AP, V_m depolarizes to a value near E_{Na} in approximately 1 msec (1/1000 of a second). The V_m then returns to the resting value in the next 1 to 2 msec. Further increases in stimulus intensity beyond the threshold level have no additional effect on the AP. If an AP is generated, its time course and amplitude

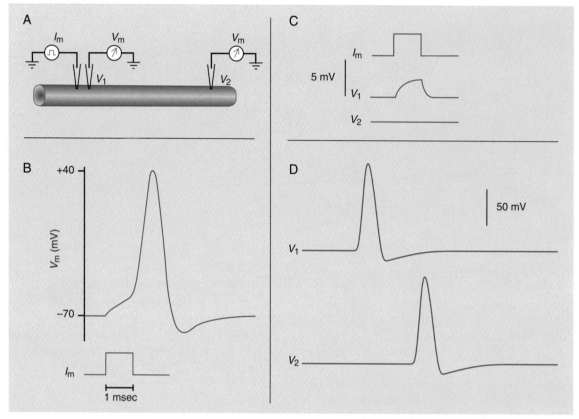

FIGURE 7-1 ■ Properties of the AP in axons. **A,** One voltage-recording electrode (V_1) is placed close to the current passing electrode (I_m), and a second voltage electrode (V_2) is placed at a distance that is at least three to four times the length constant. **B,** The time course of the AP in a squid giant axon. **C,** A subthreshold voltage change recorded at V_1 is not seen at V_2. **D,** An AP initiated at V_1 is transmitted at full amplitude to V_2.

are independent of the stimulus intensity; therefore the response is said to be **"all-or-none."** Because of the sharp, pointy appearance of an AP, it is often referred to as a "spike."

To examine the characteristics of AP propagation along the nerve axon, a second recording electrode can be placed in the axon at a position that is 3 to 4 length constants away from the stimulating electrode (Figure 7-1A). Subthreshold V_m changes are not observed at the second recording electrode (Figure 7-1C) because of the electrotonic decay of these signals caused by the passive properties of the axon (see Chapter 6). In contrast, the AP is transmitted *at full amplitude* to the second recording electrode (Figure 7-1D). Thus the AP is propagated along the

axon without decrement in size despite the passive properties of the axon.

If a pair of just-threshold stimuli is given with a long enough interval between them, both produce APs (Figure 7-2A). If the interval between stimuli is short enough, however, the second stimulus fails to evoke an AP. The nerve is said to be *refractory*. The interval of time following an AP during which a second stimulus, regardless of its amplitude, is unable to evoke a response is called the **absolute refractory period** (Figure 7-2B). The **relative refractory period** is the interval of time following an AP during which the second stimulus must be increased in intensity to evoke a second AP (Figure 7-2B).

A

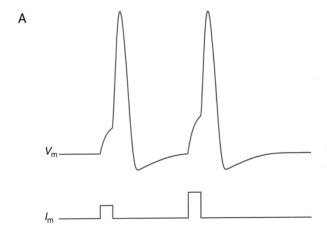

V_m

I_m

B

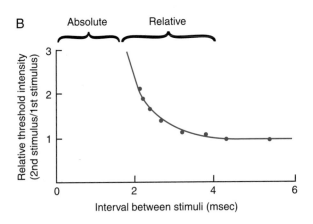

FIGURE 7-2 ■ Absolute and relative refractory periods. **A,** As the interval between two stimuli is decreased, the intensity of the second stimulus must be increased to generate an AP. **B,** The relative threshold intensity (defined as the threshold intensity of the second stimulus divided by the threshold intensity of the first stimulus) graphed as a function of the interval between stimuli. During the relative refractory period the intensity of the second stimulus must be increased to generate an AP. During the absolute refractory period an increase in stimulus intensity is ineffective: an AP cannot be generated. *(Data from Tasaki I, Takeuchi T: Pflügers Arch 245:764, 1942.)*

ION CHANNEL FUNCTION IS STUDIED WITH A VOLTAGE CLAMP

Ionic Currents Are Measured at a Constant Membrane Potential with a Voltage Clamp

In the late 1940s, Hodgkin and Huxley pioneered the study of ion channels using a technique called the **voltage clamp** to study the ionic basis of the AP in squid giant axons.* With this technique they could measure the **ionic currents** that flow across a membrane at a constant V_m. To appreciate

the advantages of a voltage clamp, consider the following parameters that have complex interdependencies during a propagated AP in an axon: current, voltage, distance, and time. As illustrated in Figure 7-1, V_m changes as a function of distance during propagation. The ionic and capacitive currents that flow during the AP must also change as a function of distance. Furthermore, both the currents and V_m change as a function of time. By eliminating some of these variables while controlling others (see Box 7-1 for details of the method), the voltage clamp simplifies the situation in the following ways. First, distance is eliminated as a variable when the voltage clamp ensures that V_m is the same over the entire membrane surface under study; this condition is called "space-clamp."

---

*Alan Hodgkin and Andrew Huxley were awarded the Nobel Prize in Physiology or Medicine in 1963 for this work.

BOX 7-1

THE VOLTAGE CLAMP IS USED TO MAINTAIN A CONSTANT MEMBRANE POTENTIAL

The voltage clamp, which uses an electronic device that allows control, or "clamping," of V_m at a desired level, is an example of a negative feedback control system. In this type of system, like the thermostat controlling the temperature in your home, a variable (temperature) is measured and compared with a "command level," or set point (the temperature setting of the thermostat). The difference between the measured variable and the set point creates an "error signal." The error signal activates an effector system (heater or air conditioner) that decreases the magnitude of the error (i.e., brings the temperature closer to the set point).

A schematic of the squid giant axon axial wire voltage clamp is shown in Figure B-1. An amplifier measures the potential difference between an intracellular electrode (V_{in}) and an extracellular electrode (V_{out}). The output of this amplifier (V_m) is the controlled variable and is compared with the command potential ($V_{command}$), or set point, by an amplifier called the control amplifier. If V_m is not equal to $V_{command}$, an error signal causes a current to flow through an axial wire that is connected to the output of the control amplifier and inserted longitudinally through the axon. The current then flows out through the membrane to a grounded electrode to complete the circuit. The current passing through the axial wire rapidly and continuously causes V_m to remain equal to $V_{command}$. One way to measure the membrane current (I_m) in the voltage clamp is simply to measure the current flowing out of the control amplifier.

An important benefit of the low-resistance axial wire is that it greatly reduces the internal, axial resistance of the axoplasm. Thus the length constant

$$\lambda = \sqrt{\frac{r_m}{r_i}}$$

is greatly increased. The result is that V_m is constant over the membrane surface under study, or in other words, V_m is "space clamped."

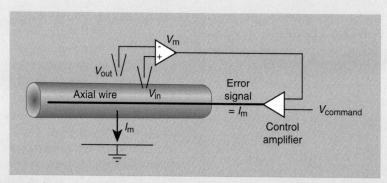

FIGURE B-1 ■ Schematic diagram of an axial wire voltage clamp. A wire is inserted longitudinally down the axon ("axial wire"). The V_m is measured as the difference between the intracellular (V_{in}) and the extracellular (V_{out}) potential. The control amplifier compares V_m to a command potential ($V_{command}$). The output of the control amplifier passes the current (I_m) that is required to hold V_m at $V_{command}$.

Second, the voltage clamp apparatus allows V_m to be held, or clamped, at a constant level. Thus

$$\frac{dV_m}{dt} = 0 \qquad [1]$$

so the capacitive current

$$I_c = C\frac{dV_m}{dt} \qquad [2]$$

is zero. Third, elimination of the capacitive current leaves only the ionic current, which can now be measured as a function of time at a constant V_m. Because the ionic current flows through open ion channels, we can investigate the functional properties of the channels by analyzing the current.

Ionic Currents Are Dependent on Voltage and Time

In a typical voltage clamp experiment V_m is changed in stepwise fashion from a negative "holding" potential (e.g., −70 mV), which is near the cell's resting potential, to some new level. When V_m is stepped to a more negative potential (e.g., −100 mV), the current shown in Figure 7-3A, is recorded. The current consists of an initial very brief spike of inward current followed by a steady (time-independent) inward current. The spike of current is capacitive current, which reflects the addition of negative charges at the inside membrane surface as V_m goes from −70 to −100 mV. The steady inward current is ionic current that flows through ion channels, called leak channels, that are open under resting conditions. The leak current is a linear function of V_m (Figure 7-3B); that is, it obeys Ohm's Law and is said to be *ohmic*. For small depolarizations from −70 mV, the membrane behaves in an analogous fashion; that is, a spike of outward capacitive current

is followed by a steady outward ionic leak current. This is consistent with the result shown in Figure 6-3, which was obtained by injecting current and measuring the change in V_m.

For larger depolarizations, however, the current pattern is strikingly different, as illustrated by the current recorded during a voltage clamp step to 0 mV (Figure 7-4A). Shortly after the spike of outward capacitive current, an inward current develops and reaches a maximum in approximately 1 msec. This inward current then declines in amplitude and is followed by an outward current that reaches a maximum and is maintained through the remainder of the voltage clamp step. This total membrane current recording contains three separable components of current (Box 7-2):

1. A linear component, similar to that shown in Figure 7-3A, contains capacitive current and ionic leak current (Figure 7-4B).
2. A time-dependent inward ionic current is carried by Na$^+$ ions flowing through **voltage-gated Na$^+$ channels** (Figure 7-4C).
3. A time-dependent outward ionic current develops more slowly than the Na$^+$ current and is carried by K$^+$ ions flowing through **voltage-gated K$^+$ channels** (Figure 7-4D).

The total current flowing through multiple channels is commonly referred to as a *macroscopic* current.

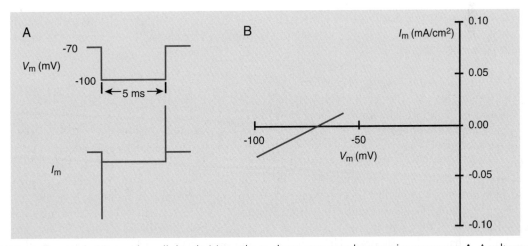

FIGURE 7-3 ■ Hyperpolarizing and small depolarizing voltage clamp steps produce passive responses. **A,** A voltage clamp step from –70 to –100 mV results in an inward current that consists of an initial spike of capacitive current, followed by a steady inward ionic current. When the clamp step ends and V_m rapidly returns to the original level, there is a spike of outward capacitive current. **B,** The magnitude of the ionic current varies as a linear function of V_m. I_m, total membrane current.

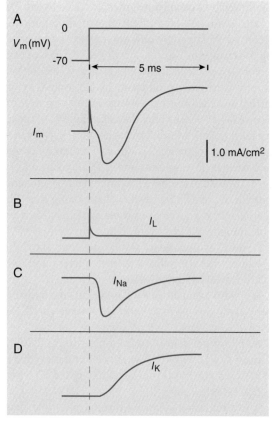

FIGURE 7-4 ■ Current recorded during a voltage clamp step to 0 mV. The total membrane current (I_m) recorded at 0 mV, **A**, contains three separable components (see Box 7-2): **B**, capacitive and ionic leakage currents (I_L), **C**, current flow through voltage-gated Na$^+$ channels (I_{Na}), and **D**, current flow through voltage-gated K$^+$ channels (I_K).

The macroscopic Na$^+$ and K$^+$ currents shown in Figure 7-4 can be described in terms of the gating (channel opening and closing) kinetics and current flow through individual channels. Because the size of the current flowing through a single open channel is constant at a constant V_m (see later), the size of the macroscopic current is proportional to the number of open channels. Thus the macroscopic Na$^+$ current (Figure 7-4C) indicates that, shortly after the depolarization, the Na$^+$ channels rapidly open. This permits Na$^+$ to flow into the cell down its electrochemical gradient (at 0 mV, the net driving force on Na$^+$ is inward and therefore the Na$^+$ current is inward in direction). The Na$^+$ current reaches a maximum in approximately

1 msec and then becomes smaller as the Na$^+$ channels close. This closure of the Na$^+$ channels during maintained depolarization is called **inactivation**. The macroscopic K$^+$ current (Figure 7-4D) indicates that the gates on K$^+$ channels open much more slowly than the Na$^+$ channel gates. Moreover, the K$^+$ channel gates stay open during the remainder of the depolarization. Thus the time-dependent characteristics of the macroscopic currents provide a measure of the kinetics of channel gating. The Na$^+$ current reaches a maximum in approximately 1 msec, a finding indicating that the Na$^+$ channels open rapidly. The K$^+$ current increases more slowly, reaching a maximum in approximately 3 to 4 msec because the K$^+$ channels open more slowly.

Voltage-Gated Channels Exhibit Voltage-Dependent Conductances

Permeability and **conductance** both provide a measure of the ease with which ions cross cell membranes (see Chapter 4). Conductance is the more appropriate measure of ease of ion movement when electrical measurements are used, such as with the voltage clamp. By analogy to the permeability ($P = J/\Delta C$), according to Ohm's Law (see Chapter 6, Equation [1]), conductance is the ratio between the rate of charge movement (current) and the potential difference across the membrane (i.e., $g = I/\Delta V$). The unit of conductance is the siemens: a 1-siemens conductor passes 1 ampere of current per volt of potential difference.

Ohm's Law can be used to calculate conductance from the Na$^+$ and K$^+$ ionic currents (I_{Na} and I_K) measured in a voltage clamp:

$$I_{Na} = g_{Na} (V_m - E_{Na}) \qquad [3]$$

$$I_K = g_K (V_m - E_K) \qquad [4]$$

The V_m is known (it is controlled by the voltage clamp). E_{Na} and E_K can be calculated from the Nernst equation. The time courses of Na$^+$ and K$^+$ conductance (g_{Na} and g_K) at 0 mV, calculated from the currents at 0 mV using Equations [3] and [4], are shown in Figure 7-5. Note that the conductances are always positive. After a brief delay, g_{Na} rapidly rises to a peak and then declines back toward zero, even though the membrane is still depolarized. The rising phase of the conductance is termed **activation**, and the declining phase is termed inactivation. The K$^+$ conductance, g_K, begins to increase (or activate) after a much longer

BOX 7-2

CURRENT COMPONENTS IN AXONS ARE SEPARATED
AND IDENTIFIED WITH IONIC SUBSTITUTIONS

The most straightforward way to separate the ionic currents is to perform ionic substitution experiments, in which permeant ions are replaced by larger, impermeant ions. In the isolated squid giant axon the axoplasm can be removed and the axon interior perfused with a solution of known composition. Thus the ionic composition of both the internal and external solutions is under experimental control. If Na^+ is replaced with the larger, impermeant cation choline, the early inward current carried by Na^+ ions is eliminated (Figure B-1B), and if the remaining current is subtracted from the total current (Figure B-1A), the isolated Na^+ current is obtained (Figure B-1C). If cesium ions (Cs^+) are then substituted for intracellular and extracellular K^+, the outward K^+ current is abolished and only linear capacitive and leakage currents remain (Figure B-1D). By subtracting the linear current (Figure B-1D) from the current shown in Figure B-1B, the isolated K^+ current is obtained (Figure B-1E).

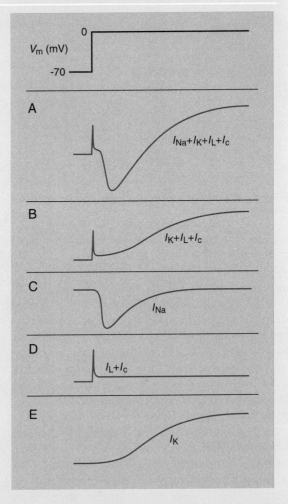

FIGURE B-1 ■ Ionic substitutions can be used to separate current components. **A,** The current recorded during a voltage clamp step to 0 mV contains Na^+ (I_{Na}), K^+ (I_K), leak (I_L), and capacitive (I_c) currents. **B,** After all the Na^+ is replaced with choline$^+$, the Na^+ current is abolished. **C,** The isolated I_{Na} is calculated by subtraction of the current shown in **B** from that in **A**. **D,** In the absence of Na^+, substitution of Cs^+ for K^+ abolishes I_K. **E,** The isolated I_K is calculated by subtraction of the current shown in **D** from that in **B**.

delay and rises more slowly than g_{Na}. After reaching a plateau, g_K remains at that level during the remainder of the depolarization (i.e., it does not inactivate).

At the molecular level, open ion channels are responsible for the conductance of the membrane, so the macroscopic conductance, g_{Na} or g_K, is proportional to the number of open channels:

$$g_{Na} = N_o \, \gamma_{Na} \qquad [5]$$

where N_o is the number of open Na^+ channels and γ_{Na} is the conductance of a single Na^+ channel. If all Na^+

channels have the same average **open probability** and they behave independently of one another, it follows that

$$g_{Na} = N_T \, p_o \, \gamma_{Na} \qquad [6]$$

where N_T is the total number of Na^+ channels and p_o is the probability that a channel is open. Because N_T and γ_{Na} are constants, it is clear that g_{Na} varies as a function of time (and voltage, as we will see later) because of variations in the probability that a Na^+ channel is open. The traces at the bottom of Figure 7-5 show how the probability of channel opening varies as a function

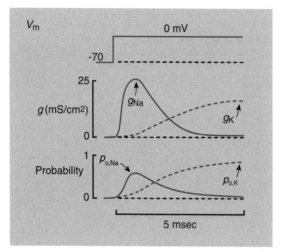

FIGURE 7-5 ■ The time course of g_{Na}, g_K, $p_{o,Na}$, and $p_{o,K}$ during a voltage clamp step to 0 mV. g_{Na} and g_K are the membrane conductances to Na^+ and K^+, respectively. $p_{o,Na}$ is the Na^+ channel open probability, and $p_{o,K}$ is the K^+ channel open probability.

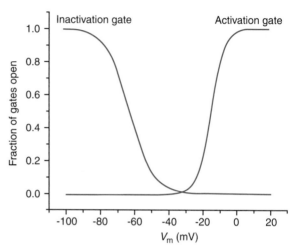

FIGURE 7-6 ■ Graph of the fraction of Na^+ channel activation and inactivation gates that are open *in the steady state,* as a function of V_m, in a squid axon. As the membrane depolarizes (i.e., V_m moves in the positive direction), the Na^+ channel activation gates open and the inactivation gates close. The fraction of open gates changes along a sigmoidal curve as V_m moves in the positive direction. *(Data from Hodgkin AL, Huxley AF: J Physiol [Lond]116:473, 1952.)*

of time during a voltage clamp step to 0 mV. The fact that Na^+ channels activate and then inactivate during maintained depolarization suggests that *two* gates are controlling the open and closed states of the channel: an activation gate and an inactivation gate (see later).

For voltage-gated ion channels the probability that a gate is in the open configuration depends on V_m. For Na^+ channels the fraction of open activation gates (i.e., p_o for the activation gate) in the steady state begins to increase at approximately -40 mV, then increases rapidly with small changes in voltage, and reaches a maximum at approximately 0 mV (Figure 7-6). This is one of the most important, fundamental properties of the *voltage-gated* Na^+ (and K^+) channels found in nerve axons and other excitable cells: the probability that a channel's activation gate will open increases as V_m is made more positive. The Na^+ channel inactivation gate has the opposite voltage dependence: as V_m is made more positive, p_o for the inactivation gate decreases (Figure 7-6). In other words, Na^+ channel inactivation gates close on depolarization.

How does a voltage-gated channel work? The answer to this question is now being clarified. The S4 region in voltage-gated channels contains a series of positive amino acids (see Chapter 5) that acts as the voltage sensor. Movement of S4, caused by depolarization,

induces a conformational change in the channel protein that causes it to go from the closed to the open conformation. Movement of the voltage sensor (S4) itself produces a current because the charged amino acid residues are moving in response to the electric field (the voltage gradient) across the membrane. This *gating current* has in fact been measured (Box 7-3).

INDIVIDUAL ION CHANNELS HAVE TWO CONDUCTANCE LEVELS

The current that flows through an individual ion channel is very small ($\sim$1 to 5 pA) and cannot be resolved with a classical voltage clamp. These small *single-channel currents* can be measured, however, by use of a **patch clamp** to isolate a small patch of membrane electrically (Figure 7-7A).* Patch clamp

---

*Erwin Neher and Bert Sakmann developed the patch clamp to measure ionic currents through single channels. They were awarded the Nobel Prize in Physiology or Medicine in 1991 for this achievement.

BOX 7-3

GATING CURRENTS DIRECTLY REFLECT MOVEMENT OF THE VOLTAGE SENSOR

Hodgkin and Huxley correctly predicted the existence of a voltage sensor in voltage-gated channels. They reasoned that the voltage sensor would move passively in response to a change in V_m. This movement of the voltage sensor would change the conformation of the channel protein and would cause the channel gate to open or close. Movement of the gating charge generates a gating current. So that the relatively small-amplitude gating current can be measured, ionic Na^+ and K^+ currents must be blocked and the spike of capacitive current that charges the membrane must be subtracted. The gating current has been measured as a brief outward current that flows before the opening of the Na^+ channels (Figure B-1).

FIGURE B-1 ■ Na^+ channel gating current is a direct measure of the movement of the voltage sensor. Following a depolarization, the gating current (I_g) is a brief outward current that flows before the opening of the Na^+ channels (note the time scale compared with Figure 7-4). The time course of Na^+ channel opening is illustrated by I_{Na}. *(Data from Armstrong CM, Bezanilla F:* Nature 242:459, 1973.)

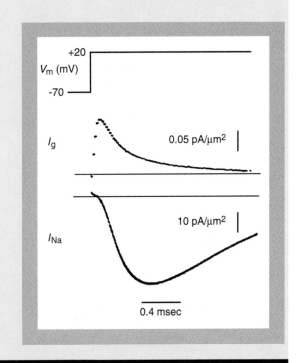

recordings reveal that a single channel has two conductance levels: zero when the channel is closed and a constant conductance, γ, when the channel is open. Thus individual ion channels gate in an all-or-none manner and pass a "pulse" of current when they are open (Figure 7-7D).

Na+ CHANNELS INACTIVATE DURING MAINTAINED DEPOLARIZATION

The time course of the Na^+ conductance change (see Figure 7-5) shows that membrane voltage has a dual effect on Na^+ channels. First, voltage causes the gates on some of the Na^+ channels to open rapidly, thus giving rise to the activation phase of g_{Na}. Later on during the depolarization, g_{Na} declines back toward zero as if the gates on the channels were closing. This dual behavior occurs because Na^+ channels have two gates.

These gates, the activation gate and the inactivation gate, together control the open and closed states of the channel. The diagrams in Figure 7-8 illustrate the configuration of the gates in the closed, open, and inactivated states of the channel. At negative voltages the activation gate is closed and the inactivation gate is open. When the membrane is depolarized, the activation gate in most Na^+ channels opens before the inactivation gate closes, thus allowing the channel to open and conduct Na^+ ions. In time the inactivation gate will close, thereby closing the channel through inactivation. On repolarization, the activation gate closes, a process called **deactivation**, and the inactivation gate reopens.

Na^+ channel inactivation is physiologically important for at least three reasons:
1. The Na^+ conductance automatically begins to turn off at the peak of the AP. This allows more rapid repolarization and potentially more rapid repetitive firing of APs.

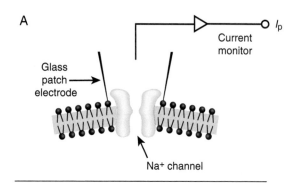

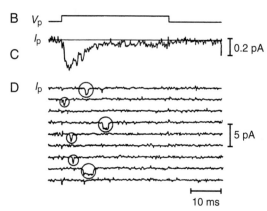

FIGURE 7-7 ■ Measurement of currents through single Na⁺ channels. **A,** A glass "patch electrode" is used to isolate, electrically, a small patch of membrane containing a single Na⁺ channel. The current flowing through the electrode (I_p), and thus through the channel, is recorded by a current monitor. **B,** Repetitive identical depolarizing voltage clamp steps (V_p) are applied to the membrane. **C,** The current summed from a large number of traces, such as those in **D. D,** Each trace is the current recorded from the patch of membrane during a single voltage clamp step. Circles mark the episodes of single channel opening. *(**B, C,** and **D** from Sigworth FJ, Neher E: Nature 287:447, 1980.)*

2. A slowly depolarizing stimulus can be ineffective in evoking an AP because the membrane can *accommodate* to a slowly rising stimulus. Accommodation occurs when the rate of rise of the stimulus is sufficiently slow, so that many Na⁺ channels inactivate before enough of them open to produce an AP.

3. The time course of recovery from inactivation is a determinant of the absolute and relative refractory periods.

A change in the inactivation gating of the skeletal muscle Na⁺ channel is responsible for the temporary weakness or paralysis experienced by patients with *hyperkalemic periodic paralysis* (Box 7-4).

ACTION POTENTIALS ARE GENERATED BY VOLTAGE-GATED Na⁺ AND K⁺ CHANNELS

The Equivalent Circuit of a Patch of Membrane Can Be Used to Describe Action Potential Generation

To explain how the activity of Na⁺ and K⁺ channels can produce an AP, we can use the equivalent electrical circuit for a patch of axonal membrane (Figure 7-9). The equivalent circuit includes a variable resistor representing voltage-gated Na⁺ channels (labeled g_{Na}), a variable resistor representing K⁺ channels (g_K), a resistor representing other, non–voltage-gated leak channels (g_L), and a capacitor for the membrane capacitance (C_m). At the resting potential the capacitor is charged to -70 mV, and in this steady state the sum of all ionic currents is zero. To change V_m, charge must flow onto, or off of, the capacitor.

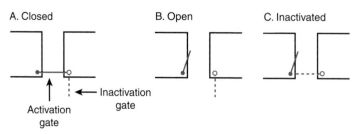

FIGURE 7-8 ■ Voltage-gated Na⁺ channels behave as if they have two gates: an activation gate and an inactivation gate. **A,** At negative membrane potentials the channel is closed, with the activation gate closed and the inactivation gate open. **B,** On depolarization the activation gate opens and the channel is open. **C,** Later during depolarization the inactivation gate closes and the channel is inactivated.

BOX 7-4

HYPERKALEMIC PERIODIC PARALYSIS RESULTS FROM GENETICALLY DEFECTIVE SKELETAL MUSCLE Na⁺ CHANNELS

Hyperkalemic periodic paralysis (HPP) is a relatively rare genetic disease that is caused by a defect in the voltage-dependent Na^+ channel isoform that is expressed in skeletal muscle. Neuronal Na^+ channels are not affected. The disease has autosomal dominant inheritance and is characterized by episodes of skeletal muscle weakness or paralysis. These episodes are preceded by a normally occurring increase in the concentration of extracellular K^+ ions, thus the name HPP. Electrophysiological experiments have shown that the resting potential of skeletal muscle fibers from HPP patients is *abnormally* depolarized in the presence of elevated extracellular K^+ (Figure B-1). Maintained depolarization of skeletal muscle causes it to become inexcitable and unable to contract. The result is weakness or paralysis.

The molecular mechanisms underlying HPP are now understood. Patients with HPP have a single point mutation in the gene coding for the skeletal muscle Na^+ channel. The mutation leads to a single amino acid substitution in the Na^+ channel. An important change

in function results: the defective Na^+ channels do not inactivate completely (Figure B-2). The noninactivating Na^+ channels give rise to the abnormal depolarization of HPP-affected skeletal muscle fibers.

The following chain of events occurs during a period of paralysis in patients with HPP: Some normal event, such as exercise, leads to an increase in extracellular K^+. This elevated extracellular K^+ level causes membrane depolarization, which activates some voltage-gated Na^+ channels. In normal individuals the Na^+ channels then rapidly inactivate, but in patients with HPP they do not. An inward current flowing through the noninactivating Na^+ channels produces an additional depolarization. This inactivates the muscle contraction apparatus, and temporary paralysis occurs.

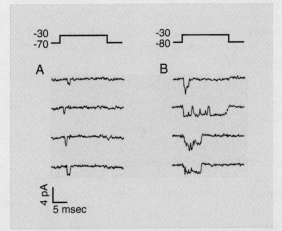

FIGURE B-2 ■ Currents through single Na^+ channels recorded from normal muscle **(A)** and skeletal muscle affected by HPP **(B)** during a depolarizing voltage clamp step (shown at the top). The Na^+ channels from normal muscle open briefly at the beginning of the depolarization and then inactivate. The Na^+ channels from HPP-affected muscle can continue to open and close throughout the depolarization; thus they do not inactivate normally. *(Data from Cannon SC, Brown RH Jr, Corey DP: Neuron 6:619, 1991.)*

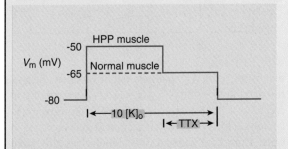

FIGURE B-1 ■ Diagram of resting V_m changes caused by an increase in $[K^+]_o$ in normal muscles and in muscle from a person with HPP. When $[K^+]_o$ is increased from 5 to 10 mM, HPP-affected muscle depolarizes more than normal muscle. Tetrodotoxin (TTX) abolishes the additional depolarization observed in HPP-affected muscle. This demonstrates that the depolarization is the result of I_{Na} through voltage-gated (TTX-sensitive) Na^+ channels.

FIGURE 7-9 ■ **A,** The equivalent circuit of a patch of axon membrane illustrates the mechanism of AP generation. A constant current source (I_m) is connected to the membrane through a switch. The resistors g_{Na} and g_K are the voltage-gated conductances of the membrane to Na^+ and K^+, respectively, and g_L is the "leakage" conductance. E_{Na} and E_K are the equilibrium potentials for Na^+ and K^+, respectively, and E_L is the reversal potential for the leakage channels. C_m, membrane capacitance. **B,** A stimulus from an external current source (or an adjacent patch of membrane) supplies outward current that depolarizes the membrane toward threshold. **C,** The depolarization opens Na^+ channels, and the resulting inward Na^+ current produces outward capacitive current that further depolarizes the membrane and causes the upstroke of the action potential. **D,** Later, the Na^+ channels inactivate and K^+ channels open. Outward current through the K^+ channels generates inward capacitive current that repolarizes the membrane.

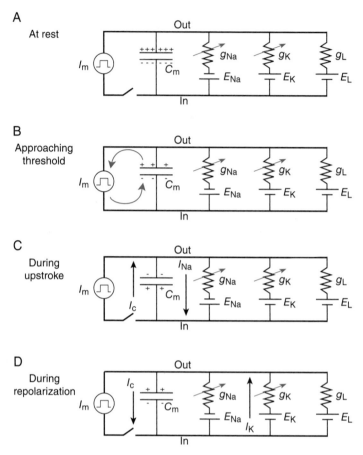

To initiate an AP, an external current source connected to the patch of membrane depolarizes the membrane just beyond threshold by adding positive charges to the inside surface of the membrane (Figure 7-9B). This depolarization causes some Na^+ channels to open quickly. Thus the membrane conductance to Na^+ ions increases and Na^+ flows into the cell down its electrochemical gradient. This is an inward I_{Na}. Current must flow in a loop, and the return pathway for the inward ionic current is through the membrane capacitor. The outward capacitive current is depolarizing, because positive charge flows onto the inside surface of the capacitor. Thus the entry of positively charged Na^+ ions depolarizes the membrane further (Figure 7-9C). This opens more Na^+ channels and permits more Na^+ to flow inward. This phase of inward Na^+ current accounts for the upstroke of the AP. After approximately 1 msec the Na^+ conductance begins to decrease (as Na^+ channels

inactivate) and K^+ channels begin to open. Because the driving force on K^+ ($V_m - E_K$) is outward, the increase in K^+ conductance gives rise to an outward K^+ current. This outward ionic current produces an inward capacitive current that repolarizes the membrane over the next 1 to 2 msec, thereby accounting for the falling phase of the AP and the undershoot (Figure 7-9D).

The Action Potential Is a Cyclical Process of Channel Opening and Closing

The cycle of events involved in the generation of an AP is shown in Figure 7-10. The rapid rising phase of the spike is the result of positive feedback; that is, membrane depolarization opens Na^+ channels (increases g_{Na}), which increases the Na^+ current, and this further depolarizes the membrane, thus opening more Na^+ channels, and so on. Once initiated, this explosive response cannot be stopped; that is, it is all-or-none.

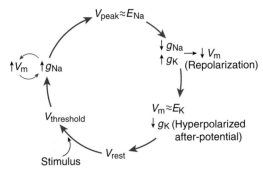

FIGURE 7-10 ■ The cycle of events in the generation of an AP in a patch of membrane. Starting at the bottom and going clockwise, a stimulus depolarizes the membrane from the resting potential (V_{rest}) toward threshold ($V_{threshold}$). If threshold is exceeded, g_{Na} increases rapidly and an inward Na^+ current develops that depolarizes V_m further. This reflects positive feedback of V_m on g_{Na}: an increase in g_{Na} causes depolarization, which further increases g_{Na}, resulting in the very rapid upstroke of the AP to its maximum value (V_{peak}). Next, Na^+ channels begin to inactivate and voltage-gated K^+ channels open. An outward K^+ current develops, causing repolarization. The V_m becomes more negative than V_{rest} (closer to E_K) because the voltage-gated K^+ channels take time to close completely. As the K^+ channels close, V_m returns to V_{rest}.

The depolarization stops when V_m approaches E_{Na} and the Na^+ current decreases for two reasons: (1) the driving force on Na^+ becomes small ($V_m \approx E_{Na}$), and (2) Na^+ channels start to inactivate. Repolarization occurs when the developing outward K^+ current exceeds the declining inward Na^+ current. The V_m approaches E_K and becomes more negative than the initial resting potential because the relative potassium permeability (P_K) is higher than at rest (see the GHK equation in Chapter 4). The P_K is higher than at rest because the voltage-gated K^+ channels take time to close. As these channels close, P_K returns to its resting level and V_m returns to the initial resting potential. If the stimulus is maintained, the cycle can be repeated so that APs can be generated repetitively.

Both Na^+ Channel Inactivation and Open Voltage-Gated K^+ Channels Contribute to the Refractory Period

During the later phases of the AP, both Na^+ channel inactivation and the activation of K^+ channels contribute to the production of the refractory period. As a

result of inactivation a fraction of the Na^+ channels are unavailable for opening in response to a depolarizing stimulus. Moreover, K^+ channels that are still open tend to "pull" V_m toward E_K. Therefore the current required to reach threshold is larger than in a resting nerve.

Pharmacological Agents That Block Na^+ or K^+ Channels, or Interfere with Na^+ Channel Inactivation, Alter the Shape of the Action Potential

Thousands of chemical agents can alter current flow through ion channels either by changing the single channel conductance or by altering the gating of the channel. Agents that block voltage-gated Na^+ or K^+ channels or that impede Na^+ channel inactivation change the shape of the AP in predictable ways. Blocking a fraction of the Na^+ channels with a low concentration of *tetrodotoxin* or *saxitoxin* decreases the magnitude of the Na^+ current (Box 7-5). This results in an AP that has a higher threshold, a slower rate of rise, and a lower peak amplitude (Figure 7-11A). A similar effect is produced by *local anesthetics* (Figure 7-11B), which also block Na^+ channels. The rate of AP repolarization is decreased by agents that interfere with Na^+ channel

---

BOX 7-5

TETRODOTOXIN AND SAXITOXIN ARE EXAMPLES OF NUMEROUS TOXINS OF ANIMAL ORIGIN THAT ALTER ION CHANNEL GATING OR CONDUCTION

Tetrodotoxin (TTX) and saxitoxin (STX) are chemically different small molecules (MW 319 and 299, respectively) that are highly selective blockers of voltage-gated Na^+ channels. Only nanomolar (10^{-9} M) concentrations are required to block the channels. TTX is a paralytic poison found in the ovaries, liver, and skin of puffer fish. The puffer fish ("fugu") is a delicacy in Japan, but occasional fatalities result from eating this fish. STX is synthesized by marine dinoflagellates (algae of the genus *Gonyaulax*). Population explosions, or "blooms," of these dinoflagellates can be recognized by their reddish color (the "red tide"). Filter-feeding shellfish can accumulate toxin to such a level that eating a single shellfish can be fatal.

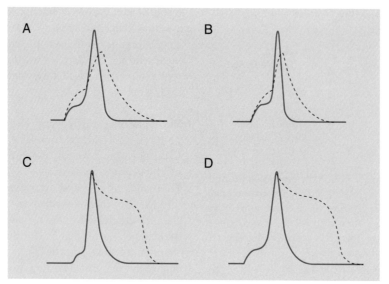

FIGURE 7-11 ■ Pharmacological agents that block Na⁺ or K⁺ channels, or interfere with Na⁺ channel inactivation, change the shape of the AP. In each panel the *solid trace* is the control and the *dashed trace* is the AP in the presence of the chemical agent. **A,** Partial block of Na⁺ channels with TTX increases the threshold for generating an AP and decreases the amplitude and the rate of rise of the AP. **B,** Partial block of the Na⁺ channels with a local anesthetic produces an effect similar to that observed with TTX. **C,** A peptide toxin isolated from sea anemones reduces the rate of Na⁺ channel inactivation and prolongs the duration of the AP. **D,** Tetraethylammonium ions block voltage-gated K⁺ channels and prolong the duration of the AP. *(**A** and **B,** Data from Narahashi T, Deguchi T, Urakawa N, Ohkubo Y:* Am J Physiol *198:934, 1959.* **C,** *Data from Rathmayer W:* Adv Cytopharmacol *3:335, 1979.* **D,** *Data from Armstrong CM, Binstock L:* J Gen Physiol *48:859, 1965.)*

inactivation, such as sea anemone toxin (Figure 7-11C), or that block voltage-gated K⁺ channels, such as tetra-ethylammonium ions (Figure 7-11D).

ACTION POTENTIAL PROPAGATION OCCURS AS A RESULT OF LOCAL CIRCUIT CURRENTS

In Nonmyelinated Axons an Action Potential Propagates as a Continuous Wave of Excitation Away from the Initiation Site

Propagation occurs because an active patch of membrane that is undergoing the upstroke of the AP can act as a source of stimulus current for the resting membrane that lies just ahead of the advancing electrical impulse. This **local circuit current** flow from the active patch supplies an outward capacitive current across the resting membrane to depolarize it toward

threshold. The process continuously and smoothly repeats itself during propagation.

Figure 7-12 shows V_m as a function of distance along a nonmyelinated axon at a specific point in time during the propagation of an AP from right to left. The local circuit currents are shown with arrows. Ahead of and behind the advancing AP, V_m is near the negative resting level. At the active patch of membrane, where the upstroke of the AP is occurring, the current is dominated by inward I_{Na}. This inward ionic current further depolarizes the membrane in the vicinity of the open Na⁺ channels, and it is *also* a current source that depolarizes the resting membrane lying ahead of the AP (this is the electrotonic spread of current described in Chapter 6; see Figure 6-7). Initially, the outward current ahead of the spike is mostly capacitive current that depolarizes the membrane toward threshold; in response, Na⁺ channels then open and generate the AP upstroke in this region of the axon. Behind the active

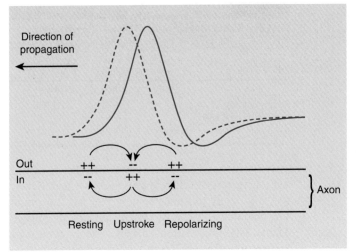

Direction of propagation

Out

In

Resting   Upstroke   Repolarizing

} Axon

FIGURE 7-12 ■ The AP propagates as a result of local circuit current flow. The *traces at the top* of the figure show V_m as a function of distance along the axon at a specific point in time during the propagation of the AP. The *dashed trace* is at a slightly later time during propagation. The *arrows* in the diagram of the axon at the *bottom* illustrate the flow of local circuit currents.

region of membrane the outward current is mostly ionic, carried by K^+, which repolarizes the membrane. A few milliseconds later, the entire V_m profile is shifted along the axon in the direction of propagation (dashed line in Figure 7-12). The AP propagates smoothly in this fashion, with an active region of membrane providing the current to stimulate the adjacent region to the threshold for spike initiation.

Under physiological conditions propagation is *unidirectional* for two reasons. First, behind the advancing wave of depolarization, the membrane is refractory because of Na^+ channel inactivation. Second, local circuit current, supplied from the active patch to membrane behind the active region, is primarily an outward ionic K^+ current that repolarizes the membrane.

Conduction Velocity Is Influenced by the Time Constant, by the Length Constant, and by Na^+ Current Amplitude and Kinetics

The **conduction velocity** of the AP can be affected by several parameters. In principle, conduction velocity can be increased either by increasing the length constant (λ) or by decreasing the time constant (τ). In nonmyelinated axons the length constant

$$\lambda = \sqrt{\frac{r_m}{r_i}} \qquad [7]$$

increases as axon diameter (d) increases because $r_m \propto 1/d$, and $r_i \propto 1/d^2$ (Box 7-6). Therefore λ is proportional to $\sqrt{d}$. Then, because λ determines the electrotonic spread of the local circuit current that initiates the AP, conduction velocity in *non*myelinated axons is proportional to $\sqrt{d}$. Increasing the λ increases the conduction velocity because an active area of membrane will be able to depolarize more distant areas of the membrane to threshold.

The membrane time constant ($\tau_m = r_m \times c_m$) is not an important determinant of variations in AP conduction velocity. During AP propagation along an axon, the stimulating current must flow through the axoplasmic resistance, r_i, before flowing out across the membrane capacitance, c_m. Therefore conduction velocity varies inversely with a different time constant, the **propagation time constant** ($\tau_p = R_i \times C_m$ where R_i is the axoplasmic resistance in ohms and C_m is the membrane capacitance in farads). A smaller τ_p results in faster conduction because V_m changes more rapidly. This factor contributes to the increase in conduction speed in myelinated axons (see next section).

Anything that changes the kinetics or magnitude of the voltage-dependent inward current will affect the conduction velocity. For example, reduction in the magnitude of the Na^+ conductance by local anesthetics, or reduction in the rate of activation of the Na^+ conductance by a decrease in temperature, will reduce the conduction velocity.

BOX 7-6
MEMBRANE RESISTANCE(r_m), AXOPLASMIC RESISTANCE(r_i), AND LENGTH CONSTANT(λ) ARE FUNCTIONS OF AXON DIAMETER

In Chapter 6, r_m was defined as R_m/circumference (see Box 6-3). Because the circumference of a cylinder of diameter d is $\pi \times d$,

$$r_m = \frac{R_m}{\pi d}$$

(i.e., $r_m \propto 1/d$). We also noted that $r_i = \rho$/cross-sectional-area. The cross-sectional area of a cylinder (i.e., the area of a circle) is $\pi d^2/4$, so

$$r_i = \frac{4\rho}{\pi d^2}$$

(i.e., $r_i \propto 1/d^2$). Thus the relationship between the length constant and axon diameter is

$$\lambda = \sqrt{\frac{r_m}{r_i}} = \sqrt{\frac{R_m/(\pi d)}{4\rho/(\pi d^2)}} = \sqrt{\frac{R_m \times d}{4\rho}}$$

In other words, the length constant is proportional to the square root of the axon diameter.

Myelination Increases Action Potential Conduction Velocity

The larger nerve fibers in vertebrate nervous systems are specialized for rapid conduction and are myelinated. In peripheral nerves the myelin is formed by Schwann cells, which wrap themselves around the axon several times to form an insulating sheath. The myelin sheath is interrupted periodically by **nodes of Ranvier** where nearly all the Na$^+$ channels are located. Table 7-1 shows representative values for conduction velocities in three different types of nerve fibers. Clearly, myelination enables much more rapid conduction. Moreover, the conduction velocity of *myelinated* fibers is directly proportional to the fiber diameter (in contrast to the dependence on $\sqrt{d}$ for *non*myelinated axons). Therefore increases in the fiber diameter, greater than approximately 1 μm, give proportionately greater increases in conduction velocity in a myelinated than in a nonmyelinated fiber (Figure 7-13). Figure 7-13 also indicates that myelination may be disadvantageous for fibers

TABLE 7-1

Conduction Velocities for Three Different Types of Nerve Fibers

FIBER TYPE	DIAMETER	CONDUCTION VELOCITY
	(μm)	(m/sec)
Aα (myelinated)	15.0	120
Nonmyelinated C fibers	0.5	1
Squid giant axon (nonmyelinated)	500.0	25

with diameters less than approximately 1 μm. Nature seems to know this, too, because very small fibers in the mammalian nervous system are not myelinated (Table 7-1). Myelination enables vertebrates to pack large numbers of *rapid* communication lines into relatively small nerve tracts, thus increasing the information-processing capability.

The high conduction velocity of myelinated fibers is the result of a very large increase in λ and a decrease in τ_p caused by the wrapping of fatty, insulating myelin around the fiber. This dramatically *increases* r_m in the myelinated (internode) regions because the current has to pass across the myelin sheath as well as the axon

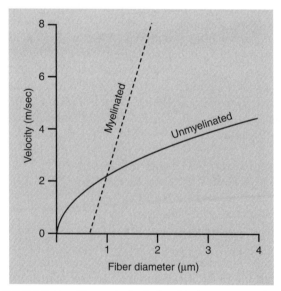

FIGURE 7-13 ■ Theoretical relationship between conduction velocity and axon diameter for both myelinated and unmyelinated axons. *(Data from Rushton WAH: J Physiol [Lond] 115:101, 1951.)*

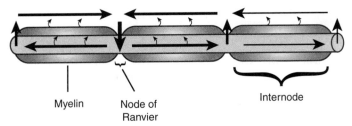

FIGURE 7-14 ■ Propagation of the AP in a myelinated axon. Most of the current that enters the axon as Na^+ current at an active node of Ranvier flows out the axon at adjacent nodes. Very little current flows out of the axon through myelin, in the internodes. As a result, propagation is very fast in the internodal region and the AP appears to skip, or jump, from node to node.

plasma membrane (Figure 7-14). In addition, c_m is *reduced* by myelination: the thicker insulation results in less charge separation per unit membrane area. The lower c_m results in a smaller propagation time constant. Then, because Na^+ channels are confined to the nodes of Ranvier, the AP effectively skips from node to node along the fiber. A spike at one node rapidly depolarizes the adjacent node, 0.2 to 2 mm away, to threshold by electrotonic spread of local circuit currents. This is known as *saltatory* (from the Latin *saltare*, meaning to dance) conduction along the fiber.

The restriction of Na^+ channels to the nodes has advantages and liabilities. Less total influx of Na^+ occurs during impulse propagation along the fiber and thereby reduces metabolic demands on the recovery mechanism (the Na^+/K^+ pump). In severe **demyelinating diseases,** such as *multiple sclerosis* (Box 7-7), conduction may not simply slow, but can fail completely. This is because there are insufficient Na^+ channels in the normally myelinated internode regions to maintain AP propagation when demyelination reduces λ and increases τ_p.

BOX 7-7

ACTION POTENTIAL PROPAGATION IS IMPAIRED IN MULTIPLE SCLEROSIS

Multiple sclerosis is one of a number of common neurological diseases that are characterized by demyelination, or the loss of the myelin sheath. Demyelination causes a range of abnormalities in AP propagation in myelinated axons. In some partially demyelinated axons the conduction velocity is decreased, and different degrees of slowing can occur among axons within the same nerve tract. In more severely demyelinated axons, complete block of AP propagation occurs.

What physiological factors contribute to the conduction abnormalities? In regions of demyelination the high-resistance, low-capacitance myelin sheath is disrupted or lost completely. This reduces conduction velocity for two reasons: the reduction in membrane resistance reduces the length constant, and the increased membrane capacitance increases the propagation time constant. These factors can account for the

slowing of conduction through demyelinated regions. There is, however, an additional factor: the differences in the distribution of voltage-gated ion channels in nodal and internodal regions of axonal membrane. Voltage-gated Na^+ channels are present at very high density ($\sim 10,000/\mu m^2$) in axon membrane at the nodes of Ranvier, in contrast to a very low density (0 to $25/\mu m^2$) in the internodal regions. There is an opposite distribution of voltage-gated K^+ channels: they are nearly absent at the nodes and present in relatively high density at the internodes. This ion channel distribution helps to explain the loss of AP propagation through severely demyelinated regions. The Na^+ channels are not present in high enough density in the internodes to support AP propagation following demyelination. In addition, the presence of K^+ channels in the internodes tends to hold V_m close to E_K.

SUMMARY

1. An AP is a rapid and transient depolarization of V_m.
2. During an AP in neurons and other excitable cells, the membrane rapidly increases its permeability to Na^+ ions. This permits Na^+ to flow into the cell and make the inside potential more positive. The Na^+ permeability then falls and the K^+ permeability rises. This allows K^+ to flow out of the cell and return V_m to the resting level. The membrane permeabilities to Na^+ and K^+ are controlled by voltage-gated Na^+ and K^+ channels, respectively.
3. Ion channel properties are studied with a voltage clamp, which allows ionic current to be measured as a function of time at a constant V_m. Because ionic currents flow through open ion channels, many of the functional properties of channels can be investigated by analysis of the ionic current.
4. Depolarizing voltage clamp steps generate three separable components of current in nerve axons:
   a. A voltage-independent component, which contains capacitive current and linear ionic leakage current.
   b. Na^+ current flowing through voltage-gated Na^+ channels.
   c. K^+ current flowing through voltage-gated K^+ channels.
5. Voltage-gated Na^+ channels open early during depolarization and then close by inactivating later during depolarization. Voltage-gated K^+ channels open more slowly than Na^+ channels and stay open during the depolarization.
6. The open probability for voltage-gated Na^+ and K^+ channels is both voltage- and time-dependent.
7. Single ion channels are studied with a patch clamp, which is a variation of the voltage clamp technique. Many single ion channels have two conductance levels: zero when the channel is closed and a constant conductance, γ, when the channel is open.
8. Voltage-gated Na^+ channels have two gates: an activation and an inactivation gate. During depolarization, rapid opening of the activation gate opens the channel. Closure of the more slowly moving inactivation gate then closes the channel by inactivation. Both gates must be open for the channel to conduct Na^+ ions. Closure of either gate closes the channel.

9. AP propagation along a nonmyelinated axon occurs by local circuit current flow. Conduction velocity is influenced by the length constant, λ, the propagation time constant, τ_p, and the Na^+ current, I_{Na}.
10. In myelinated axons, Na^+ channels are located at the nodes of Ranvier. Propagation occurs by local circuit current flow, but the conduction velocity is much faster than in nonmyelinated axons for two reasons. The internodal region has a large length constant because of the large membrane resistance and has a short propagation time constant because of the low membrane capacitance.

KEY WORDS AND CONCEPTS

- Action potential (AP)
- Threshold
- All-or-none
- Absolute and relative refractory periods
- Voltage clamp
- Ionic currents
- Voltage-gated Na^+ channels
- Voltage-gated K^+ channels
- Inactivation
- Conductance
- Activation and deactivation
- Channel open probability
- Patch clamp
- Propagation
- Local circuit current
- Conduction velocity
- Propagation time constant (τ_p)
- Nodes of Ranvier
- Demyelinating diseases (e.g., multiple sclerosis)

STUDY PROBLEMS

1. Voltage-gated Na^+ channels in a nerve axon are being studied by recording macroscopic Na^+ currents with a voltage clamp. Assume that all the Na^+ channels in the membrane are opened with

a voltage clamp step to $+20$ mV, that the peak magnitude of the Na^+ current is -2 mA/cm^2 during this step to $+20$ mV, and that $E_{Na} = +60$ mV.

a. Estimate the maximum Na^+ conductance ($g_{Na,max}$) of this membrane.

b. Describe $g_{Na,max}$ in terms of single Na^+ channels.

c. During a voltage clamp step to -20 mV we also measure a Na^+ current with a peak magnitude of -2 mA/cm^2. Estimate the probability that a Na^+ channel is open at -20 mV.

d. What voltage-dependent property of Na^+ channels is illustrated by the calculations in a to c?

2. a. An AP is initiated in an axon by passing outward current across the membrane using an external current source. Describe the sequence of events that occurs at the site of initiation of the AP. Include the role of voltage-gated Na^+ channels, voltage-gated K^+ channels, I_{Na}, I_K, I_C, and inactivation.

b. The total membrane current (I_m) flowing across the axonal membrane is monitored at a point on the axon several centimeters away from the site of initiation as the AP propagates past that point. The current consists of an initial phase of outward current, followed by an inward current and finally another phase of outward current. Explain the origin of these currents.

c. Does the AP propagate in both directions away from the initiation site? Why?

3. List three factors that can alter the conduction velocity of an AP in a nerve axon. Describe the change caused by each factor, and explain how it occurs.

BIBLIOGRAPHY

Armstrong CM. Voltage-dependent ion channels and their gating, *Physiol Rev* 72:S5, 1992.

Armstrong CM, Bezanilla F. Charge movement associated with the opening and closing of the activation gates of the Na channels, *J Gen Physiol* 63:533, 1974.

Armstrong CM, Binstock L. Anomalous rectification in the squid giant axon injected with tetraethylammonium chloride, *J Gen Physiol* 48:859, 1965.

Armstrong CM, Hille B. Voltage-gated ion channels and electrical excitability, *Neuron* 20:371, 1998.

Cannon SC, Brown RH Jr, Corey DP. A sodium channel defect in hyperkalemic periodic paralysis: potassium-induced failure of inactivation, *Neuron* 6:619, 1991.

Hille B. *Ionic channels of excitable membranes*, ed 3, Sunderland, MA, 2001, Sinauer.

Hodgkin AL. Chapter 4. In *The conduction of the nervous impulse*, Springfield, IL, 1964, Charles C Thomas.

Hodgkin AL, Huxley AF. A quantitative description of membrane current and its application to conduction and excitation in nerve, *J Physiol (Lond)* 117:500, 1952.

Hodgkin AL, Huxley AF, Katz B. Measurements of current-voltage relations in the membrane of the giant axon of Loligo, *J Physiol (Lond)* 116:424, 1952.

Narahashi T, Deguchi T, Urakawa N, et al. Stabilization and rectification of muscle fiber membrane by tetrodotoxin, *Am J Physiol* 198:934, 1959.

Rathmayer W. Sea anemone toxins: tools in the study of excitable membranes, *Adv Cytopharmacol* 3:335, 1979.

Rushton WAH. A theory of the effects of fibre size in medullated nerve, *J Physiol (Lond)* 115:101, 1951.

Sigworth FJ, Neher E: Single Na^+ channel currents observed in cultured rat muscle cells, *Nature* 287:447, 1980.

8

ION CHANNEL DIVERSITY

OBJECTIVES

1. Compare and contrast the properties of voltage-gated Ca^{2+} and Na^+ channels.

2. Understand the mechanism of action of Ca^{2+} antagonist drugs, and describe their use as therapeutic agents.

3. Understand the properties of TRP channels.

4. Describe the role of A-type K^+ channels and Ca^{2+}-activated K^+ channels in regulating the AP firing pattern in a bursting neuron.

5. Describe the properties of ATP-sensitive K^+ channels, and explain their role in glucose-induced insulin release from pancreatic β-cells.

6. Describe the role of hERG K^+ channels in repolarizing the cardiac AP.

7. Understand how the activation of β-adrenergic receptors modulates Ca^{2+} channel activity and how this modulation enhances force development in the heart.

VARIOUS TYPES OF ION CHANNELS HELP TO REGULATE CELLULAR PROCESSES

Ion channels are found in all cells. The properties of voltage-gated Na^+ and K^+ channels in squid giant axons are described in Chapter 7. These channels are essential for generating the AP in nerve axons and skeletal muscle cells. Many other types of ion channels also play important roles in regulating membrane electrical activity. Table 8-1 lists some examples. Each of the ion channel types is defined by a combination of specific physiological, pharmacological, and structural characteristics. As an example, the characteristics of voltage-gated Na^+ channels in nerve axons include, in part, Na^+ selectivity, voltage-dependent activation, fast inactivation, and high-affinity block by tetrodotoxin. In this chapter we focus on the properties of several other physiologically important channels. These examples illustrate

the diversity of ion channel types, the variety of roles they play in normal cell function, and their pathophysiology. Ion channels that open after the binding of a neurotransmitter or other chemical activator, called ligand-gated ion channels, are discussed in Chapters 12 and 13.

VOLTAGE-GATED Ca^{2+} CHANNELS CONTRIBUTE TO ELECTRICAL ACTIVITY AND MEDIATE Ca^{2+} ENTRY INTO CELLS

Ion channels that selectively allow Ca^{2+} ions to permeate are of vital importance in the normal functioning of nearly all cells. All excitable cells have Ca^{2+} channels. In some cells, **voltage-gated Ca^{2+} channels** can generate APs in the absence of Na^+ channels. However, in many cells that express both voltage-gated Na^+ and Ca^{2+} channels, Ca^{2+} channel activity

TABLE 8-1

Examples of Three Classes of Ion Channels

CHANNEL TYPE	LOCATION	FUNCTIONS
Voltage-Gated Channels		
Na$^+$ channel	Axon, skeletal muscle	Upstroke of the AP
K$^+$ channels	Axon, skeletal muscle	AP repolarization
Ca^{2+} channels (many types exist)	Heart, nerve terminals, endocrine cells	Inject Ca^{2+} into cells
Ligand-Gated Channels		
Nicotinic ACh receptor channel	Neuromuscular junction, neurons	Synaptic transmission
Glutamate receptor channels	Neurons	Synaptic transmission
GABA receptor channel	Neurons	Synaptic transmission
Ca^{2+}-activated K$^+$ channel	Almost all excitable cells	Regulate burst length
Other Channels		
ATP-sensitive K$^+$ channel	Pancreas, heart, smooth muscle	Glucose sensor in β-cells
NC$_{Ca-ATP}$	Astrocytes	Cytotoxic edema
Inward rectifier K$^+$ channel	Heart, brain, skeletal muscle	Permit long depolarizations
Store-operated channels	Nearly all cells	Regulation of [Ca^{2+}]$_i$
Receptor-operated channels	Nearly all cells	Regulation of [Ca^{2+}]$_i$
Two-pore K$^+$ channel	All cells	Regulate resting potential

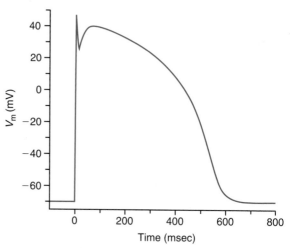

FIGURE 8-1 ■ An AP recorded from a human cardiac muscle cell. The duration of the AP in ventricular muscle cells is hundreds of times longer than the AP in a nerve axon (see Figure 7-1). Ca^{2+} influx through voltage-gated Ca^{2+} channels contributes to the long duration of these APs. (*Data from Piacentino III V, Weber CR, Chen X et al.: Circ Res 92:651, 2003.*)

modifies the shape of the AP. Ca^{2+} channels contribute to the APs generated in the heart (Figure 8-1), some endocrine cells (e.g., pancreatic β-cells), and specific regions of neurons (e.g., dendrites and nerve terminals).

Ca^{2+} channels influence a wide range of cellular activities by allowing Ca^{2+} ions to enter the cell. The extracellular free Ca^{2+} concentration is ~1 mM (10^{-3} M), whereas [Ca^{2+}]$_i$ is normally ~0.0001 mM (10^{-7} M). This means that E_{Ca} is approximately +120 mV. Therefore, when a Ca^{2+} channel opens, Ca^{2+} ions flow into the cell down their electrochemical gradient. Because [Ca^{2+}]$_i$ is normally very low, a relatively small influx of Ca^{2+} can produce a significant increase in [Ca^{2+}]$_i$. [Ca^{2+}]$_i$ ions directly regulate secretion, contraction, energy metabolism, ion channel activity, and numerous other processes.

Several distinct types of Ca^{2+} channels have been identified based on their physiological and pharmacological properties (Box 8-1 and Table 8-2). Indeed, several Ca^{2+} channel mutations have been identified that are associated with hereditary diseases. One example is X-linked congenital stationary night blindness. This is caused by loss-of-function mutations in the L-type **Ca^{2+} channel subtype** that is expressed in photoreceptors (Box 8-2).

Ca^{2+} Currents Can Be Recorded with a Voltage Clamp

Voltage-gated Ca^{2+} channels can be studied with the voltage clamp. Whole-cell currents recorded from a neuronal cell body with a voltage clamp are shown in Figure 8-2. With K$^+$ channels blocked, a depolarizing voltage clamp step evokes an inward current that has both a fast transient component and a more sustained component. The fast transient component of the current is generated by Na$^+$ channels. When external Na$^+$ is removed, the Na$^+$ channels no longer conduct inward current (because there are no Na$^+$ ions to flow into the cell), and the remaining inward current is

BOX 8-1

TYPES OF VOLTAGE-GATED Ca²⁺ CHANNELS

Multiple types of voltage-gated Ca^{2+} channels have been identified on the basis of physiological and pharmacological criteria. These include the following.

L Type

The L type is the major type of Ca^{2+} channel in muscle cells (cardiac, smooth, and skeletal). L-type Ca^{2+} current is characterized by a relatively high voltage for activation, fast deactivation, slow voltage-dependent inactivation, a large single-channel conductance, modulation by cAMP-dependent phosphorylation, and inhibition by dihydropyridines, phenylalkylamines, and benzothiazepines. These Ca^{2+} channels are called L-type channels because of their *large* conductance (25 pS) and *long*-lasting single-channel open times.

T Type

In comparison with L-type Ca^{2+} currents, T-type currents activate at more negative voltages, have fast voltage-dependent inactivation, have slow deactivation, have a small single-channel conductance (8 pS), and are insensitive to the drugs that block L-type channels. They are named T type for their *t*iny single channel conductance and their *t*ransient kinetics.

N Type

N-type Ca^{2+} currents were first identified by their intermediate voltage-dependence and intermediate inactivation kinetics. The channels are insensitive to the L-type Ca^{2+} channel blockers but are blocked selectively by a cone snail peptide toxin, ω-conotoxin GVIA. They were called N type because they were *n*either L type *n*or T type.

P/Q Type

P-type Ca^{2+} currents were first recorded in cerebellar Purkinje neurons and are blocked selectively by the spider toxin, ω-agatoxin IVA. Q-type Ca^{2+} currents were first recorded in cerebellar granule cells and have a lower affinity for ω-agatoxin IVA.

TABLE 8-2

Voltage-Gated Ca²⁺ Channel Subtypes

Ca²⁺ CURRENT TYPE	LOCATION	SPECIFIC BLOCKER	FUNCTION
L	Cardiac muscle, endocrine cells	Dihydropyridines	Excitation-contraction coupling, hormone secretion
P/Q	Nerve terminals	ω-Agatoxin IVA	Neurotransmitter release
N	Nerve terminals	ω-Conotoxin-GVIA	Neurotransmitter release
R	Neurons	SNX482	Neurotransmitter release
T	Heart, neurons	Mibefradil	Pacemaker currents

carried by Ca^{2+} ions flowing into the cell through voltage-gated Ca^{2+} channels (Figure 8-2). The Ca^{2+} channels activate more slowly than Na^+ channels. Furthermore, the magnitude of the Ca^{2+} current (I_{Ca}) decreases very slowly over tens of milliseconds. This indicates that Ca^{2+} channels inactivate slowly (much more slowly than Na^+ channels). Inactivation also is incomplete, so that Ca^{2+} current can continue to flow during maintained depolarization. The maximum amplitude of the Ca^{2+} current ($\sim$100 μA/cm² of membrane area) is usually smaller than the Na^+ or K^+ currents in axons.

The relationship between Ca^{2+} channel open probability and V_m is similar to that for voltage-gated Na^+ channels, except that the channels activate over a more depolarized voltage range. In some cell types, voltage-gated Ca^{2+} channels can generate APs in a manner similar to Na^+ channels. A depolarizing stimulus opens Ca^{2+} channels, and Ca^{2+} ions flow into the cell down their electrochemical gradient. This Ca^{2+} current further depolarizes the cell and opens more channels. The positive feedback of membrane depolarization on channel opening produces the AP upstroke. Compared with axonal APs generated by Na^+ channels, pure Ca^{2+} APs have slower upstrokes and longer durations because the Ca^{2+} channels activate more slowly than Na^+ channels, and inactivation is slow and incomplete.

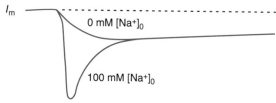

FIGURE 8-2 ■ Ca^{2+} and Na^+ currents recorded from the same neuron during a voltage clamp step (V_m) from –50 to –7.5 mV. With 100 mM $[Na^+]_o$ there are two current components: (1) a fast, transient inward current that is generated by voltage-gated Na^+ channels, and (2) a steady inward current that remains after removal of the external Na^+ (0 mM $[Na^+]_o$). Voltage-gated Ca^{2+} channels generate the second component. (*Data from Kostyuk PG, Kristhal OA, Shakhovalov YA:* J Physiol [Lond] *270:545, 1977.*)

Ca^{2+} Channel Blockers Are Useful Therapeutic Agents

Various drugs reduce Ca^{2+} currents by blocking voltage-gated Ca^{2+} channels. These **Ca^{2+} antagonists** include phenylalkylamines (e.g., verapamil and its methoxy derivative known as D600 or gallopamil), benzothiazepines (e.g., diltiazem), and the dihydropyridine derivatives (e.g., nifedipine, nimodipine, and nicardipine). At physiological pH, D600 and diltiazem are protonated and therefore positively charged, whereas nifedipine is uncharged. These drugs selectively block L-type Ca^{2+} channels (see Box 8-1).

Ca^{2+} channel blockers are widely used as therapeutic agents in the management of coronary artery disease, hypertension, and cardiac arrhythmias (Box 8-3). Some properties of the block produced by Ca^{2+} antagonists are shown in Figure 8-3. The data in Figure 8-3A illustrate the effect of repetitive depolarizations on the development of block by the charged compound D600. Very little reduction in I_{Ca} occurs in response to the first stimulus after D600 is added (Figure 8-3A). Subsequent depolarizations, however, produce smaller and smaller Ca^{2+} currents. This behavior is called **use-dependent block** of the Ca^{2+} channels. In contrast, uncharged drugs, such as nifedipine, block Ca^{2+} channels without requiring depolarization; that is, they produce a significant degree of *tonic* block and very little use-dependent block. Diltiazem is intermediate between D600 and nifedipine in its use dependence. Use-dependent block by charged drugs is explained by the idea that the activation gate in the channel must open before charged drug molecules can gain access to the blocking site (Figure 8-3B).

BOX 8-3

Ca²⁺ ANTAGONIST DRUGS ARE USED AS THERAPEUTIC AGENTS

The force generated by contracting cardiac or vascular smooth muscle cells depends on $[Ca^{2+}]_i$. A major pathway for Ca^{2+} entry into these cells is through voltage-gated Ca^{2+} channels. Thus, block of Ca^{2+} channels with a Ca^{2+} antagonist drug decreases Ca^{2+} entry into the cell and thereby decreases the force of contraction. The relaxation of vascular smooth muscle results in arterial vasodilation. This effect underlies the rationale for using Ca^{2+} channel blockers to treat hypertension (high blood pressure). Chronic hypertension is the result of increased vascular resistance. Ca^{2+} channel blockers reduce blood pressure by relaxing arteriolar smooth muscle and decreasing vascular resistance. Ca^{2+} antagonists are also used in the treatment of angina pectoris (chest pain arising from inadequate blood supply to heart muscle). The benefits may result from coronary artery dilation (and consequently increased blood supply) or from decreased myocardial oxygen consumption that is secondary to decreased cardiac contractility.

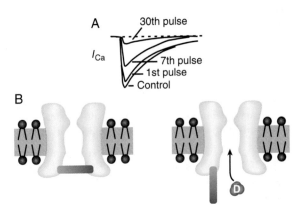

FIGURE 8-3 ■ Ca^{2+} channel block by D600. **A,** The I_{Ca} were recorded during 120-msec voltage clamp steps to +20 mV. The trace-labeled control was recorded before the addition of D600. The remaining traces were obtained in the presence of D600 by use of a series of voltage clamp pulses (to +20 mV) delivered at the rate of three pulses per minute. Little block occurs in the absence of pulses, but blockade becomes progressively larger with repeated depolarizations. This type of channel inhibition is called use-dependent block. **B,** Use-dependent block by D600 will occur if the drug cannot gain access to its binding site when the channel activation gate is closed. When the gate opens, D600 *(D)* can enter the inner vestibule of the channel, bind to its receptor, and block the channel. (**A,** *Data from Lee KS, Tsien RW:* Nature *302:790, 1983.*)

MANY MEMBERS OF THE TRANSIENT RECEPTOR POTENTIAL SUPERFAMILY OF CHANNELS MEDIATE Ca²⁺ ENTRY

The first member of the transient receptor potential (TRP) superfamily of ion channels was the *Drosophila* TRP channel, which mediates a transient electrical response to light in photoreceptors. Six families of mammalian TRP channel proteins have been identified based on sequence homology: TRP canonical (TRPC), TRP vanilloid (TRPV), TRP melastatin (TRPM), TRP polycystin (TRPP), TRP mucolipin (TRPML), and TRP ankyrin (TRPA). Three of these, TRPV1, TRPM8, and TRPA1, are involved in detection of hot and cold and in chemosensation (Box 8-4). All TRP proteins contain six transmembrane domains; these proteins assemble as homotetramers or heterotetramers to form nonselective cation channels. TRP channel subtypes vary widely in their permeability

to monovalent and divalent cations. Many of these channels conduct Ca^{2+} and thus contribute to rises in $[Ca^{2+}]_i$.

Some Members of the TRPC Family are Receptor-Operated Channels

Based on sequence homology, members of the mammalian TRPC family can be divided into four subfamilies: TRPC1, TRPC2, TRPC3/6/7, and TRPC4/5. TRPC channels are all nonselective cation channels with different permeabilities to Ca^{2+} and Na^+.

All channels of the TRPC family are opened by activation of phospholipase C (PLC), which is itself activated by ligand binding to cell surface receptors. Therefore some members of the TRPC family are referred to as **receptor-operated channel** (ROCs). Activated PLC cleaves phospholipids to yield diacylglycerol (DAG) and other signaling molecules (see Chapter 13). DAG appears to be the physiological

SOME TRP CHANNELS ARE INVOLVED IN CHEMOSENSATION AND IN THE DETECTION OF NOXIOUS TEMPERATURE CHANGES

Members of the TRP superfamily of channels are involved in the detection of hot and cold. Interestingly, the same channels respond to chemicals (e.g., capsaicin and menthol) that produce hot or cold sensations.

TRPV1 Channels Are Involved in the Detection of Heat
The TRPV family is made up of four mammalian subfamilies: TRPV1/2, TRPV3, TRPV4, and TRPV5/6. TRPV1 channels were first identified in primary somatosensory neurons in dorsal root and trigeminal ganglia. They are ligand-gated nonselective cation channels. TRPV1 channels contribute to the sensation of heat, either in the form of increased temperature or by the presence of chemicals that produce a hot feeling. The channel is activated by elevated temperatures ($\geq 43°C$) or by capsaicin, which is the pungent essence of hot chili peppers.

TRPM8 Channels Are Activated by Cold or by Cooling Chemicals
There are four subfamilies of TRPM channels: TRPM1/3, TRPM2/8, TRPM4/5, and TRPM6/7. TRPM8 is expressed in very high levels in sensory neurons. TRPM8 channels function as PM Ca^{2+} channels. The channel is activated by low temperature (8°C to 28°C) or by chemicals, such as menthol, that evoke a cooling sensation.

TRPA1 Channels Are Activated by Pungent Chemicals
The TRPA family contains only a single member, TRPA1, which is expressed in somatosensory neurons in dorsal root and trigeminal ganglia. TRPA1 is activated by garlic and by isothiocyanate compounds, which include the pungent ingredients in wasabi, horseradish, and mustard oil.

---

activator of TRPC3/6/7. In smooth muscle cells, for example, DAG activates TRPC3 and TRPC6 channels, which mediate Ca^{2+} entry and depolarization, leading to vasoconstriction (see Chapters 13 and 15). Some TRPC channels also appear to be regulated by the level of filling of $[Ca^{2+}]_i$ stores and thus represent **store-operated channels** (SOCs).

K⁺-SELECTIVE CHANNELS ARE THE MOST DIVERSE TYPE OF CHANNEL

K^+-selective ion channels are found in all cells. They are very diverse in their activity, structure, and distribution. For example, K^+ channels are important in regulating cell volume, insulin secretion, heart rate, neuronal excitability, neuronal firing patterns, and epithelial transport (see Chapter 11).

The first K^+ channel gene was cloned from the *Shaker* mutant of the fruit fly, *Drosophila*, in 1987. Since then, more than 200 genes (including >50 human genes) have been identified that encode various types of K^+ channels (Table 8-3 and Box 8-5). Genetic advances have led to the identification of naturally

occurring K^+ channel mutations that are associated with a variety of diseases.

Neuronal K⁺ Channel Diversity Contributes to the Regulation of Action Potential Firing Patterns

One of the most important ways that information is encoded in the nervous system is in the temporal pattern of APs. The simplest form of encoding involves variations in AP frequency. Indeed, some cells generate complex patterns of APs, such as the *high-frequency bursts* shown in Figure 8-4. In most neurons the cell body is responsible for initiating the AP. The cell body and dendrites integrate a large number of electrical inputs that either depolarize or hyperpolarize the cell; when threshold is reached, an AP is generated. Two populations of ion channels in axons, voltage-gated Na^+ channels and one type of voltage-gated K^+ channel, are responsible for AP propagation (see Chapter 7). Through these channels the axon can faithfully transmit to the nerve terminals an AP initiated at the cell body. In contrast, the cell body often contains many other types of ion channels that contribute significantly to the regulation of AP firing patterns. We now consider two

TABLE 8-3

K$^+$ Channel Types and Associated Heritable Diseases

TYPE	NOMENCLATURE	GENETIC DISORDER
Voltage-gated (Shaker)	Kv1.1-Kv1.7	Episodic ataxia*
Shab, Shaw, Shal	Kv2.1-Kv6.1	
Human *ether-a-go-go*	hERG	Long QT syndrome†
KvLQT1	KCNQ1	Long QT syndrome
Inward rectifier	Kir1.1-Kir7.1	Bartter's syndrome (see Chapter 11)
Large conductance Ca^{2+} activated	Slo, BK$_{Ca}$	
Small conductance Ca^{2+} activated	SK1-SK3	
Two-pore	TWIK1, TREK, TASK, TRAAK	
Sulfonylurea receptor	SUR1	Persistent hyperinsulinemic hypoglycemia of infancy

*Episodic ataxia is an autosomal dominant disease characterized by hyperexcitability in motor neurons. Exercise may provoke attacks of atactic (uncoordinated) walking and jerking movements that can last minutes.
†Long QT syndrome is a group of disorders characterized by a long QT interval in the electrocardiogram that indicates delayed repolarization of the ventricles. It can be associated with lightheadedness or sudden death caused by cardiac arrhythmias.

such channels: rapidly inactivating voltage-gated K$^+$ channels and Ca^{2+}-activated K$^+$ channels.

Rapidly Inactivating Voltage-Gated K$^+$ Channels Cause Delays in Action Potential Generation

Many neurons contain a population of K$^+$ channels that generate *transient* outward K$^+$ currents (Figure 8-5). For historical reasons these currents are called A-currents (I_A); we will therefore refer to the channels that produce these currents as **K$_A$ channels.** In response to a depolarizing voltage clamp step, K$_A$ channels activate relatively rapidly, in a few milliseconds, and then inactivate over tens of milliseconds during the maintained depolarization (Figure 8-5). The overall time course of these transient K$^+$ currents resembles that of voltage-dependent Na$^+$ currents, although the absolute rate of K$_A$ channel gating is approximately 10 times slower than Na$^+$ channel gating.

BOX 8-5

K$^+$ CHANNELS ARE STRUCTURALLY AND FUNCTIONALLY DIVERSE

More than 200 genes have been cloned that encode the pore-forming subunit of various types of K$^+$ channels. Each functional K$^+$ channel is composed of four such subunits. All the channels have a homologous pore-forming region that selectively allows passage of K$^+$ ions. As a result of other structural homologies, the channels can be classified into three families based on the number of hydrophobic transmembrane α-helical segments that are present (see Chapter 5):
1. Channels with six transmembrane α-helical segments. Included in this family are the voltage-gated K$^+$ channels (the Kv channels in Table 8-2), human *ether-a-go-go* (hERG), Ca^{2+}-activated K$^+$ channels, and KCNQ channels. Voltage-gated K$^+$ channels in nerve axons contribute to the repolarization of the action potential (see Chapter 7).
2. Channels with two transmembrane α-helical segments. The inward rectifier K$^+$ channels (Kir1.1-Kir7.1) all have two transmembrane segments with a pore-forming loop between them. The inward rectifiers have a much larger conductance for inward current than for outward current. An example is the K$_{ATP}$ channel in pancreatic β-cells. This channel is composed of four Kir6.2 inward rectifier subunits and four SUR1 sulfonylurea receptors.
3. Channels with four transmembrane α-helical segments and two pore domains. TWIK, TREK, TASK, and TRAAK are genes that form channels containing two pore regions. These channels are believed to play a role in setting the resting V_m.

Steady-state activation and inactivation curves for K$_A$ channels are shown in Figure 8-6. K$_A$ channels inactivate over a relatively negative range of V_m. The resting potential of many neurons can vary from approximately −50 to −75 mV. Over this range of V_m, the change in the number of K$_A$ channels that are inactivated is significant (Figure 8-6). If the cell's resting potential is −50 mV, most K$_A$ channels are inactivated. If the resting potential is more negative, fewer K$_A$ channels are inactivated, so more

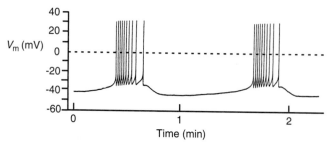

FIGURE 8-4 ■ Spontaneous bursts of APs generated by a neuron. The V_m in this cell alternates between a quiet period near −40 mV and an active period containing a high-frequency burst of approximately 10 APs. During the quiet period the cell slowly depolarizes toward the threshold for initiation of an AP. *(Data from Hille B: Ionic channels of excitable membranes, ed 3, Sunderland, Mass, 1991, Sinauer.)*

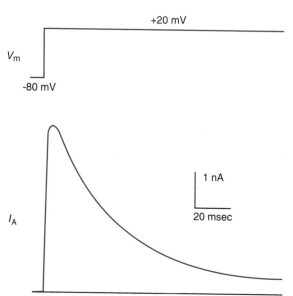

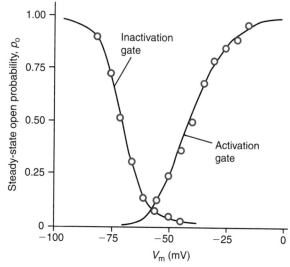

FIGURE 8-5 ■ An outward K$^+$ current through K$_A$ channels (I_A) during a depolarizing voltage clamp step. The channels open, or activate, in a few milliseconds following the depolarization. Over the next 100 to 200 msec, the current decreases in size as the K$_A$ channels inactivate.

FIGURE 8-6 ■ Steady-state activation and inactivation curves for K$_A$ channels. The curves show the probability that the activation or inactivation gate is open in the steady state. K$_A$ channels activate and inactivate over a relatively negative voltage range when compared with the Na$^+$ and K$^+$ channels in nerve axons (see Figure 7-6). *(Data from Hille B: Ionic channels of excitable membranes, ed 3, Sunderland, Mass, 1991, Sinauer.)*

noninactivated channels are available to regulate subsequent electrical activity.

Neurons often fire APs repetitively, and by doing so they can encode information in the spike frequency. K$_A$ channels are found in neuronal cell bodies where they help to regulate the rate of repetitive AP firing.

The outward current through open K$_A$ channels produces a delay in the approach toward AP threshold. This can be demonstrated by pharmacological blocking of K$_A$ channels. Figure 8-7 illustrates the effect of blocking K$_A$ channels on AP generation. The neuron is stimulated by injection of a constant depolarizing

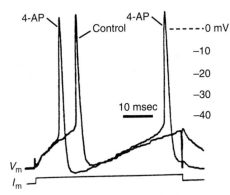

FIGURE 8-7 ■ K$_A$ channels regulate the frequency of APs in some neurons. This figure shows the response of a hippocampal neuron to a 50-msec depolarizing current pulse. The control response shows a single AP generated during the stimulus. In the presence of 5 mM 4-aminopyridine (4-AP), which blocks K$_A$ channels, the threshold is reached earlier, the latency to the first AP is reduced, and a second AP is generated during the stimulus. (*Data from Segal M, Rogawski MA, Barker JL: J Neurosci 4:604, 1984.*)

current. The depolarization causes the cell to fire an AP. After the AP, the maintained stimulation again causes the cell to depolarize. K$_A$ channels now open in response to this depolarization. The increase in outward I_A opposes the inward stimulus current, and the rate of depolarization is slowed. Only a single AP is generated during the 50-msec stimulus under control conditions (Figure 8-7). When the K$_A$ channels are blocked with 4-aminopyridine, the reduced I_A is less effective in opposing the inward stimulus current. As a result, depolarization is faster and a second AP is generated before the end of the stimulus. Thus K$_A$ channel activity can regulate the interval between APs in a burst.

Ca^{2+}-Activated K$^+$ Channels Are Opened by Intracellular Ca^{2+}

K$^+$-selective channels that are opened, at least in part, by intracellular Ca^{2+} are found in almost all excitable cells. There are three subtypes of **Ca^{2+}-activated K$^+$ (K$_{Ca}$) channels**: large, intermediate, and small conductance channels. Here we consider only the large conductance (**BK$_{Ca}$**) **channel**, whose single-channel conductance is one of the largest (>200 pS). The

probability that a BK$_{Ca}$ channel is open depends not only on [Ca^{2+}]$_i$, but also on V_m (Figure 8-8). At −50 mV, in the presence of 1 μM [Ca^{2+}]$_i$, the channels are closed almost all of the time. If [Ca^{2+}]$_i$ is raised to 100 μM and V_m is held constant at −50 mV, channel openings increase dramatically (Figure 8-8B). If [Ca^{2+}]$_i$ is held at 1 μM, changing V_m from −50 mV to +50 mV opens nearly all the BK$_{Ca}$ channels (Figure 8-8A). Thus the open probability of BK$_{Ca}$ channels depends on both V_m and [Ca^{2+}]$_i$. The increase in [Ca^{2+}]$_i$ that potentiates or triggers the opening of BK$_{Ca}$ channels could result from Ca^{2+} entry through Ca^{2+} channels near the BK$_{Ca}$ channels or from Ca^{2+} release by intracellular stores. In arteriolar smooth muscle cells, Ca^{2+} release from the sarcoplasmic reticulum opens BK$_{Ca}$ channels in the plasma membrane and causes membrane hyperpolarization, relaxation of the smooth muscle, and dilation of the arteriole.

In cells that generate bursts of APs, such as certain neurons and pancreatic β-cells, K$_{Ca}$ channels play a role in regulating the burst pattern. During a burst of APs, Ca^{2+} ions enter the cell through voltage-gated Ca^{2+} channels that open during the APs. This Ca^{2+} influx can produce a significant increase in [Ca^{2+}]$_i$. The increase in [Ca^{2+}]$_i$ activates K$_{Ca}$ channels, and the outward K$^+$ current through these channels hyperpolarizes the cell and terminates the burst.

ATP–Sensitive K$^+$ Channels Are Involved in Glucose-Induced Insulin Secretion from Pancreatic β-Cells

The β-cells in the islets of Langerhans in the endocrine pancreas play a critical role in the regulation of the plasma glucose concentration. The β-cell secretes the hormone insulin in response to increased plasma levels of glucose. This normally occurs immediately after a meal, when glucose and other nutrients are absorbed from the gastrointestinal tract. Insulin stimulates the uptake, metabolism, and storage of glucose in muscle and fat cells. Thus glucose and insulin are involved in a negative feedback system that regulates the plasma glucose concentration. When the plasma glucose level rises, insulin is secreted by the pancreas. Insulin then stimulates the uptake and storage of glucose, and as a result, the plasma glucose level falls.

The ATP–sensitive K$^+$ (**K$_{ATP}$**) **channel** in the pancreatic β-cell plays an important role in linking the glucose

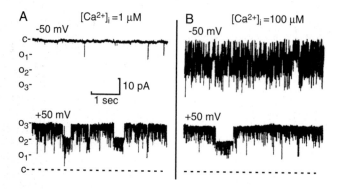

FIGURE 8-8 ■ BK_{Ca} channels are opened by elevation of $[Ca^{2+}]_i$ and by depolarization. These single-channel patch clamp recordings were obtained from an isolated patch of skeletal muscle membrane in which the former inside (cytosolic) surface of the membrane is exposed to the bathing solution. The K^+ concentrations on the two sides of the membrane are the same (so that $E_K = 0$ mV). This patch contained 3 BK_{Ca} channels, and the current levels corresponding to all channels closed (c) or to one, two, or three channels open (o_1, o_2, and o_3, respectively) are indicated at the left-hand side of the figure. **A,** With 1 μM Ca^{2+} at the cytosolic surface of the membrane, the BK_{Ca} channels are closed most of the time at –50 mV and are open most of the time at +50 mV. **B,** With 100 μM Ca^{2+} at the cytosolic surface of the membrane, channel openings are frequent even at –50 mV. Note that the current is inward at –50 mV and outward at +50 mV because of the difference in the driving force ($V_m - E_K$) at these two membrane potentials. *(Data from Barrett JN, Magleby KL, Pallotta BS: J Physiol [Lond] 331:211, 1982.)*

concentration, through cellular metabolism, to β-cell electrical activity and insulin secretion. The K_{ATP} channel is blocked by a high concentration of intracellular ATP ($[ATP]_i$). When the plasma glucose concentration rises, glucose is transported into the β-cell (by the GLUT-2 transporter; see Chapter 10), and glucose metabolism results in an increase in $[ATP]_i$. This closes K_{ATP} channels and thus reduces membrane K^+ permeability. As a result of the reduced K^+ permeability, the β-cell depolarizes and generates bursts of APs (Figure 8-9). Voltage-gated Ca^{2+} channels are activated during this glucose-induced electrical activity, and Ca^{2+} ions enter the cell through the open Ca^{2+} channels. The resulting increase in $[Ca^{2+}]_i$ activates insulin secretion.

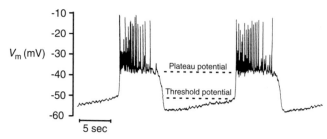

FIGURE 8-9 ■ Glucose-induced bursts of APs in a pancreatic β-cell. In the presence of elevated extracellular glucose levels ($[glucose]_o > 7$ mM), β-cells generate bursts of APs. Starting from a negative V_m of approximately –55 mV, V_m slowly depolarizes. When threshold is reached, the cell rapidly depolarizes to a plateau at approximately –35 mV. At the plateau a burst of small-amplitude APs is generated. After a few seconds the burst terminates, the membrane repolarizes, and the cycle repeats. The rate of depolarization (from the resting potential) and the frequency and duration of the bursts, which trigger insulin secretion, are directly related to $[glucose]_o$. *(Data from Ashcroft FM, Rorsman P: Prog Biophys Mol Biol 54:87, 1989.)*

Thus the β-cell K_{ATP} channel plays a critical role in glucose-induced insulin secretion and consequently the regulation of the plasma glucose concentration. Sulfonylureas, such as glibenclamide, selectively block K_{ATP} channels and thereby enhance insulin secretion. For this reason sulfonylureas are used in the treatment of **type 2 diabetes mellitus** (Box 8-6). Loss-of-function mutations of the K_{ATP} channel cause a metabolic disease, familial persistent hyperinsulinemic hypoglycemia of infancy (Box 8-7).

A Ca^{2+}-activated nonselective cation channel, NC_{Ca-ATP}, is found in astrocytes and has many properties similar to K_{ATP} channels. The NC_{Ca-ATP} channel is blocked by intracellular ATP and by the drug glibenclamide. During ischemia, opening of the channel as a result of ATP depletion causes cytotoxic edema.

A Voltage-Gated K$^+$ Channel Helps to Repolarize the Cardiac Action Potential

The *ether-a-go-go* (EAG) channels are a family of voltage-gated K$^+$ channels. One member of the family, the human EAG-related gene (hERG) K$^+$ channel, is crucial for the repolarization of the cardiac AP. Like voltage-gated Na$^+$ channels, hERG channels have voltage-dependent activation and inactivation gates. However, the kinetic properties of the hERG channel gates are markedly different from Na$^+$ channel kinetics. Na$^+$ channel activation and deactivation are fast processes, whereas inactivation is slow (see Chapter 7): thus most Na$^+$ channels open transiently on depolarization and then inactivate and remain closed (inactivated) on repolarization. In contrast, hERG K$^+$ channels have slow activation and deactivation kinetics and fast inactivation. On depolarization, hERG K$^+$ channels slowly activate but rapidly inactivate, so that few channels open during depolarization. During repolarization, however, the hERG channels rapidly recover from inactivation, and because the activation gate is still open (deactivation is slow), the channels open and conduct outward K$^+$ current. This outward current contributes significantly to the repolarization of the cardiac AP. Mutations in hERG K$^+$ channels can result in prolonged cardiac APs and cardiac arrhythmias (Box 8-8).

ION CHANNEL ACTIVITY CAN BE REGULATED BY SECOND-MESSENGER PATHWAYS

The activity of many ion channels can be modulated by a variety of intracellular second messengers. Second messengers are diffusible intracellular molecules that undergo a concentration change as a result

---

BOX 8-6

NON–INSULIN-DEPENDENT DIABETES MELLITUS (TYPE 2 DIABETES) CAN BE TREATED WITH SULFONYLUREA DRUGS

Diabetes mellitus is a disease characterized by insulin deficiency or by decreased receptor sensitivity to insulin. Diabetes is a heterogeneous disease, but all forms have in common the diagnostic features of hyperglycemia, polyuria, and polydipsia. Hyperglycemia, or high plasma glucose concentration, occurs because glucose uptake and metabolism in peripheral tissues are reduced. Polyuria, or excessive urine production, occurs because the kidneys cannot reabsorb all the filtered glucose and the glucose in the urine acts as an osmotic diuretic. This diuresis causes dehydration and increased thirst, or polydipsia.

The two main types of diabetes are insulin-dependent diabetes mellitus (IDDM) and non–insulin-dependent diabetes mellitus (NIDDM). NIDDM, which is also called type 2 diabetes, affects approximately 14 million people in the United States and is usually associated with obesity. In contrast to patients with IDDM, those with NIDDM do not usually require exogenous insulin. This form of the disease is often associated with decreased receptor sensitivity to insulin and decreased insulin secretion from β-cells. Sulfonylurea drugs (e.g., tolbutamide, glibenclamide) stimulate insulin secretion and are commonly used in the treatment of NIDDM. The sulfonylurea receptor is part of the K_{ATP} channel, and binding of the drug closes the channel. This leads to depolarization, bursts of action potentials, and enhanced insulin secretion.

BOX 8-7

FAMILIAL PERSISTENT HYPERINSULINEMIC HYPOGLYCEMIA OF INFANCY RESULTS FROM LOSS-OF-FUNCTION MUTATIONS IN ATP-SENSITIVE K$^+$ CHANNELS

Familial hyperinsulinism is a rare autosomal recessive disorder characterized by severe hypoglycemia caused by continuous, unregulated insulin secretion. Symptoms of the disease, which is referred to as persistent hyperinsulinemic hypoglycemia of infancy (PHHI), include lethargy, poor feeding, high-pitched cry, irritability, and convulsions. If the disease is untreated, neurological damage and mental retardation can occur. Milder forms of the disease are often successfully treated with intravenous glucose or with drugs that inhibit insulin secretion (e.g., diazoxide, which opens K$_{ATP}$ channels). More severe forms necessitate near-total pancreatectomy to control

the hypoglycemia. After pancreatectomy, growth and development are normal but insulin-dependent diabetes sometimes develops because of the loss of too many β-cells.

Mutations of the K$_{ATP}$ channel are implicated in PHHI. Several distinct mutations have been reported, all of which result in loss of function of the K$_{ATP}$ channels in pancreatic β-cells. As a result, the β-cells generate bursts of action potentials continuously, even when the plasma glucose concentration is very low (hypoglycemia). Thus insulin is secreted continuously.

BOX 8-8

MUTATIONS IN hERG K$^+$ CHANNELS CAN RESULT IN LONG QT SYNDROME

The QT interval in the electrocardiogram (ECG; Figure B-1A) is a measure of the duration of the cardiac AP. Mutations in genes that code for cardiac ion channels can cause a prolongation of the cardiac AP and the disorder called long QT syndrome. The increased QT interval may result in a life-threatening arrhythmia, a form of ventricular tachycardia referred to as *torsade de pointes*, which may progress to ventricular fibrillation and cardiac arrest (Figure B-1B). Loss-of-function mutations in hERG K$^+$ channels can produce long QT syndrome. In these mutations, the decreased outward K$^+$ current through hERG channels delays repolarization and thus increases cardiac AP duration.

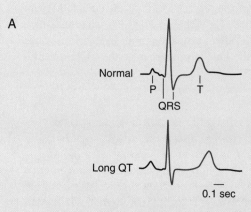

FIGURE B-1 ■ ECGs illustrating long QT and cardiac arrhythmias. **A,** The normal trace illustrates the waves of the ECG: the P wave reflects depolarization of the atria, the QRS complex indicates depolarization of the ventricles, and the T wave reflects ventricular repolarization. Thus the interval from the beginning of the Q wave to the end of the T wave, the QT interval, is a measure of the duration of the ventricular AP. The trace labeled "Long QT" illustrates an abnormally long QT interval. **B,** Long QT intervals can progress to *torsades de pointes*, a type of ventricular tachycardia, or even to ventricular fibrillation and cardiac arrest. *(Data from Keating MT, Sanguinetti MC: Cell 104:569, 2001.)*

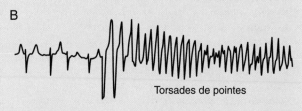

Torsades de pointes

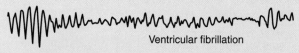

Ventricular fibrillation

of neurotransmitter or hormone action. The cyclic adenosine monophosphate (cAMP) system is a typical second-messenger pathway. The membrane-bound enzyme adenylyl cyclase converts ATP to cAMP and is activated by transmitter binding to a coupled receptor (e.g., the β-adrenergic receptor activates adenylyl cyclase). The main effect of cAMP in most cells is to activate the cAMP-dependent protein kinase (PKA). The activated kinase can then phosphorylate ion channels or other proteins and thereby change their activity to produce a cellular response.

β-Adrenergic Receptor Activation Modulates L-Type Ca²⁺ Channels in Cardiac Muscle

Heart muscle cell membranes contain several hormone and neurotransmitter receptors that play a role in regulating the activity of the heart. One such receptor, the β-adrenergic receptor, is activated by circulating epinephrine or by norepinephrine released from sympathetic nerve endings. Activation of cardiac β-adrenergic receptors causes an increase in the duration of the cardiac AP and increased cardiac contractility (i.e., an increase in the force generated by contracting heart muscle; see Chapter 13). These effects result from enhanced Ca^{2+} entry into the cell through voltage-gated L-type Ca^{2+} channels. Voltage clamp experiments show that the amplitude of cardiac I_{Ca} is increased in the presence of β-adrenergic agonists.

The enhanced I_{Ca} prolongs the duration of the cardiac AP. This, in turn, allows the Ca^{2+} channels to stay open longer, resulting in a larger influx of Ca^{2+} ions. The increase in $[Ca^{2+}]_i$ causes a more complete activation of the contractile machinery and a larger resultant force.

The β-adrenergic enhancement of I_{Ca} in the heart is mediated by the second messenger cAMP. The β-adrenergic receptor is coupled to the membrane-bound enzyme adenylate cyclase, which catalyzes the conversion of ATP to cAMP. β-Adrenergic receptor activation increases the activity of adenylyl cyclase, and this results in an increase in the cytosolic cAMP concentration, $[cAMP]_i$. The increase in $[cAMP]_i$ activates PKA, which then phosphorylates the Ca^{2+} channel. Two factors contribute to the increase in I_{Ca} that results from channel phosphorylation: (1) the number of *functional* Ca^{2+} channels that can be activated by depolarization increases and (2) the probability of channel opening at a given voltage increases. Thus, after activation of β-adrenergic receptors, the number of functional Ca^{2+} channels in the cardiac cell membrane is greater, and, when the membrane is depolarized (e.g., during an AP), a larger fraction of the functional channels is activated. This allows a larger influx of Ca^{2+} ions, which triggers a stronger activation of the contractile apparatus and greater force development. This increased development of force is called a **positive inotropic effect.**

SUMMARY

1. Ion channel types are characterized by their selectivity and by their structures and pharmacology.
2. The various types of ion channels play specific, critical roles in normal cell function, and channel defects can have serious pathophysiological consequences.
3. Voltage-gated Ca^{2+} channels can generate the upstroke of APs. In addition, Ca^{2+} channels can influence a large variety of cellular activities because Ca^{2+} ions regulate many cellular processes.
4. Several distinct types of Ca^{2+} channels can be distinguished on the basis of their physiological and pharmacological properties.
5. Ca^{2+} antagonist drugs reduce Ca^{2+} entry into the cell by blocking voltage-gated Ca^{2+} channels. These Ca^{2+} channel blockers are widely used as therapeutic agents in the management of cardiac arrhythmias, coronary artery disease, and hypertension.
6. TRPC channels are receptor-operated or store-operated cation channels.
7. K^+-selective ion channels, which are found in all cells, are diverse in their activity, structure, and distribution. Neuronal K^+ channel diversity contributes to the regulation of AP firing patterns.

8. Many neurons express K_A channels, which generate transient outward currents. These channels activate relatively rapidly during depolarization and then inactivate. In neurons that generate bursts of APs, the length of the interval between APs in the burst is regulated by the activity of K_A channels.

9. The three subtypes of Ca^{2+}-activated K^+ (K_{Ca}) channels (large, intermediate, and small conductance) are all opened by intracellular Ca^{2+}. The BK_{Ca} channels have one of the largest single-channel conductances. The current through K_{Ca} channels helps to repolarize individual APs and can also play a role in terminating a burst of APs.

10. K_{ATP} channels in the pancreatic β-cell, which are blocked by high $[ATP]_i$, are involved in glucose-induced insulin secretion. In response to a rise in plasma glucose concentration, glucose metabolism by the β-cell leads to an increase in $[ATP]_i$, which blocks K_{ATP} channels. Consequently, the β-cell depolarizes, thereby opening voltage-gated Ca^{2+} channels and allowing Ca^{2+} ions to enter the cell and activate insulin secretion.

11. Sulfonylureas, which block K_{ATP} channels and thereby enhance insulin secretion, are used in the treatment of type 2 diabetes.

12. The hERG K^+ channels are voltage-gated channels with slow activation and deactivation and fast inactivation. These channels contribute to repolarization of the cardiac AP.

13. Ion channel activity can be regulated by second-messenger pathways. For example, L-type Ca^{2+} channel activity in cardiac muscle is regulated by β-adrenergic receptors. The activation of these β-adrenergic receptors by epinephrine or norepinephrine causes phosphorylation of the Ca^{2+} channels. This increases the activity of the L-type Ca^{2+} channels, enhances Ca^{2+} entry, and increases force development in the heart.

KEY WORDS AND CONCEPTS

- Voltage-gated Ca^{2+} channels
- Ca^{2+} channel subtypes (e.g., L-type, T-type)
- Ca^{2+} antagonists (Ca^{2+} channel blockers)
- Use-dependent block
- TRP channels
- Receptor-operated channels (ROCs)
- Store-operated channels (SOCs)
- K_A channels
- Ca^{2+}-activated K^+ channels
- BK_{Ca} channels
- K_{ATP} channels
- Sulfonylureas
- Type 2 diabetes mellitus
- hERG K^+ channels
- Positive inotropic effect

STUDY PROBLEMS

1. In Chapter 4 we discussed the fact that the ionic flux that occurs during a nerve AP does not *significantly* alter the ion concentration in the cell (see Box 4-3). Why, then, can Ca^{2+} entry through Ca^{2+} channels significantly increase the $[Ca^{2+}]_i$ concentration? Support your answer with a calculation of the change in $[Ca^{2+}]_i$ that would occur under the following conditions: an average I_{Ca} of 20×10^{-12} amps flows for 100 msec into a spherical cell that has a diameter of 10 μm.

2. A certain type of neuron has a resting V_m of –75 mV, and it generates a single burst of APs in response to a brief depolarizing stimulus. The time interval between APs in the burst is regulated in part by the activity of K_A channels. If the resting potential of the cell were to become depolarized (e.g., to –65 mV), what would you expect to happen to the frequency of APs in the burst? Explain your answer.

3. The bursting pattern of electrical activity in pancreatic β-cells is controlled by the following populations of ion channels: L-type Ca^{2+} channels, BK_{Ca} channels, K_{ATP} channels, and voltage-gated K^+ channels. Propose some mechanisms involving ion channels (in addition to sulfonylurea block of K_{ATP} channels) that could be used to enhance insulin secretion from β-cells in patients with type 2 diabetes.

BIBLIOGRAPHY

Ashcroft FM, Rorsman P: Electrophysiology of the pancreatic β-cell, *Prog Biophys Mol Biol* 54:87, 1989.

Barrett EF, Magleby KL, Pallotta BS: Properties of single calcium-activated potassium channels in cultured rat muscle, *J Physiol (Lond)* 331:211, 1982.

Berne RM, Levy MN: *Cardiovascular physiology*, ed 8, St Louis, MO, 2001, Mosby.

Catterall WA: Structure and regulation of voltage-gated Ca^{2+} channels, *Annu Rev Cell Dev Biol* 16:521, 2000.

Chen M, Dong Y, Simard JM: Functional coupling between sulfonyl-urea receptor type 1 and a nonselective cation channel in reactive astrocytes from adult rat brain, *J Neurosci* 23:8568, 2003.

Connor JA, Stevens CF: Prediction of repetitive firing behaviour from voltage clamp data on an isolated neurone soma, *J Physiol (Lond)* 213:31, 1971.

Hille B: *Ionic channels of excitable membranes*, ed 3, Sunderland, MA, 2001, Sinauer.

Kostyuk PG, Kristhal OA, Shakhovalov YA: Separation of sodium and calcium currents in the somatic membrane of mollusc neurones, *J Physiol (Lond)* 270:545, 1977.

Lee KS, Tsien RW: Mechanism of calcium channel blockade by vera-pamil, D600, diltiazem, and nitrendipine in single dialysed heart cells, *Nature* 302:790, 1983.

Lehmann-Horn F, Jurkat-Rott K: Voltage-gated ion channels and hereditary disease, *Physiol Rev* 79:1317, 1999.

Modell SM, Lehmann MH: The long QT syndrome family of cardiac ion channelopathies: a HuGE review, *Genet Med* 8:143, 2006.

Ohno K, Wang H-L, Milone M, et al: Congenital myasthenic syndrome caused by decreased agonist binding affinity due to a mutation in the acetylcholine receptor e subunit, *Neuron* 17:157, 1996.

Pedersen SF, Owsianik G, Nilius B: TRP channels: an overview, *Cell Calcium* 38:233, 2005.

Segal M, Rogawski MA, Barker JL: A transient potassium conductance regulates the excitability of cultured hippocampal and spinal neurons, *J Neurosci* 4:604, 1984.

Shieh C-C, Coghlan M, Sullivan JP, et al: Potassium channels: molecular defects, diseases, and therapeutic opportunities, *Pharm Rev* 52:557, 2000.

Solute Transport

9 ELECTROCHEMICAL POTENTIAL ENERGY AND TRANSPORT PROCESSES

OBJECTIVES

1. Understand that concentration gradients and electrical potential gradients store chemical and electrical potential energy, respectively.

2. Understand that electrochemical potential energy drives all transport processes.

3. Use the concept of electrochemical potential energy to analyze transport processes.

ELECTROCHEMICAL POTENTIAL ENERGY DRIVES ALL TRANSPORT PROCESSES

In Chapter 2, by examining the permeability of biological membranes to various solutes, we concluded that, with the exception of simple, small, and typically lipid-soluble molecules (e.g., O_2, CO_2, ethanol), most biologically important solutes (e.g., sugars, amino acids, inorganic ions) cannot readily traverse cellular membranes. Therefore special transport mechanisms are required to move these impermeant solutes from one side of a membrane to the other. An important class of special transport mechanisms—the ion channels—has already been discussed in Chapters 5 to 8. There we observed that differences in ion concentrations and electrical potential across a membrane can drive the movement of ions through channels. In this chapter we introduce the concepts of chemical and electrical potential energy, which are stored in concentration gradients and electrical potential gradients, respectively. We also demonstrate

that electrochemical potential energy drives all solute transport processes.

The Relationship Between Force and Potential Energy Is Revealed by Examining Gravity

Experience tells us that the gravitational **force** acts on an object at any height so that, when the object is released, it is pulled toward the ground. Because force is mass (m) times acceleration (a), the gravitational force (F_G) must be:

$$F_G = -ma_G \qquad [1]$$

where a_G is the acceleration caused by gravity and the minus sign indicates that the force is directed downward (in the negative y direction).

Lifting an object of mass, m, from the ground to some height, y, requires an investment of energy. The amount of energy invested in lifting an object is directly proportional to the mass of the object, as well as the height to which the object is lifted. One way to

103

conceptualize this is to say that the object has greater **potential energy** when it is at a greater height from the surface of the Earth. These considerations are neatly summarized in the definition of gravitational potential energy (PE_G):

$$PE_G = ma_G \, y \qquad [2]$$

Implicit in this equation is the fact that ground level is the reference point against which gravitational potential energy is measured; that is, at ground level ($y = 0$), $PE_G = 0$. From this definition, it is clear that a change in height, Δy, causes a corresponding change in gravitational potential energy, $\Delta PE_G = ma_G \Delta y$. In other words, a gradient of gravitational potential energy exists in the y direction. The gravitational potential energy gradient is $\Delta PE_G / \Delta y$, or, when written as a derivative, dPE_G / dy:

$$\frac{dPE_G}{dy} = ma_G = -F_G \qquad [3]$$

We interpret this equation as saying that a gradient in gravitational potential energy gives rise to the gravitational force, which moves an object *down* the potential energy gradient, from some height toward ground level. This example provides an important and general insight: a **gradient of potential energy** gives rise to a force that will tend to move material *down* the potential energy gradient.

A Gradient in Chemical Potential Energy Gives Rise to a Chemical Force That Drives the Movement of Molecules

Fick's First Law of Diffusion (see Chapter 2) describes how a concentration gradient drives the diffusion of molecules down the concentration gradient. This diffusive movement of molecules may be viewed as being driven by a "chemical force." Because a potential energy gradient gives rise to a force, if a concentration gradient can give rise to a chemical force, the concentration gradient must also embody a gradient of **chemical potential energy**. The chemical potential energy is commonly represented by the symbol μ. For one mole of any solute, S, at concentration [S], the chemical potential energy has the form (Box 9-1)

$$\mu_S = \mu_S^0 + RT \ln[S] \qquad [4]$$

where μ_S^0 is the chemical potential energy of the solute at the reference concentration of 1 M; thus, at [S] = 1 M, $\mu_S = \mu_S^0$. Equation [4] tells us that solute molecules located in a region of high concentration are at a higher chemical potential energy than the same molecules located in a region of low concentration.

An Ion Can Have Both Electrical and Chemical Potential Energy

All molecules have chemical potential energy. Ions, because they carry electric charge, can also have **electrical potential energy**. At any electrical potential, V, the electrical potential energy associated with a single ion carrying z charges (e.g., $z = +2$ for Ca^{2+}; $z = -1$ for Cl^-) is zeV, where e is the magnitude of a single electric charge ($e = 1.602 \times 10^{-19}$ coulomb). For a mole of such ions, the electrical potential energy should be multiplied by Avogadro's number ($N_A = 6.022 \times 10^{23}$/mol). The magnitude of one mole of elementary charges is given a special name, the *Faraday* (symbol F), and has the value 96,485 coulombs/mol. Thus the electrical potential energy for one mole of ions, each carrying z charges, is zFV.

Because an ionic solute, S^z, can be at a particular concentration and electrical potential, its potential energy is the sum of both chemical and electrical components. The complete expression for the **electrochemical potential energy** of one mole of an ionic solute, S^z, at concentration, $[S^z]$, and at an electrical potential, V, is:

$$\mu_{S^z} = (\mu_{S^z}^0 + RT \ln[S^z]) + zFV \qquad [5]$$

Again, although μ_{S^z} has the units of energy, it is more commonly referred to as the **electrochemical potential**. Box 9-2 shows examples of expressions for the electrochemical potential of two common solutes.

The Nernst Equation Is a Simple Manifestation of the Electrochemical Potential

To illustrate the usefulness of the concept of electrochemical potential, we revisit a scenario first analyzed in Chapter 4: a cell with $[K^+]_i = 140$ mM and $[K^+]_o = 5$ mM, and whose PM is permeable *only* to K^+ (Figure 9-1). The K^+ concentration gradient would initially drive a net movement of K^+ ions out of the cell, thus causing the electrical potential inside

BOX 9-1

A CONCENTRATION GRADIENT STORES CHEMICAL POTENTIAL ENERGY, WHICH DRIVES THE MOVEMENT OF MOLECULES

In diffusion, a gradient in the concentration of a solute ([S]), drives a net flux, J, of the solute, as described by Fick's First Law (see Chapter 2):

$$J = -D\frac{d[S]}{dx} \qquad [B1]$$

Alternatively, we can view the diffusive movement as the result of a chemical force pushing on the solute molecules. It is quite reasonable that the velocity of the molecule, v, should be proportional to the magnitude of the chemical force, F_c:

$$v = uF_c \qquad [B2]$$

where u is the *mobility coefficient* of the molecule. We can determine how the flux is related to the molecular velocity by examining Figure B-1, which shows molecules moving at velocity, v, along a cylinder of solution. The flux, J, is simply the number of moles of molecules passing through the cross-sectional area, A, per unit time. Because the molecules are moving at velocity, v, in a time period, Δt, they would move a distance $v\Delta t$, which is the distance between the two shaded areas. Thus any

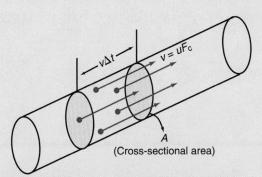

(Cross-sectional area)

FIGURE B-1 ■ Movement of molecules being driven by a "chemical force." Molecules are moving at average velocity, v, under the influence of a chemical force, F_c. In a time period, Δt, each molecule will move a distance equal to $v \times \Delta t$. Two surfaces of cross-sectional area, A, and separated by a distance, $v \times \Delta t$, are shown. All molecules located between these two surfaces are expected to move past the back surface after a time period, Δt, has elapsed.

molecule initially located between these two areas would pass through the area, A, during the time interval, Δt. The volume contained between the two shaded areas is $A \times (v\Delta t)$. The number of moles of solute in this volume is just $[S] \times A \times (v\Delta t)$. However, the flux is just this number of moles of solute passing through the area, A, in the time interval, Δt, or:

$$J = \frac{\text{moles of } S/A}{\Delta t} = \frac{[S] \times A \times (v\Delta t)/A}{\Delta t} \qquad [B3]$$
$$= [S]v = [S]uF_c$$

We can verify that the concentration, $[S]$, times the velocity, v, is indeed a flux by checking the units. Because $[S]$ has units of moles per volume (e.g., moles/cm^3), and v is expressed as distance per unit time (e.g., cm/sec), $[S] \times v$ must have units of moles per unit area per unit time, such as (moles/cm^2)/sec. These units imply the movement of a certain amount of material through a certain area in a certain amount of time, which is indeed the definition of flux.

The flux driven by the chemical force is the same flux as that specified by Fick's First Law, so the expressions for J in Equations [B1] and [B3] can be equated. Making use of the relationship between the ionic mobility, u, and the diffusion coefficient, D, $u = D/RT$ (see Box 4-1), and solving for the chemical force, yields:

$$F_c = -RT\frac{1}{[S]}\frac{d[S]}{dx} = -RT\frac{d\ln[S]}{dx} \qquad [B4]$$

The last step of Equation [B4] required us to recognize the derivative of the natural logarithm function (see Appendix B).

We already know that the chemical force arises from a gradient in chemical potential energy, μ_S, and tends to push molecules *down* the gradient. In other words:

$$\frac{d\mu_S}{dx} = -F_c = RT\frac{d\ln[S]}{dx} \qquad [B5]$$

By applying the integration techniques of Appendix B, we find that the chemical potential energy must have the form:

$$\mu_S = constant + RT\ln[S] \qquad [B6]$$

Continued

BOX 9-1

A CONCENTRATION GRADIENT STORES CHEMICAL POTENTIAL ENERGY, WHICH DRIVES THE MOVEMENT OF MOLECULES—cont'd

where *constant* is the integration constant, which must be determined independently. Examining Equation [B6] shows that when [S] = 1 M, μ_S = *constant*. In other words, the constant is the chemical potential energy when the solute is at a concentration of 1 M. This special value is given the special symbol, μ_S^0. Therefore the

complete expression for the chemical potential energy of a mole of solute, S, at concentration, [S], is:

$$\mu_S = \mu_S^0 + RT \ln[S] \qquad [B7]$$

Although μ_S has the units of energy, it is more commonly referred to as the *chemical potential*.

BOX 9-2

MATHEMATICAL EXPRESSIONS FOR THE ELECTROCHEMICAL POTENTIAL OF SOLUTES INSIDE AND OUTSIDE A CELL

The general expression for the electrochemical potential of a solute is given by Equation [B1]:

$$\mu_{S^z} = (\mu_{S^z}^0 + RT \ln[S^z]) + zFV \qquad [B1]$$

For a neutral solute like glucose (G), which carries no electrical charge ($z = 0$), its potential energy should not depend on the membrane potential (V_m) of the cell. Therefore the electrochemical potentials of glucose inside and outside the cell are given by the simple expressions:

$$\mu_{G,i} = \mu_G^0 + RT \ln[G]_i \qquad [B2]$$

and

$$\mu_{G,o} = \mu_G^0 + RT \ln[G]_o \qquad [B3]$$

respectively.

For an ionic solute such as the sulfate ion, SO_4^{2-}, which carries a charge ($z = -2$), its potential energy is expected to depend on V_m. Because V_m is the electrical potential inside relative to that outside (defined to be zero), the electrochemical potential for sulfate inside the cell is given by:

$$\mu_{SO_4^{2-},i} = \left(\mu_{SO_4^{2-}}^0 + RT \ln\left[SO_4^{2-}\right]_i\right) + (-2)FV_m \qquad [B4]$$
$$= \mu_{SO_4^{2-}}^0 + RT \ln\left[SO_4^{2-}\right]_i - 2FV_m$$

and that for sulfate outside the cell is given by:

$$\mu_{SO_4^{2-},o} = \left(\mu_{SO_4^{2-}}^0 + RT \ln\left[SO_4^{2-}\right]_o\right) + (-2)F(0) \qquad [B5]$$
$$= \mu_{SO_4^{2-}}^0 + RT \ln\left[SO_4^{2-}\right]_o$$

the cell to become negative relative to the outside. The developing negative V_m of the cell tends to resist further efflux of K^+ ions. When equilibrium is reached, a stable unchanging V_m will be established and no net flux of K^+, either into or out of the cell, will occur. In Chapter 4, by analyzing the fluxes, we concluded that the V_m attained at equilibrium is the K^+ equilibrium potential (E_K) and is given by the Nernst equation.

We now analyze the same situation by focusing on the meaning of *equilibrium*, without giving any attention to dynamic processes such as fluxes. When equilibrium is reached, no net movement of K^+ ions occurs between the inside and the outside of the cell. We know that a difference in electrochemical potential between two locations (i.e., a gradient) gives rise to the electrochemical force driving the movement of ions. Therefore the only way to have zero net K^+ flux is to

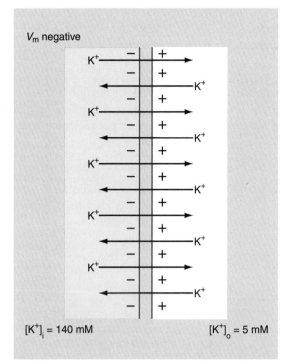

V_m negative

$[K^+]_i = 140$ mM

$[K^+]_o = 5$ mM

FIGURE 9-1 ■ Equilibration of K^+ across a plasma membrane (PM) that is permeable only to K^+. A PM separates the cytosol, with $[K^+]_i = 140$ mM, from the extracellular solution, where $[K^+]_o = 5$ mM. If the PM is permeable only to K^+, the K^+ concentration gradient will drive net movement of K^+ out of the cell, thus causing the electrical potential inside the cell to become negative relative to the outside. When equilibrium is reached, the negative V_m established is just sufficient to prevent further net flux of K^+ across the PM.

have no net electrochemical force. This is possible only if there is no longer any difference in electrochemical potential for K^+ between the inside and outside of the cell. In other words, at equilibrium, $\mu_{K^+,i}$ must be equal to $\mu_{K^+,o}$. The expression for $\mu_{K^+,i}$ is:

$$\mu_{K^+,i} = \mu_{K^+}^0 + RT\ln[K^+]_i + (+1)FV_m \qquad [6]$$

and the expression for $\mu_{K^+,o}$ is:

$$\mu_{K^+,o} = \mu_{K^+}^0 + RT\ln[K^+]_o + (+1)F(0)$$
$$= \mu_{K^+}^0 + RT\ln[K^+]_o \qquad [7]$$

In writing these two expressions, we have used the same convention that was introduced in Chapter 4.

That is, the intracellular electrical potential, V_m, is measured relative to the outside electrical potential, which is defined to be zero. Setting $\mu_{K^+,i} = \mu_{K^+,o}$ and rearranging algebraically (Box 9-3) gives:

$$V_{m,eq} = \frac{RT}{(+1)F} \ln \frac{[K^+]_o}{[K^+]_i} = E_K \qquad [8]$$

which we recognize as the Nernst equation. The V_m reached at equilibrium is indeed the K^+ equilibrium potential, E_K. This example illustrates the simplicity of using the electrochemical potential in analyzing equilibrium situations.

BOX 9-3
THE NERNST EQUATION IS A NATURAL CONSEQUENCE OF THE CONCEPT OF ELECTROCHEMICAL POTENTIAL

At equilibrium, the electrochemical potential of K^+ is equalized across the plasma membrane (PM); that is:

$$\mu_{K^+,i} = \mu_{K^+,o} \qquad [B1]$$

We can thus equate the expressions for $\mu_{K^+,i}$ and $\mu_{K^+,o}$ (Equations [6] and [7] in text).

$$\mu_{K^+}^0 + RT\ln[K^+]_i + (+1)FV_m \qquad [B2]$$
$$= \mu_{K^+}^0 + RT\ln[K^+]_o$$

The standard electrochemical potential of K^+, $\mu_{K^+}^0$, occurring on both sides of the equation, can be eliminated to give:

$$RT\ln[K^+]_i + FV_m = RT\ln[K^+]_o \qquad [B3]$$

Rearranging yields:

$$V_m = \frac{RT}{F}(\ln[K^+]_o - \ln[K^+]_i) \qquad [B4]$$

Knowing that the difference of two logarithms is the same as the logarithm of a quotient (see Appendix B), we can rewrite Equation [B4] as:

$$V_m = \frac{RT}{F} \ln \frac{[K^+]_o}{[K^+]_i} \qquad [B5]$$

which is the Nernst equation for K^+.

How to Use the Electrochemical Potential to Analyze Transport Processes

The scenario we just discussed, involving K^+ efflux and attainment of E_K, is the simplest example of a cellular *transport process*, that is, a process whereby solute on one side of a membrane is moved to the other side. The transport of K^+ can be represented as the "reaction" scheme:

$$K^+_{in} \rightleftharpoons K^+_{out}$$

To obtain the equation that describes the process *at equilibrium*, we simply equated the K^+ electrochemical potentials on the two sides of the reaction scheme.

A slightly more complex transport process is chloride-bicarbonate (Cl^-/HCO_3^-) exchange, which is mediated by an integral membrane protein in the PM of erythrocytes. The process involves the coordinated movement of one Cl^- ion into the cell and one HCO_3^- ion out of the cell, or vice versa. The Cl^-/HCO_3^- exchange process can be represented by the reaction scheme:

$$Cl^-_{out} + HCO^-_{3\,in} \rightleftharpoons Cl^-_{in} + HCO^-_{3\,out}$$

To obtain the equation that describes how all the concentrations and the V_m are related *at equilibrium*, we equate the total electrochemical potentials on the two sides of the reaction scheme:

$$\mu_{Cl^-,o} + \mu_{HCO_3^-,i} = \mu_{Cl^-,i} + \mu_{HCO_3^-,o} \qquad [9]$$

Then, substitution of the full expressions for the electrochemical potentials and algebraic rearrangement yield (Box 9-4):

$$\frac{[HCO_3^-]_i}{[HCO_3^-]_o} = \frac{[Cl^-]_i}{[Cl^-]_o} \qquad [10]$$

BOX 9-4
THE EQUILIBRIUM TRANSPORT EQUATION FOR CHLORIDE-BICARBONATE EXCHANGE

The chloride-bicarbonate (Cl^-/HCO_3^-) exchange process can be represented as:

$$Cl^-_{out} + HCO^-_{3\,in} \rightleftharpoons Cl^-_{in} + HCO^-_{3\,out}$$

At equilibrium the following balance in electrochemical potentials must be achieved:

$$\mu_{Cl^-,o} + \mu_{HCO_3^-,i} = \mu_{Cl^-,i} + \mu_{HCO_3^-,o} \qquad [B1]$$

The electrochemical potentials for Cl^- and HCO_3^- inside and outside the cells are given by the following equations:

$$\mu_{Cl^-,i} = \mu^0_{Cl^-} + RT \ln[Cl^-]_i + (-1)FV_m \qquad [B2]$$

$$\mu_{Cl^-,o} = \mu^0_{Cl^-} + RT \ln[Cl^-]_o \qquad [B3]$$

$$\mu_{HCO_3^-,i} = \mu^0_{HCO_3^-} + RT \ln[HCO_3^-]_i + (-1)FV_m \qquad [B4]$$

$$\mu_{HCO_3^-,o} = \mu^0_{HCO_3^-} + RT \ln[HCO_3^-]_o \qquad [B5]$$

Substituting the expressions [B2] to [B5] into Equation [B1] gives:

$$(\mu^0_{Cl^-} + RT \ln[Cl^-]_o) + (\mu^0_{HCO_3^-} + RT \ln[HCO_3^-]_i$$
$$- 1FV_m) = (\mu^0_{Cl^-} + RT \ln[Cl^-]_i - 1FV_m)$$
$$+ (\mu^0_{HCO_3^-} + RT \ln[HCO_3^-]_o) \qquad [B6]$$

All terms involving μ^0 and V_m cancel to give:

$$RT \ln[Cl^-]_o + RT \ln[HCO_3^-]_i$$
$$= RT \ln[Cl^-]_i + RT \ln[HCO_3^-]_o \qquad [B7]$$

Dividing both sides by RT and rearranging gives:

$$\ln[HCO_3^-]_i - \ln[HCO_3^-]_o$$
$$= \ln[Cl^-]_i - \ln[Cl^-]_o \qquad [B8]$$

Noting that a difference of two logarithms is the same as the logarithm of a quotient (see Appendix B), we can rewrite Equation [B8]:

$$\frac{[HCO_3^-]_i}{[HCO_3^-]_o} = \frac{[Cl^-]_i}{[Cl^-]_o} \qquad [B9]$$

Equation [B9] relates the intracellular and extracellular Cl^- and HCO_3^- concentrations when the Cl^-/HCO_3^- exchange process has reached equilibrium (i.e., when no further net movement of Cl^- and HCO_3^- can occur).

We see that V_m does not appear in this equation. This is not unexpected, because for every Cl^- ion transported in one direction, an HCO_3^- is transported in the opposite direction by the exchanger protein. The net transport of electrical charges is zero; that is, Cl^-/HCO_3^- exchange is *electroneutral*. Therefore the transport process should affect only the concentrations of Cl^- and HCO_3^-, without affecting V_m. It is thus entirely reasonable that the equilibrium concentrations of Cl^- and HCO_3^- are *not* related to V_m. Although the operation of the Cl^-/HCO_3^- exchanger is relatively simple, it has fundamental physiological importance in allowing the circulatory system to transport CO_2 generated metabolically in the body tissues to the lungs, where the gas can be expelled from the body (Box 9-5).

We have seen that the electrochemical potential is a useful concept for analyzing processes that transport solutes across membranes. In Chapter 10 we use the same approach to analyze several physiologically important transport processes of greater complexity.

BOX 9-5

THE CHLORIDE-BICARBONATE EXCHANGER IS ESSENTIAL FOR REMOVING CARBON DIOXIDE FROM THE BODY

In the course of 24 hours an average person expires more than 800 liters of CO_2 gas. This represents more than 30 moles of CO_2 ($\sim$1400 g, or $\sim$3 lb, by weight). This amount of CO_2, generated metabolically within tissues of the body, must be transported by the circulation to the lungs to be expelled. Although CO_2 is approximately 20 times more soluble than O_2 in plasma, its solubility is still far too low to allow its efficient transport as a simple dissolved gas in plasma. Just as hemoglobin is an evolutionary adaptation for transporting sparingly soluble O_2, a special mechanism has evolved to enable the transport of large amounts of CO_2 through the circulation.

CO_2 produced in the tissues travels by diffusion into capillaries and into the red blood cells (RBCs) that course through the capillaries (Figure B-1). RBCs contain high concentrations of carbonic anhydrase (CA), an enzyme that catalyzes (accelerates) the reversible reaction of CO_2 with water to form carbonic acid (H_2CO_3):

$$CO_2 + H_2O \underset{\text{anhydrase}}{\overset{\text{Carbonic}}{\rightleftarrows}} H_2CO_3$$

Without catalysis, hydration of CO_2 is a very slow process; each CO_2 molecule takes many seconds to react with water. As one of the fastest enzymes on Earth, CA greatly speeds up the hydration process. Each molecule of CA can catalyze the reaction of 1 million CO_2 molecules per second.

As soon as the H_2CO_3 molecules are formed, they dissociate into protons (H^+) and bicarbonate ions (HCO_3^-):

$$H_2CO_3 \rightleftarrows H^+ + HCO_3^-$$

Most of the H^+ ions are immediately bound to hemoglobin, whose concentration within an RBC is in excess of 5 mM (i.e., hemoglobin is an effective buffer for H^+ ions). Abundant Cl^-/HCO_3^- exchangers in the RBC PM rapidly transport HCO_3^- ions out of the RBC into the plasma with the concomitant transport of Cl^- ions into the RBC. In the tissues the net result of the coordinated action of CA and the Cl^-/HCO_3^- exchanger is to convert poorly soluble CO_2 into two highly soluble ions, H^+ and HCO_3^-, which are then kept in physically separate compartments (H^+ in the RBC and HCO_3^- out in the plasma), so that they cannot recombine to regenerate CO_2. These separated moieties are transported as an ensemble by the circulation toward the lungs.

As blood reaches the lungs, the processes that occurred in the tissues are reversed (Figure B-1). The Cl^-/HCO_3^- exchanger now mediates movement of Cl^- out of, and HCO_3^- back into, the RBC. The HCO_3^- ions can thus combine with H^+ ions in the RBC to reform H_2CO_3, which is rapidly *dehydrated* by CA to yield simple, dissolved CO_2. The CO_2 formed in the RBC then diffuses from the alveolar capillaries into the alveolar air space, where the CO_2 concentration (or partial pressure, $p{CO_2}$) is low.

In summary, in the tissues the higher concentration of CO_2 drives CO_2 entry into RBCs, and the subsequent increase in $[HCO_3^-]_i$ drives the Cl^-/HCO_3^- exchanger to operate in the Cl^--entry/HCO_3^--exit mode. In the alveolar capillaries the CO_2 concentration gradient is reversed (low CO_2 concentration in the alveolar space) and drives CO_2 from the RBCs into the alveolus. This leads to CA catalyzing net dehydration of H_2CO_3. The

Continued

BOX 9-5

THE CHLORIDE-BICARBONATE EXCHANGER IS ESSENTIAL
FOR REMOVING CARBON DIOXIDE FROM THE BODY—cont'd

In the Tissues

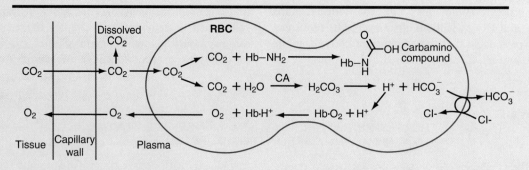

In the Lungs

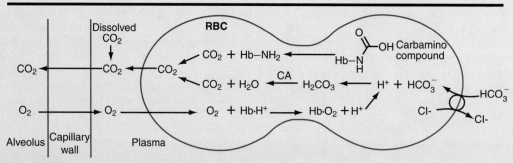

FIGURE B-1 ■ Transport of CO_2 by the circulation. CO_2 generated by metabolism in tissues is transported in three forms in the circulation: as HCO_3^- in the plasma, as bound carbamino compounds formed with hemoglobin, and as dissolved CO_2 gas. CO_2 diffusing into the RBC is rapidly converted to H^+ and HCO_3^- ions. Through the action of the Cl^-/HCO_3^- exchanger, the HCO_3^- is extruded out of the RBC in exchange for Cl^- from the plasma. The H^+ ions bind to oxyhemoglobin ($Hb \cdot O_2$); this favors release of O_2 from hemoglobin and makes O_2 available to the tissues. Of the CO_2 produced in tissues, the greatest fraction (~70%) is transported as HCO_3^-. A smaller fraction (~20%) is carried by proteins in the blood (principally hemoglobin) in the form of carbamino compounds. Relatively little (~10%) is transported as simple dissolved CO_2. In the lungs the processes that occur in the tissues are reversed, thus allowing CO_2 to be released to, and O_2 to be taken up from, the alveolar air space. $Hb \cdot H^+$ is protonated hemoglobin; $Hb-NH_2$ represents a side-chain amino group on hemoglobin; CA is carbonic anhydrase.

consequent decrease in $[HCO_3^-]_i$ drives the exchanger to operate in the Cl^--exit/HCO_3^--entry mode.

The importance of the coordinated action of CA and the Cl^-/HCO_3^- exchanger in CO_2 transport is reflected by the fact that 70% of the CO_2 produced in the body is moved into the lungs by this mechanism, whereas only 10% is carried in the plasma as simple dissolved CO_2. The remaining 20% is transported in protein-bound forms, which are commonly referred to as carbamino compounds (Figure B-1).

SUMMARY

1. The *chemical potential energy* of a substance is a function of its concentration. A molecule in a region of high concentration is at a higher chemical potential energy than the same molecule in a region of low concentration.

2. A difference in chemical potential energy between two regions gives rise to a "chemical force" that drives the movement of molecules down the potential energy gradient, that is, from a location of higher potential energy toward a location of lower potential energy.

3. An ion, being electrically charged, can also have *electrical potential energy*. The combined chemical and electrical potential energy is referred to as the *electrochemical potential*.

4. A difference in electrochemical potential between two regions drives the transport of ions and molecules down the electrochemical potential gradient.

KEY WORDS AND CONCEPTS

- Force
- Potential energy
- Gradient of potential energy
- Chemical potential energy
- Electrical potential energy
- Electrochemical potential energy (electrochemical potential)

STUDY PROBLEMS

1. A cell has a membrane potential, V_m.
   a. The glucose concentration inside the cell is $[G]_i$. Write the mathematical expression representing the electrochemical potential of glucose inside the cell.
   b. The Ca^{2+} concentrations inside and outside the cell are $[Ca^{2+}]_i$ and $[Ca^{2+}]_o$, respectively. Write the mathematical expressions for the electrochemical potential of Ca^{2+} inside and outside the cell.

BIBLIOGRAPHY

Atkins PW: *Physical chemistry*, ed 5, New York, NY, 1994, WH Freeman.

10

PASSIVE SOLUTE TRANSPORT

OBJECTIVES

1. Explain how the distribution of lipids and proteins in the cell membrane influences the membrane permeability to hydrophobic and hydrophilic solutes and ions.

2. Differentiate the following mechanisms based on the source of energy driving the process and the necessity for an integral cell membrane protein: diffusion, mediated (facilitated) transport, and secondary active transport.

3. Explain how the transport rates of certain molecules and ions are accelerated by specific integral membrane proteins ("carrier" and "channel" molecules).

4. Understand how coupling of solute transport enables one solute to be transported against its electrochemical gradient by using energy stored in the electrochemical gradient of the coupled solute.

5. Explain the two-step process involved in the net transport of certain solutes across epithelia.

DIFFUSION ACROSS BIOLOGICAL MEMBRANES IS LIMITED BY LIPID SOLUBILITY

We have just learned that all transport processes are driven by electrical or chemical gradients (see Chapter 9). The question is, "How are substances (solutes and the solvent, water) actually transported across biological membranes?" Bear in mind that some substances must be concentrated in cells, or rapidly taken up, because they are essential for cell function. Other substances must be excluded or extruded from cells. For example, too much Na^+ (or any other solute) in cells would be an excessive osmotic burden and cause cells to swell (see the discussion of the Donnan effect in Chapter 4).

Biological membranes are primarily lipid bilayers (see Chapter 1), which are poorly permeable to **polar** and hydrophilic solutes; these substances bear a net charge or have internal charge separation

(i.e., they are uncharged molecules that behave like electric dipoles in which the positive and negative charge are separated). The ability of polar substances to pass across lipid membranes tends to be inversely proportional to molecular weight: the larger the molecule, the lower the permeability. Moreover, although pure phospholipid bilayers are modestly permeable to water, which is a polar solvent (see Chapter 1), the presence of cholesterol greatly reduces water permeability. Consequently, the phospholipid-cholesterol bilayer that constitutes the PM in many types of cells is not sufficiently permeable to water for physiological needs.

Consider the problem of moving a polar solute, such as glucose, across a biological membrane. Glucose crosses lipid bilayers extremely slowly (see Chapter 2). In other words, the bilayer is an effective barrier to the transport of these substances. Yet glucose is an essential fuel for cells.

CHANNEL, CARRIER, AND PUMP PROTEINS MEDIATE TRANSPORT ACROSS BIOLOGICAL MEMBRANES

To mediate and regulate the transfer of water and polar solutes, biological membranes contain integral proteins called channels, carriers, and "pumps" (Figure 10-1). More than one third of all the genes in the human genome code for membrane proteins, and approximately half of these genes (i.e., one sixth of the genome) code for transport proteins.

Transport Through Channels Is Relatively Fast

Channels in biological membranes (see Chapter 7) are proteins with central pores that open to both the extracellular fluid and the cytoplasm simultaneously (Figure 10-1A). The narrowest region within the pore, called the selectivity filter, determines which substance(s), ions, water, etc., may pass through the channel. For example, channels in the family called **aquaporins**[*] are selectively permeable to water. Some other channels (see Chapters 7 and 8) are highly selective for Na^+ ions, for K^+ ions, or for Ca^{2+} ions, whereas still others are less selective and may, for example, be permeable to both Na^+ and K^+ (e.g., the nicotinic acetylcholine channel; see Chapter 13).

A characteristic feature of channels is their relatively high permeability. Typical **turnover numbers** for ion channels (i.e., the maximum number of ions

---

[*]Peter Agre shared the 2003 Nobel Prize in Chemistry for his discovery of aquaporins and his determination of their structure and function.

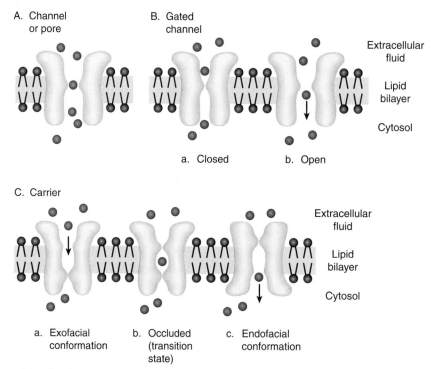

A. Channel or pore
B. Gated channel
a. Closed
b. Open
Extracellular fluid
Lipid bilayer
Cytosol

C. Carrier
a. Exofacial conformation
b. Occluded (transition state)
c. Endofacial conformation
Extracellular fluid
Lipid bilayer
Cytosol

FIGURE 10-1 ■ Models of a channel or pore (**A**), a gated channel (**B**), and a carrier (**C**). The gated channel is shown in, closed *(a)* and open *(b)* configurations. The carrier is shown in the exofacial configuration *(a)* with the solute binding site open to the extracellular fluid, the "occluded" (transition state) configuration with the bound solute inaccessible to either fluid *(b),* and the endofacial configuration with the solute binding site open to the cytosol *(c).* Note that the "simple" carrier must also be able to switch between the exofacial and endofacial configurations in the absence of bound solute to effect net transport of the solute down its electrochemical gradient.

that can pass through a channel in 1 second) are approximately 10^6 to 10^8. For example, approximately 30 million K^+ ions can pass through a single K^+ channel in 1 second (Table 10-1).

Channel Density Controls the Membrane Permeability to a Substance

Aquaporins facilitate water flow between the extracellular fluid and the cytoplasm to maintain osmotic balance (equilibrium) in cells that require high water permeability (e.g., skeletal muscle and some epithelia). The relative permeability to water depends on the number of channels (aquaporin tetramers) present, per unit area, in the PM (i.e., the channel density), and hormones may regulate this. For example, the hormone *vasopressin* increases water permeability in the principal cells of the renal cortical collecting duct by stimulating the production of cAMP. In turn, cAMP promotes the insertion of aquaporin-2 (AQP-2) molecules into the apical membrane of the epithelial cells. As a result, the osmotic driving force can speed water reabsorption from the renal tubule lumen. Defects in this hormonal mechanism, such as loss-of-function mutations in the vasopressin receptor or the AQP-2 molecule, lead to *diabetes insipidus* (excretion of a high volume of dilute urine). This occurs because AQP-2 molecules cannot be inserted into the apical membrane of the renal cortical collecting duct. Therefore, even though the osmotic force may favor water reabsorption, the intrinsically low water permeability

in the kidney cortical collecting ducts minimizes water reabsorption.

The Rate of Transport Through Open Channels Depends on the Net Driving Force

In contrast to the aforementioned situation, impaired insulin secretion or reduced sensitivity to insulin leads to high glucose content in the proximal renal tubule fluid. The resulting osmotic force reduces the reabsorption of water even though the epithelial cell apical membranes may contain a large number of AQP-2 molecules. This results in *diabetes mellitus* (excretion of a high volume of sweet urine with an osmolality approaching that of plasma). In this case the abnormally high osmotic pressure in the kidney tubule lumen reduces the osmotic driving force for water reabsorption from the lumen even though apical membrane water permeability may be high (see Chapter 11).

Transport of Substances Through Some Channels Is Controlled by "Gating" the Opening and Closing of the Channels

Another mechanism of regulating channel permeability is by channel gating (Figure 10-1B), which may be regulated by voltage or by ligands such as Ca^{2+} or ATP (see Chapters 7 and 8). Still, other channels are gated by pressure and membrane deformation, as exemplified by the mechanosensitive monovalent cation channels in some sensory nerve terminals.

TABLE 10-1
Relative Transport Rates for Various Types of Transporters

TRANSPORTER	TURNOVER NUMBER* (PER SEC)
K^+ channel	30,000,000
Valinomycin (carrier)	30,000
Glucose carrier (GLUT-1)	3,000
Na^+/Ca^{2+} exchanger	2,000
Ca^{2+} pump (SERCA)	200
Na^+ pump	150

*Rate of cycling of the transporter molecule, except for ion channels such as the K^+ channel, where the turnover number indicates the maximum number of ions transported in 1 second under physiological conditions. Thus each Na^+ pump molecule cycles 150 times per second and transports 450 Na^+ (and 300 K^+) per second.

CARRIERS ARE INTEGRAL MEMBRANE PROTEINS THAT OPEN TO ONLY ONE SIDE OF THE MEMBRANE AT A TIME

In contrast to channels, the solute binding sites in carriers undergo spontaneous *conformational changes* and thus have *alternating access* to the two sides of the membrane. A solute binds to the carrier at one side of the membrane; then, as a result of a conformational change, a "gate" closes and the solute is transiently **occluded** (the **transition state**; Figure 10-1C). Then, through a further conformational change in the protein, the "gate" on the opposite side opens so that the solute can dissociate from the carrier at this side of the membrane (Figure 10-1C, c). We use the terms exofacial and endofacial to denote the transporter conformations in which the solute binding sites face the extracellular fluid and cytoplasm, respectively.

Carriers Facilitate Transport Through Membranes

Simple carriers are integral membrane proteins that bind and transport a single solute species across a membrane. They move the bound solute down its electrochemical gradient (i.e., down concentration gradients and, in the case of charged solutes, voltage gradients). The rate of transport is substantially slower than that mediated by channels (Table 10-1) because substrate binding and carrier conformation changes take time. Nevertheless, transport is much faster than would be expected for diffusion in the absence of such carriers. In other words, carriers simply speed up (facilitate) processes that would normally occur, albeit much more slowly, by diffusion; hence this process is sometimes called **facilitated diffusion** (Box 10-1). Thus *transport by the simple carrier cannot be used to generate a **steady-state** electrochemical gradient* because the generation of such a gradient requires the expenditure of energy. This distinguishes **passive transport** from *active transport*, which is discussed later and in Chapter 11.

Sugars Are Transported by Carriers Several solutes are transported by solute-selective simple carriers. A good example is the **glucose carrier** that mediates glucose transport across the human red blood cell (RBC) PM. This protein, GLUT-1, belongs to a family of sugar transporters (Box 10-2). GLUT-1 has 12 membrane-spanning helices, 5 of which are *amphiphilic*: each has a hydrophobic and a hydrophilic surface. The hydrophobic regions interact with the surrounding lipids in the bilayer. In contrast, the hydrophilic surfaces of the 5 helices face one another and form a central, water-filled transmembrane channel or pore (Figure 10-2). This is the general structure of many transporter molecules.

The need for carrier-mediated glucose transport is exemplified by the consequences of mutations in the human GLUT-1 gene. GLUT-1 is also expressed in the brain, where it mediates glucose transport across the blood-brain barrier and glucose uptake by glial cells. Individuals with a mutated GLUT-1 gene have an abnormally low cerebrospinal fluid glucose concentration (approximately half normal) despite a normal blood glucose level. The reduced brain glucose causes a devastating neurological syndrome (Box 10-3).

Transport by Carriers Exhibits Kinetic Properties Similar to Those of Enzyme Catalysis

Carrier-Mediated Transport Is Saturable and Vectorial Because limited numbers of carrier molecules are present in the membrane, raising the solute

---

BOX 10-1

THE MECHANISM OF CARRIER-MEDIATED TRANSPORT AS EXEMPLIFIED BY THE GLUT-1 GLUCOSE TRANSPORTER

Carriers such as the GLUT-1 sugar transporter spontaneously change conformation, whether or not glucose is bound, so that the glucose binding site is open (accessible) either to the extracellular fluid or to the cytoplasm. Consider, as an example, the situation in which the glucose concentration in the extracellular fluid is high and the concentration in the cytoplasm is low. Glucose molecules will then bind to the GLUT-1 binding sites with higher probability (because of the higher glucose concentration) when the sites are facing the extracellular fluid than when the sites are facing the cytoplasm. Glucose that is bound at the outside will tend to dissociate when the carrier conformation changes and the sites face the cytoplasm with its low glucose concentration. The result will be a net transport of glucose from the extracellular fluid to the cytoplasm. When the concentrations of glucose are equal on the two sides of the membrane, no *net* transport will occur because the probability of glucose binding at the internal and at the external faces of the membrane will then be the same. Large movements of glucose (fluxes) in both directions may take place under the latter circumstances, but they will be equal in magnitude. Thus, in the steady state, the simple carrier cannot be used to transport a solute against a concentration gradient.

BOX 10-2
THE FAMILY OF SUGAR TRANSPORTERS

GLUT-1 is one of five homologous human sugar transporters (GLUT-1 to GLUT-5). Each is the product of a different gene, and each has a different tissue distribution and is regulated differently. Moreover, these transporters are members of one family of a superfamily of solute transporters, the *major facilitator superfamily*, with more than 1000 members and some well-conserved sequence motifs. The superfamily already includes as many as 34 separate families. Each family has a family-specific signature sequence and specificity for a single class of substrates. Some other families within this superfamily include the monocarboxylic acid transporters that transport pyruvate and lactate, for example, and the anion/cation cotransporters that transport sodium and phosphate simultaneously.

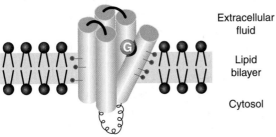

Extracellular fluid

Lipid bilayer

Cytosol

FIGURE 10-2 ■ Model of a simple glucose carrier (e.g., GLUT-1). The model shows the five transmembrane helices that surround the hydrophilic central pore, to which the glucose binds. One of the helices is tilted so that the glucose binding site (which is thereby made visible in this rendition) is accessible to the extracellular fluid only. Also diagrammed in this model are some of the hydrophobic residues on the amphipathic helices that face the surrounding phospholipids.

BOX 10-3
THE DEFECTIVE GLUCOSE TRANSPORTER PROTEIN SYNDROME

More than 400 persons with a defective GLUT-1 glucose transporter have been identified. Individuals with this rare defect have infantile seizures (convulsions), starting at age 3 to 4 months. They also have microcephaly (small head size) and developmental delays. Laboratory examination reveals *hypoglycorrhachia* (i.e., a very low glucose concentration, approximately half normal) in the cerebrospinal fluid (CSF) and a low CSF/blood glucose ratio (< 0.5). The rate of glucose uptake by the red blood cells (RBCs) from these patients is also less than 50% of normal; this can be used as a diagnostic test. These manifestations can be explained by mutations in the GLUT-1 gene that cause malfunction of the expressed protein (a reduced turnover rate or maximum velocity of transport). GLUT-1 normally is expressed in many cells, including RBCs and epithelial cells of the choroid plexus and ependyma, as well as in blood vessel endothelial cells; it facilitates the transport of glucose from the blood to the CSF. GLUT-1 is also expressed in glia, where the transporter is concentrated in foot processes that surround neuronal synapses. Thus, in patients with defective GLUT-1, the glucose concentration within glial cells may be particularly low. This may be rate-limiting for cellular energy metabolism and brain function.

enzymes because both of them speed up spontaneous processes. The product of a carrier's reaction, however, is the *vectorial* (directional) movement of the substrate from one side of the membrane to the other, rather than the chemical alteration of the substrate.

Carrier-Mediated Transport Exhibits Substrate Specificity The binding site on a transporter protein recognizes certain solutes, but not others. GLUT-1, for example, transports D-glucose and D-galactose, but not L-glucose (the transporter is *stereospecific*) or maltose or ribose.

Carrier-Mediated Transport Can Be Inhibited Either Competitively or Noncompetitively GLUT-1 transports both D-galactose and D-glucose. By competing for sugar binding sites on the carriers, D-galactose can be expected to inhibit the transport

(e.g., sugar) concentration sufficiently will saturate all the carrier molecules so that no more sugar can be bound. This limits the maximal rate of transport (Figure 10-3). The transport follows *Michaelis-Menten (saturation) kinetics*, which also governs the rate of enzymatic reactions. Indeed, carriers are comparable to

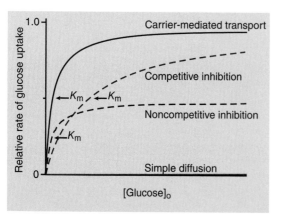

FIGURE 10-3 ■ Relative rate of glucose uptake into human red blood cells (mediated by the glucose transporter GLUT-1) graphed as a function of the glucose concentration in the medium ($[Glucose]_o$). The curves show the uptake under control conditions ("carrier-mediated transport") and in the presence of a competitive inhibitor (e.g., galactose) and a noncompetitive inhibitor (e.g., phloretin). GLUT-1 uptake is half-maximally activated by 1.6 mM glucose (= K_m; indicated by *arrows* on the control carrier-mediated transport and noncompetitive inhibitor curves). The competitive inhibitor increased the apparent K_m for glucose fourfold. The maximum rate of glucose uptake is 0.6 mmol/sec at 20°C. A saturating concentration of noncompetitive inhibitor reduces the rate to that of simple diffusion, which is vanishingly low.

of D-glucose and vice versa (competitive inhibition). The presence of galactose reduces the apparent affinity of GLUT-1 for glucose but does not affect the maximum velocity of glucose transport at saturating glucose concentrations (Figure 10-3). In contrast, certain molecules such as phloretin and dinitrofluorobenzene are noncompetitive inhibitors of GLUT-1. These inhibitors do not affect the affinity of the carrier for glucose, but reduce the maximum transport velocity (Figure 10-3).

Simple Carriers Exhibit Reversibility and Countertransport Carriers facilitate the movement of solutes in *both* directions across the membrane (i.e., they are "reversible"). The direction of *net* movement is, in general, determined by the electrochemical gradient of the transported substance. Consider, however, a situation in which RBCs are equilibrated with glucose (i.e.,

equilibrium is achieved, in which external and internal concentrations are equal and there is no net driving force). If a high concentration of galactose is added to the extracellular solution, the carriers will mediate a *net efflux* of glucose (i.e., they will move glucose outward). Initially, carriers in the endofacial conformation bind only glucose (the only solute available). When these carriers change to the exofacial conformation, some of the bound glucose will be displaced by galactose, which will then be transported inward—this is glucose-galactose exchange. The glucose concentration inside the RBCs will *temporarily* fall below that outside the cells until the galactose concentration inside the cells rises to equal that outside (i.e., when the galactose concentration *gradient* falls to zero). During this brief period (i.e., under *non–steady-state conditions*), a countertransport of glucose in exchange for galactose takes place. This demonstrates that the carriers alternately open to the two membrane surfaces.

Carrier-Mediated Transport Can Be Regulated The glucose carrier isoform found in adipocytes (fat cells) and in skeletal and cardiac muscle cells, GLUT-4, is regulated by insulin. In the absence of insulin stimulation, most GLUT-4 molecules reside in intracellular vesicular membranes. Insulin stimulation promotes the fusion of these vesicles with the PM, thereby increasing the density of GLUT-4 molecules in the PM and accelerating glucose transport. As a result, insulin shortens the time required for the intracellular glucose concentration to reach that in the extracellular fluid. This is advantageous after a meal, when the blood sugar concentration rises and insulin secretion is increased. The enhanced rate of glucose entry enables faster glycogen synthesis in muscle and adipocytes. Thus insulin facilitates glucose storage (as glycogen) but does *not* enable GLUT-4 to concentrate free glucose in the cells.

β-Adrenergic agonists such as epinephrine and isoproterenol inhibit GLUT-4 mediated glucose uptake in skeletal muscle, apparently by stimulating glycogenolysis. Glycogenolysis causes the glucose concentration within the cells to rise, inhibiting further net entry of glucose. In other words, this inhibition is the result of a change in the glucose concentration gradient rather than a change in the GLUT-4 turnover (or cycling rate).

COUPLING THE TRANSPORT OF ONE SOLUTE TO THE "DOWNHILL" TRANSPORT OF ANOTHER SOLUTE ENABLES CARRIERS TO MOVE THE COTRANSPORTED OR COUNTERTRANSPORTED SOLUTE "UPHILL" AGAINST AN ELECTROCHEMICAL GRADIENT

Coupled transport that does not directly use ATP hydrolysis is sometimes called **secondary active transport**. The reason is that the movement of one substance down its electrochemical energy gradient (i.e., "downhill" movement; see Chapter 9) can be used to concentrate another substance (i.e., move it "uphill" against a concentration or voltage gradient). For example, the carrier-mediated transport of various solutes is coupled to Na^+ transport, using energy from the Na^+ electrochemical gradient that is maintained by the sodium pump (see Chapter 11). The Na^+ may be either cotransported or countertransported with the coupled solute. **Cotransport** is also referred to as **symport** because both solutes move simultaneously in the same direction; **countertransport** is also called **antiport** or **exchange** because the two coupled solutes move in opposite directions.

Na$^+$/H$^+$ Exchange Is an Example of Na$^+$-Coupled Countertransport

Intracellular pH regulation in all cells depends on several transport mechanisms. Among these are the Cl^-/base (e.g., Cl^-/HCO_3^-, and Cl^-/OH^-) exchangers, which can mediate net proton (H^+) influx or efflux, and the Na^+-HCO_3^- cotransporter, which mediates acid efflux. One other type of H^+ transport system, present in virtually all cells, is the Na^+/H^+ exchanger (**sodium/proton exchanger** or NHE) that mediates the electroneutral exchange (i.e., there is no net charge transfer) of 1 Na^+ for 1 H^+. Five molecular isoforms of NHE are expressed in mammalian cells in a tissue-specific manner. NHE1, for example, is expressed in cardiac muscle and in kidney and intestinal epithelia. NHE1 has 12 transmembrane segments and a large C-terminal cytoplasmic tail containing several sites that are involved in regulating the

exchanger activity (e.g., by phosphorylation and by Ca^{2+}-calmodulin).

NHE rapidly extrudes protons from cells when intracellular pH (pH$_i$) falls to less than the normal value of approximately 7.2. Enough energy is available in the Na^+ electrochemical gradient to transport sufficient H^+ out of most cells to raise pH$_i$ to more than 8.0 (Box 10-4), but this pH is nonphysiological. To prevent pH$_i$ from rising too much, intracellular H^+ regulates the NHE at an H^+ binding site that is distinct from the H^+ transport site. This H^+ regulatory site has a steep pH dependence and activates transport when protons are produced and pH$_i$ falls. Then, as protons are extruded and pH$_i$ approaches 7.2, NHE activity is rapidly downregulated. This behavior is different from that of many other Na^+-coupled transport systems (discussed later), which operate close to electrochemical equilibrium.

Na$^+$ IS COTRANSPORTED WITH A VARIETY OF SOLUTES SUCH AS GLUCOSE AND AMINO ACIDS

The **Na$^+$-glucose cotransporters**, SGLTs, some isoforms of which are found in the brush border (apical or luminal) membranes of intestinal and renal epithelial cells, are good examples of carriers that are obliged to cotransport two solutes simultaneously. These transporters (Figure 10-4) are members of a family of more than 35 Na^+-coupled cotransporters. Other members include several Na^+–amino acid cotransporters (e.g., the Na^+-alanine cotransporter), the Na^+-K^+-$2Cl^-$ cotransporter, various neurotransmitter transporters (e.g., the Na^+-norepinephrine, Na^+-dopamine, and Na^+-serotonin cotransporters), and the **Na$^+$– I$^-$ cotransporter.**

Several widely prescribed antidepressant drugs such as fluoxetine (Prozac) act by selectively inhibiting presynaptic serotonin (5-hydroxytryptamine [5-HT]) reuptake by Na^+-serotonin cotransport. This enhances and prolongs the activation of postsynaptic serotonergic neurons by 5-HT.

In thyroid glands, the Na^+-I^- cotransporter (sometimes called the Na^+-I^- symporter) is used to concentrate I^-, a trace element (i.e., one that is

BOX 10-4

THE SODIUM/PROTON EXCHANGER MEDIATES THE ELECTRONEUTRAL EXTRUSION OF PROTONS FROM CELLS

The transport process mediated by the Na^+/H^+ exchanger (NHE) with a $1\ Na^+:1\ H^+$ coupling ratio is described by the following equation:

$$Na^+_{out} + H^+_{in} \rightleftharpoons Na^+_{in} + H^+_{out} \qquad [B1]$$

To determine how the Na^+ and H^+ concentrations and the V_m are related at equilibrium, we equate the total electrochemical potential on the two sides of the membrane (see Chapter 9):

$$\mu_{Na^+,out} + \mu_{H^+,in} = \mu_{Na^+,in} + \mu_{H^+,out} \qquad [B2]$$

Expansion of this equation (see Chapter 9) yields

$$\begin{aligned}
&\mu^0_{Na^+} + RT \ln[Na^+]_o + (+1)(0)F \\
&+ \mu^0_{H^+} + RT \ln[H^+]_i + (+1)V_m F \\
&= \mu^0_{Na^+} + RT \ln[Na^+]_i + (+1)V_m F \\
&+ \mu^0_{H^+} + RT \ln[H^+]_o + (+1)(0)F
\end{aligned} \qquad [B3]$$

Note that the electrical terms drop out of this equation because the coupled exchange of $1\ Na^+$ for $1\ H^+$ is electroneutral, as is also the case for Cl^-/HCO_3^- exchange (see Chapter 9).

Algebraic rearrangement and division by RT (see Box 9-4) then yield the expression:

$$\begin{aligned}
&RT \ln[H^+]_i - RT \ln[H^+]_o \\
&= RT \ln[Na^+]_i - RT \ln[Na^+]_o
\end{aligned} \qquad [B4]$$

or (see Box 9-4):

$$\frac{[H^+]_i}{[H^+]_o} = \frac{[Na^+]_i}{[Na^+]_o} \qquad [B5]$$

Now, solving for $[H^+]_i$, we have:

$$[H^+]_i = \frac{[Na^+]_i}{[Na^+]_o} \times [H^+]_o \qquad [B6]$$

Then, if $[Na^+]_i = 15\ mM$ and $[Na^+]_o = 150\ mM$, at equilibrium the $[H^+]_i$ should be 10-fold lower than $[H^+]_o\ (= 10^{-7.4}\ M)$. Alternatively, if $[H^+]$ is expressed as pH (the negative logarithm of $[H^+]$), intracellular pH (pH_i) should be 1.0 pH unit larger than extracellular pH ($pH_o = 7.4$). In other words, there is sufficient potential energy in the Na^+ electrochemical gradient to lower $[H^+]_i$ by a factor of 10, to approximately $10^{-8.4}\ M$ (i.e., raise pH_i to 8.4), through NHE. However, the fact that pH_i is normally approximately 7.2 and not 8.4 indicates that NHE-mediated proton extrusion is inactivated (as a result of the dissociation of H^+ from the regulatory site) before pH_i reaches the equilibrium value predicted by Equation [B6].

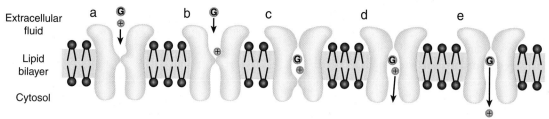

FIGURE 10-4 ■ Mechanism of glucose (G) transport by a Na^+–glucose cotransporter (SGLT-2) with a 1:1 coupling ratio. Note the ordered binding and release of the transported solutes: in the exofacial conformation (*a* and *b*), Na^+ goes on first (*a*), followed by glucose (*b*). In the endofacial conformation (*d* and *e*), the sequence is reversed, and glucose comes off last (*e*). *c* is the occluded conformation with both substrates bound. Not shown is the occluded conformation in which neither substrate is bound; this is essential to effect *net* solute transport by the cotransporter (see text).

present at very low abundance). This process promotes the iodination of tyrosine; iodinated tyrosine is then used to form the thyroid hormones, which are iodothyronines (Box 10-5).

How Does the Electrochemical Gradient for One Solute Affect the Gradient for a Cotransported Solute?

The answer to this question is based on the knowledge that both solutes must be bound to the carrier at the same side of the membrane before the carrier can alter its conformation to translocate either solute. We can visualize this by considering the steps involved in the SGLT-2 carrier cycle (Figure 10-5). This carrier isoform is expressed in the *apical* (brush border) membranes of kidney proximal tubule epithelial cells. It cotransports 1 Na^+ ion and 1 glucose molecule (Figure 10-5B). With unloaded SGLT-2 in the exofacial conformation, the carrier readily binds Na^+ because of the high concentration of Na^+ in the renal tubular lumen. The binding of Na^+ increases the affinity for lumenal glucose, which also then binds to the carrier (Figure 10-5A). The binding of both solutes permits a spontaneous conformational change to the endofacial configuration. Then, because of the

BOX 10-5
SODIUM-IODIDE COTRANSPORT: PHYSIOLOGY AND PATHOPHYSIOLOGY

The sodium-iodide (Na^+-I^-) cotransporter (NIC), which is expressed in thyroid follicular cells, the stomach, lactating mammary gland, and several other cell types, cotransports 1 I^- with 2 Na^+ ions. Thus I^-, an ion normally present in trace amounts, may be concentrated as much as 1000-fold within these cells (see Boxes 10-6 and 10-7). The I^- trapped in the thyroid is used for the synthesis of the thyroid hormones *thyroxine* and *triiodothyronine*. The ability of the NIC to concentrate I^- in the thyroid is used to concentrate radioactive ^{131}I in cancerous thyroid cells, both to detect the spread of the cancer and, after surgery, to destroy the remaining cancer cells by radiation. Moreover, inherited defects in the NIC result in defective I^- trapping and thus in congenital *hypothyroidism* (low thyroid hormone levels).

low concentration of Na^+ in the cytoplasm, as well as the negative V_m (i.e., cytoplasm negative to tubule lumen), the bound Na^+ dissociates readily. This lowers the binding affinity for glucose so that it, too, is discharged into the cytoplasm. The conformational change between exofacial and endofacial configurations can take place only when *both* Na^+ and glucose are bound to the carrier or when *neither* solute is bound. Thus little glucose should exit from the cells through this carrier because the low $[Na^+]_i$ makes Na^+ binding in the endofacial conformation unlikely.

Energetic Consequences of Cotransport If the transport of 1 molecule of glucose is tightly coupled to the transport of 1 Na^+ ion by SGLT-2 (Figure 10-5B), the energy dissipated by the downhill movement of Na^+ can be stored in the glucose gradient. Of course, as Na^+ moves into the cell through SGLT-2 at the *apical membrane*, the Na^+ pump extrudes Na^+ across the *basolateral membrane* (see Chapter 11). This maintains the Na^+ gradients across *both* the apical and basolateral membranes. In this coupled transport system no net gain or loss of energy can occur, so the energy built up in the glucose gradient must equal the energy that is available from the Na^+ gradient (Box 10-6). This type of transport is sometimes called *secondary active transport*. The Na^+ pump (see Chapter 11) uses the energy from ATP to build up and *maintain* the Na^+ electrochemical gradient (*primary active transport*), which is used, in turn, to drive the secondary transport of other solutes.

A more quantitative, thermodynamic treatment is presented in Box 10-6. It shows that the SGLT-2 transporter should be able to concentrate glucose 100-fold within the cell when the Na^+ concentration ratio ($[Na^+]_o/[Na^+]_i$) is 10 and the membrane potential across the apical membrane is −62 mV. This is important in the kidneys, where we need to reabsorb as much glucose from the lumen of the renal tubule as possible.

Glucose Uptake Efficiency Can Be Increased by a Change in the Na^+-Glucose Coupling Ratio

SGLT-2, with an Na^+-glucose coupling ratio of 1:1, is expressed in the early portion of kidney proximal convoluted tubules. The late portion of the proximal convoluted tubules contains a different Na^+-glucose

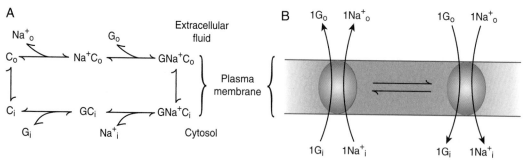

FIGURE 10-5 ■ State diagram (**A**) of net transport reactions (**B**) mediated by the SGLT-2 Na^+–glucose (G) cotransporter (**C**). Subscripts "o" and "i" refer to the extracellular fluid or exofacial configuration of the carrier and the cytosol or endofacial configuration of the carrier, respectively. Note that the carrier can switch between exofacial and endofacial conformations only when it is unloaded (C_o and C_i) or when both Na^+ and G are bound (GNa^+C_o and GNa^+C_i). The transporter can move 1 Na^+ ion and 1 glucose molecule either out of the cell (**B,** *left*) or into the cell (**B,** *right*).

cotransporter isoform, SGLT-1, with an Na^+-glucose coupling ratio of 2:1. SGLT-2 markedly lowers the glucose concentration in the renal tubular fluid. Then SGLT-1, which can concentrate glucose 100-fold more than can SGLT-2 (Box 10-7), further minimizes the loss of glucose in the urine by mediating uptake of most of the residual glucose in the late proximal tubule fluid.

NET TRANSPORT OF SOME SOLUTES ACROSS EPITHELIA IS EFFECTED BY COUPLING TWO TRANSPORT PROCESSES IN SERIES

We have just seen that SGLT-2 and SGLT-1 are capable of markedly concentrating glucose in epithelial cells. If the transported glucose actually accumulated in cells, however, the osmotic stress would cause the cells to gain water and swell (see Chapter 3). Moreover, the object of glucose transport across the apical membranes of the epithelial cells is *not* to concentrate the glucose in these cells. Rather, the object is to transfer glucose from the renal tubule or intestinal lumen, *across the epithelium* to the interstitial space, to transport the glucose into the blood to circulate to other cells in the body. To accomplish this net intestinal absorption (or renal reabsorption) of glucose, GLUT-2, a simple glucose carrier (and an isoform of GLUT-1), is expressed in the basolateral membranes of these epithelial cells. Thus the net transepithelial transport occurs in a two-step process

(Figure 10-6): glucose is transported into these epithelial cells, across their apical membranes, by SGLT-2 or SGLT-1. The glucose diffuses through the cytoplasm to the basolateral membrane and is then transported into the interstitial space by GLUT-2, so that it can be carried, in the blood plasma, to all the other cells in the body. Consequently, no large buildup of glucose occurs in the cytoplasm of the epithelial cells; thus osmotic pressure does not increase and the cells do not swell.

Osmotic problems do not arise in neurons as a result of Na^+-coupled neurotransmitter reuptake for a different reason: the *total* amount of neurotransmitter that must be reaccumulated at the end of an action potential is very small relative to the total cell volume. Cell volume changes therefore are negligible.

Various Inherited Defects of Glucose Transport Have Been Identified

SGLT-1 is also expressed in intestinal tract (jejunum) epithelial cells, where it is responsible for glucose and galactose absorption. Mutations in SGLT-1 result in glucose-galactose malabsorption, a rare syndrome that is manifested as severe, potentially fatal diarrhea (Box 10-8). These mutations in SGLT-1 cause only mild *glycosuria* (glucose in the urine) because most of the renal glucose uptake in the proximal tubules is mediated by SGLT-2. Genetically defective SGLT-2 is associated with a much more marked glycosuria (in this situation, caused by a renal tubular defect rather than by a high blood glucose level and excessive

BOX 10-6

THE ENERGETICS OF COUPLED COTRANSPORT IS EXEMPLIFIED BY THE 1 Na^+:1 GLUCOSE COTRANSPORTER (SGLT-2)

The transport of 1 molecule of glucose (G) is tightly coupled to the transport of 1 sodium (Na^+) ion through the SGLT-2 cotransporter across the apical membrane of a renal proximal tubule cell. Accordingly, for this transporter, the coupled transport reaction is:

$$G_{in} + Na^+_{in} \rightleftharpoons G_{out} + Na^+_{out} \qquad [B1]$$

This transport reaction is somewhat more complex than NHE (see Box 10-4) because Na^+-G cotransport involves the net movement of charge when 1 Na^+ and 1 G move across the membrane. The driving forces for this coupled transport are the electrochemical potential energies for Na^+ on the two sides of the membrane and the chemical potential energies for G (because $z = 0$; see Chapter 9). Equating the potential energies on the two sides of transport equation [B1], we have:

$$\mu_{G,in} + \mu_{Na^+,in} = \mu_{G,out} + \mu_{Na^+,out} \qquad [B2]$$

Expanding this equation (see Chapter 9), with $z = +1$ for Na^+, gives us:

$$\mu_G^0 + RT\ln[G]_i + \mu_{Na^+}^0 + RT\ln[Na^+]_i + (+1)V_mF$$
$$= \mu_G^0 + RT\ln[G]_o + \mu_{Na^+}^0 + RT\ln[Na^+]_o + (+1)(0)F \qquad [B3]$$

Rearrangement then yields:

$$RT\ln[G]_i - RT\ln[G]_o =$$
$$RT\ln[Na^+]_o - RT\ln[Na^+]_i - V_mF \qquad [B4]$$

Dividing by RT, we have:

$$\ln[G]_i - \ln[G]_o = \ln[Na^+]_o - \ln[Na^+]_i - V_mF/RT \qquad [B5]$$

Then, recalling that, for any solute, S, $(\ln[S]_i - \ln[S]_o) = \ln([S]_i/[S]_o)$ (see Appendix B), we can rewrite the preceding equation as:

$$\ln\frac{[G]_i}{[G]_o} = \ln\frac{[Na^+]_o}{[Na^+]_i} - zV_mF/RT \qquad [B6]$$

Taking antilogarithms (see Appendix B), we get:

$$\frac{[G]_i}{[G]_o} = \frac{[Na^+]_o}{[Na^+]_i}e^{\frac{-V_mF}{RT}} \qquad [B7]$$

Thus if $[Na^+]_o = 150$ mM, $[Na^+]_i = 15$ mM, and $V_m = -62$ mV, because $RT/F = 26.7$ mV at 37°C, the expected maximal glucose gradient, $[G]_i/[G]_o$, is:

$$\frac{[G]_i}{[G]_o} = \frac{150}{15} \times e^{2.32} = 10 \times 10 = 100 \qquad [B8]$$

In other words, with a typical Na^+ concentration gradient ($[Na^+]_o/[Na^+]_i = 10$), $[G]_i = 100 \times [G]_o$; that is, the SGLT-2 carrier should be able to concentrate glucose 100-fold within the cell. (Note the importance of V_m when net transfer of charge occurs during the transport cycle.)

---

glucose in the glomerular filtrate, as occurs in diabetes mellitus). SGLT-2 is the main carrier responsible for glucose reabsorption in the kidney, but it has little role in intestinal glucose uptake, whereas SGLT-1 is most important in the intestine.

Fanconi-Bickel syndrome is a rare inherited disorder that results from loss-of-function mutations in GLUT-2, which is expressed in the intestine, kidney, liver, and pancreas. The syndrome is manifested by glucose intolerance and resting hypoglycemia (low blood glucose level). Glucose absorption in the intestine and kidneys is impaired, and the liver exhibits excessive (pathological) glycogen storage because it cannot export the glucose it produces by *gluconeogenesis.*

Na^+ IS EXCHANGED FOR SOLUTES SUCH AS Ca^{2+} AND H^+ BY COUNTERTRANSPORT MECHANISMS

Now let's consider carriers in which the inward (downhill) transport of Na^+ is tightly coupled to the outward transport of other solutes. Two good examples of this second, countertransported solute are protons (Na^+/H^+ exchange, or NHE) and Ca^{2+} ions (**Na^+/Ca^{2+} exchange**). As we have already learned, the NHE is one of several transporters (the Cl^-/HCO_3^- exchanger is another) involved in maintaining intracellular pH in various types

THE ENERGETICS OF A 2 Na⁺:1 GLUCOSE CARRIER (SGLT-1) ILLUSTRATES THE POWER OF THE EXPONENTS IN THE TRANSPORT EQUATION

Because the Na^+-glucose cotransporter, SGLT-1, has a coupling ratio of 2 Na^+:1 glucose, the transport process in this case is (contrast with Box 10-6):

$$G_{in} + 2\,Na^+_{in} \rightleftharpoons G_{out} + 2\,Na^+_{out} \qquad [B1]$$

Rewriting this equation in terms of the electrochemical potential energies, we have:

$$\mu_{G,\,in} + 2\,\mu_{Na^+,in} = \mu_{G,\,out} + 2\,\mu_{Na^+,out} \qquad [B2]$$

Note that, when more than one ion or molecule of a single species is transported during a transporter cycle, the electrochemical potential energy for that species must be multiplied by the number of ions or molecules transported. Expanding this equation (see Box 10-6) now gives us:

$$\mu_G^0 + RT\ln[G]_i + 2\mu_{Na^+}^0 + 2RT\ln[Na^+]_i$$
$$+ 2(+1)V_mF = \mu_G^0 + RT\ln[G]_o + 2\mu_{Na^+}^0 \qquad [B3]$$
$$+ 2RT\ln[Na^+]_o + 2(+1)(0)F$$

Rearrangement, and division by RT, then yields:

$$\ln[G]_i - \ln[G]_o =$$
$$2\ln[Na^+]_o - 2\ln[Na^+]_i - 2V_mF/RT \qquad [B4]$$

Taking antilogarithms (see Box 10-6), and remembering that the antilogarithm of $n\ln X$ is X^n (see Appendix B), gives:

$$\frac{[G]_i}{[G]_o} = \left(\frac{[Na^+]_o}{[Na^+]_i}\right)^2 e^{\frac{-2V_mF}{RT}} \qquad [B5]$$

Thus if $[Na^+]_o = 150$ mM, $[Na^+]_i = 15$ mM, and $V_m = -62$ mV, because $RT/F = 26.7$ mV, the expected maximal glucose concentration gradient, $[G]_i/[G]_o$ is:

$$\frac{[G]_i}{[G]_o} = \left(\frac{150}{15}\right)^2 \times e^{4.64} = 10^2 \times 100 = 10,000 \qquad [B6]$$

In other words, with the same Na^+ electrochemical gradient as is used in the preceding example (see Box 10-6), SGLT-1 should be able to concentrate glucose 10,000-fold within the cell.

---

of cells. The NHE also plays a major role in the reabsorption of Na^+ and excretion of protons in the kidney. The Na^+/Ca^{2+} exchanger, NCX, another important countertransporter, helps to maintain intracellular Ca^{2+} balance. This is critical because of the central role of Ca^{2+} ions in cell signaling.

Na⁺/Ca²⁺ Exchange Is an Example of Coupled Countertransport

The NCX is found in the PM of a large variety of cell types. These include all types of muscle (cardiac, skeletal, and smooth muscle), neurons, and intestinal and renal epithelial cells (where it is prevalent in the basolateral membranes of cells involved in Ca^{2+} absorption or resorption). In these cells the NCX coupling ratio (stoichiometry) is 3 Na^+:1 Ca^{2+} and there is a net movement of one positive charge during each cycle. In other words, the NCX, which is reversible, can mediate

the entry of 3 Na^+ ions in exchange for 1 exiting Ca^{2+} ion, or it can mediate the entry of 1 Ca^{2+} ion in exchange for 3 exiting Na^+ ions (Figure 10-7). With this coupling ratio, and with an Na^+ concentration ratio ($[Na^+]_o/[Na^+]_i$) of 10:1 and a V_m of -62 mV, the Na^+ electrochemical potential difference ($\mu_{Na^+,out} - \mu_{Na^+,in}$) provides sufficient energy to maintain a Ca^{2+} concentration ratio ($[Ca^{2+}]_o/[Ca^{2+}]_i$) of 10,000:1 (Box 10-9). Then, because the free (unbound, ionized) Ca^{2+} concentration in blood plasma ($[Ca^{2+}]_o$) is approximately 1 mM (0.001 M = 10^{-3} M), we would expect the free Ca^{2+} concentration in the cytoplasm ($[Ca^{2+}]_i$) to be approximately 10^{-7} M, or 100 nM (Box 10-9). In fact, this is essentially correct. In most resting cells, $[Ca^{2+}]_i$ is approximately 100 nM, or approximately 1/10,000th of the Ca^{2+} concentration in the extracellular fluid. Thus the exchanger usually operates close to electrochemical equilibrium (Box 10-9).

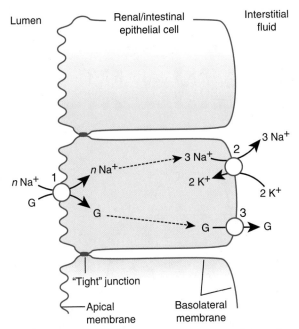

Lumen · Renal/intestinal epithelial cell · Interstitial fluid

FIGURE 10-6 ■ Renal proximal tubule or small intestine epithelial cell illustrating the two-step sequential transport of glucose (G) across the apical and basolateral membranes. G is cotransported, with Na^+, across the epithelial cell apical membrane by the Na^+-G cotransporter SGLT-2 or SGLT-1 *(1)*; n Na^+ ions are transported with 1 G, where $n = 2$ for SGLT-1 and 1 for SGLT-2. Na^+ and G then diffuse, in the cytosol, to the basolateral membrane where Na^+ is extruded by the Na^+ pump *(2)* (see Chapter 11), and G is transported into the interstitial fluid by the simple G carrier GLUT-2 *(3)*.

Na^+/Ca^{2+} Exchange Is Influenced by Changes in the Membrane Potential

As discussed in Chapters 7 and 8, the V_m of excitable cells undergoes marked changes when the cells are activated. This has special significance for Ca^{2+} transport mediated by the NCX in excitable cells because the exchanger is influenced by V_m. (Box 10-9). A noteworthy example is cardiac muscle, in which the AP has a relatively long duration (see Box 10-9, Figure B-1). Because $[Na^+]_o$, $[Na^+]_i$, and $[Ca^{2+}]_o$ do not change significantly during the AP, but $[Ca^{2+}]_i$ does, the direction of the exchanger-mediated Ca^{2+} movement is determined by the changes in V_m and $[Ca^{2+}]_i$. As indicated in Box 10-9, with a resting V_m of -62 mV, there is no net NCX-mediated movement of Ca^{2+}

BOX 10-8
GLUCOSE-GALACTOSE MALABSORPTION

Glucose-galactose malabsorption is a rare disorder of sugar transport. The disease is manifested, beginning in neonatal life, as severe watery, acidic diarrhea that is brought on by ingestion of lactose [milk sugar, 4-(β-D-galactosido)-D-glucose], which is hydrolyzed to glucose and galactose in the intestinal lumen. The disease can be fatal within a few weeks if lactose and glucose (or sucrose, which is hydrolyzed to glucose and fructose) are not removed from the diet. The cause of the disease is a mutational defect in the intestinal brush border SGLT-1 Na^+-glucose cotransporter that virtually abolishes the absorption of glucose and galactose in the intestine. The diarrhea results from retention of these sugars (and Na^+) in the intestinal lumen. These solutes exert an osmotic effect; thus, not only is fluid absorption reduced, but also fluid is drawn from the plasma into the intestinal lumen. For this reason, this type of diarrhea is referred to as osmotic diarrhea.

under the (resting) conditions shown. During the cardiac AP, however, when the membrane depolarizes (Box 10-9, Figure B-1), Ca^{2+} is driven *into* the cells by the NCX. This helps to initiate and maintain cardiac contraction. Conversely, early in diastole (the relaxation phase of the cardiac contraction cycle), when the membrane repolarizes while $[Ca^{2+}]_i$ is still elevated, Ca^{2+} is driven *out* of the cells by the NCX. This promotes cardiac relaxation and recovery.

Na^+/Ca^{2+} Exchange Is Regulated by Several Different Mechanisms

As is the case for many transport systems, the kinetic properties of the NCX are regulated in a tissue-specific manner. This enables the NCX to accommodate to physiological demands. For example, NCX activity may be increased when the exchanger is phosphorylated or when it is activated by phosphatidylinositol bisphosphate (PIP_2) or even by intracellular Ca^{2+}, which acts at a regulatory site distinct from the site involved in ion translocation. For example, even with a high $[Na^+]_i$, the NCX will not transport Ca^{2+} *into* cells if the $[Ca^{2+}]_i$ concentration is low (i.e., ≤100 nM).

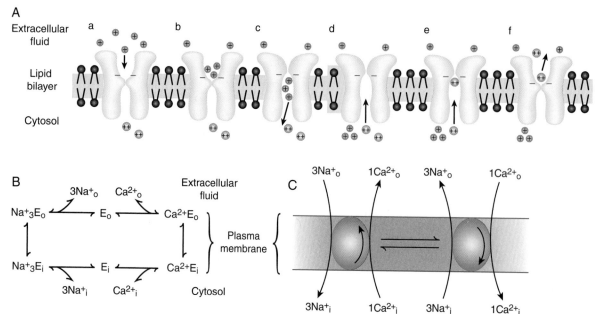

FIGURE 10-7 ■ A, Mechanism of Ca^{2+} transport by the Na^+/Ca^{2+} exchanger (NCX). In the exofacial configuration *(a)*, the NCX can bind either 3 Na^+ *(b)* or 1 Ca^{2+} (not shown). Then, after a transition through an occluded state (not shown) to the endofacial configuration *(c)*, one solute species can dissociate (e.g., Na^+, as in the transition from *c* to *d*) and either Na^+ (not shown) or Ca^{2+} *(e)* can bind. After another conformation change (through the occluded state) to the exofacial conformation, the bound solute (Ca^{2+} in this case) can be discharged to the extracellular fluid *(f)*. State diagram of (**B**) and net transport reactions (**C**) mediated by the NCX (E). Subscripts "o" and "i" refer to the extracellular fluid or exofacial configuration of the exchanger and the cytosol or endofacial configuration of the NCX, respectively. Note that the carrier can switch between exofacial and endofacial conformations only when it is loaded ($Na^+_3E_o$, $Na^+_3E_i$, $Ca^{2+}E_o$, or $Ca^{2+}E_i$); the unloaded carrier does not undergo conformational change (i.e., between E_o and E_i). As shown in **C,** the NCX can either move 3 Na^+ ions into the cell in exchange for 1 exiting Ca^{2+} ion *(left)* or 3 Na^+ ions out of the cell in exchange for 1 entering Ca^{2+} ion *(right)*.

Intracellular Ca^{2+} Plays Many Important Physiological Roles

A low resting $[Ca^{2+}]_i$ is physiologically important because small increases in $[Ca^{2+}]_i$ serve numerous essential second-messenger functions.[*] An increase in $[Ca^{2+}]_i$ triggers contraction in all types of muscle (see Section V); it activates neurotransmitter release at nerve endings (see Chapter 12) and many other secretory processes; it plays a central role in visual and auditory signal transduction; it also controls the

fertilization of the ovum and cell division. Moreover, many pathophysiological processes are associated with deranged Ca^{2+} homeostasis, and Ca^{2+} overload usually leads to cell death. Thus cellular regulation of $[Ca^{2+}]_i$ is extremely important, and the NCX is one of the critical mechanisms involved in this regulation (see Chapter 11).

MULTIPLE TRANSPORT SYSTEMS CAN BE FUNCTIONALLY COUPLED

The same principles of coupled transport (cotransport and countertransport) apply to the transport of numerous other solutes. As exemplified by Cl^-/HCO_3^- exchange (see Chapter 9), some coupled

---

[*]In the 1950s, the Nobel Laureate, Otto Loewi, made the prescient and oft-cited statement, "Ja, Kalzium, das ist alles!" (Calcium is everything!).

BOX 10-9

THE ENERGETICS OF COUPLED COUNTERTRANSPORT IS EXEMPLIFIED BY Na^+/Ca^{2+} EXCHANGE

The cardiac/neuronal plasma membrane Na^+/Ca^{2+} exchanger (NCX) transports 3 Na^+ ions in exchange for 1 Ca^{2+}. Thus the transport equation can be written as:

$$Ca^{2+}_{in} + 3\ Na^+_{out} \rightleftharpoons Ca^{2+}_{out} + 3\ Na^+_{in} \quad [B1]$$

Rewriting this equation in terms of the electrochemical potential energies on the two sides of the PM, we have (see Box 10-7):

$$\mu_{Ca^{2+},in} + 3\ \mu_{Na^+,out} = \mu_{Ca^{2+},out} + 3\ \mu_{Na^+,in} \quad [B2]$$

Expanding this equation (remembering that $z = +2$ for Ca^{2+}) gives:

$$\begin{aligned}
&\mu^0_{Ca^{2+}} + RT\ln[Ca^{2+}]_i + (+2)V_m F \\
&+ 3\ \mu^0_{Na^+} + 3\ RT\ln[Na^+]_o + 3(+1)0F \\
&= \mu^0_{Ca^{2+}} + RT\ln[Ca^{2+}]_o + (+2)0F \\
&+ 3\ \mu^0_{Na^+} + 3\ RT\ln[Na^+]_i + 3(+1)V_m F
\end{aligned} \quad [B3]$$

Rearrangement and division by RT yields:

$$\ln[Ca^{2+}]_i - \ln[Ca^{2+}]_o =$$
$$3\ln[Na^+]_i - 3\ln[Na^+]_o + \frac{V_m F}{RT} \quad [B4]$$

Taking antilogarithms (and remembering that the antilogarithm of $3\ln X$ is X^3) yields:

$$\frac{[Ca^{2+}]_i}{[Ca^{2+}]_o} = \left(\frac{[Na^+]_i}{[Na^+]_o}\right)^3 e^{\frac{V_m F}{RT}} \quad [B5]$$

Thus with $[Na^+]_o = 150$ mM, $[Na^+]_i = 15$ mM, and $V_m = -62$ mV, because $RT/F = 26.7$ mV, we get:

$$\frac{[Ca^{2+}]_i}{[Ca^{2+}]_o} = \left(\frac{[Na^+]_i}{[Na^+]_o}\right)^3 e^{-2.32} = (0.1)^3 \times 0.1 = 0.0001 \quad [B6]$$

Thus $[Ca^{2+}]_i = 0.0001[Ca^{2+}]_o$. Then, because the free Ca^{2+} concentration in blood plasma ($[Ca^{2+}]_o$) is approximately 1.0 mM (0.001 M), we have:

$$[Ca^{2+}]_i = 0.0001 \times 0.001\ M = 0.0000001\ M$$
$$= 100 \times 10^{-9}\ M\ \text{or}\ 100\ nM$$

The equation indicates that $[Ca^{2+}]_i$ should be approximately 100 nM, which is just about what is actually observed in most cells.

Equation B4 can also be rearranged to solve for the V_m at which there is no net NCX-mediated Na^+ or Ca^{2+} transport. This V_m is known as the NCX *reversal potential*, $E_{Na/Ca}$ (Figure 6-3):

$$V_m = E_{Na/Ca} = 3\frac{RT}{F}\ln\frac{[Na^+]_o}{[Na^+]_i} - \frac{RT}{F}\ln\frac{[Ca^{2+}]_o}{[Ca^{2+}]_i} \quad [B7]$$

We recognize (Chapter 4, Box 4-1, Equation B13) that:

$$\frac{RT}{F}\ln\frac{[Na^+]_o}{[Na^+]_i} = E_{Na}\ \text{ and }\ \frac{RT}{2F}\ln\frac{[Ca^{2+}]_o}{[Ca^{2+}]_i} = E_{Ca}$$

Therefore $E_{Na/Ca}$ is related to the equilibrium potentials for the two transported ions:

$$E_{Na/Ca} = 3E_{Na} - 2E_{Ca} \quad [B8]$$

In cardiac muscle cells, $[Ca^{2+}]_i$, and, thus, E_{Ca}, changes during the action potential (Figure B-1A), with consequent changes in $E_{Na/Ca}$. Knowing all the ion concentrations enables us to use Equation B7 to calculate $E_{Na/Ca}$, which changes throughout the AP (Figure B-1B).

The driving force on the NCX-mediated current, $V_m - E_{Na/Ca}$ (Figure B-1C), enables us to understand how the NCX operates during the cardiac AP. Before the AP, $V_m - E_{Na/Ca}$ is negative. Thus an inward current and net efflux of Ca^{2+} help to keep $[Ca^{2+}]_i$ low (i.e., during each NCX cycle, 3 Na^+ enter and 1 Ca^{2+} exits, resulting in a net entry of 1 positive charge). During the AP upstroke, $V_m - E_{Na/Ca}$ becomes positive and large, and the NCX mediates outward current and net Ca^{2+} entry. During the AP plateau (at ~50 to 250 milliseconds), $V_m - E_{Na/Ca}$ is very small and there is little NCX-mediated net flux of Ca^{2+}. Repolarization (after ~250 milliseconds) makes $V_m - E_{Na/Ca}$ negative and large, driving Ca^{2+} out of the myocytes. Importantly, the high $[Ca^{2+}]_i$ during the AP also exerts a regulatory role: it activates the NCX so that efflux through the exchanger is particularly rapid during repolarization.

Continued

THE ENERGETICS OF COUPLED COUNTERTRANSPORT IS EXEMPLIFIED BY Na^+/Ca^{2+} EXCHANGE—cont'd

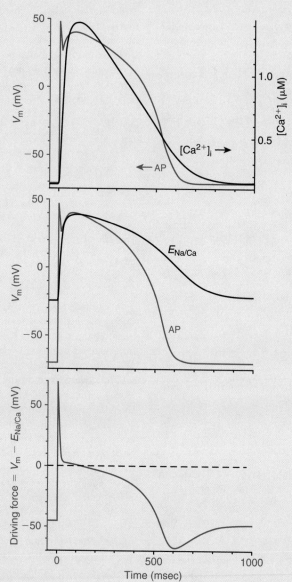

FIGURE B-1 ■ The relationship between V_m and $[Ca^{2+}]_i$ (top), the NCX reversal potential ($E_{Na/Ca}$) *(middle)*, and the electrochemical driving force on the NCX ($V_m - E_{Na/Ca}$) *(bottom)* during a cardiac action potential (AP). Note that the NCX reverses direction to favor Ca^{2+} influx during the upstroke of the AP. The NCX reverses direction again during the AP plateau, to favor Ca^{2+} efflux during repolarization. *(Recalculated from data in Weber CR, Piacentino V 3rd, Houser SR, Bers DM: Circulation 108:2224, 2003.)*

solute transport systems do not use Na^+. Nevertheless, like most Na^+-coupled systems, the RBC and kidney Cl^-/HCO_3^- exchanger (also known as anion exchanger type 1, or AE1) operates near equilibrium so that it can use the gradient of either one of the transported ions to move the other.

AE1 mediates the exchange of the anion of a weak acid (H_2CO_3 [carbonic acid]) for the anion of a strong acid (HCl [hydrochloric acid]). In the distal convoluted tubule of the kidney, AE1 is used to reabsorb HCO_3^- and extrude Cl^- into the tubular lumen, thereby acidifying the urine. Consequently, hereditary defects in AE1 are associated with *renal tubular acidosis* because the kidneys cannot excrete sufficient acid.

Tertiary Active Transport

Figure 10-8 shows how a metabolic intermediate, α-ketoglutarate ($αKG^{2-}$), is coupled to the countertransport of organic anions (OA^-) in the basolateral membrane of the kidney *proximal* tubule cells. This transporter, the multispecific OA transporter-1, or OAT-1, is a major route of drug excretion. Examples of OA^- transported by this system are urate, *p*-aminohippurate (PAH^-, an agent used to measure renal plasma flow), the penicillins, and salicylates (e.g., acetylsalicylic acid, or aspirin, is excreted primarily as salicylurate). Mitochondrial metabolism provides the $αKG^{2-}$ that is transported out of the cell in exchange for entering OA^-. OAT-1 is selective for $αKG^{2-}$: other dicarboxylic acids such as succinate, fumarate, and malate cannot replace the $αKG^{2-}$. So that $αKG^{2-}$ will not be wasted (and cellular $αKG^{2-}$ levels will be maintained), $αKG^{2-}$ is transported back into the cells across the basolateral membrane by an Na^+/dicarboxylate cotransporter with a high affinity for dicarboxylates (NaDC-3); it couples the transport of 1 $αKG^{2-}$ to 3 Na^+. Indeed, the Na^+ pump indirectly promotes the transport of OA^- into the cells by inducing a gradient for $αKG^{2-}$ (so-called **tertiary active transport**). The OA^- does not accumulate in the cell; it leaves the cell across the apical membrane by a different transporter. The net effect is secretion of the OA^- while Na^+ and $αKG^{2-}$ are recycled (Figure 10-8). This is particularly important for rapid clearance of xenobiotics ("foreign" chemicals) from the body.

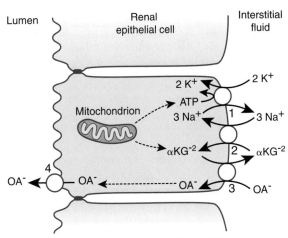

FIGURE 10-8 ■ Renal tubule epithelial cell illustrating the two-step sequential transport of organic anions (OA^-) across the basolateral and apical membranes. To move from the interstitial fluid into the cell cytosol, OA^- is transported across the basolateral membrane by an OA^-/α-ketoglutarate ($αKG^{-2}$) exchanger, OAT-1 (3). The OA^- then diffuses through the cytosol to the apical membrane; it is next transported into the renal tubular lumen by another organic acid carrier (4). The $αKG^{-2}$ is a Krebs cycle intermediate produced in mitochondria; it is reclaimed from the interstitial fluid by the basolateral $Na^+/αKG^{-2}$ cotransporter, NaDC-3, which couples the transport of 3 Na^+ to 1 $αKG^{-2}$ (2). The Na^+ electrochemical gradient that drives this cotransporter is maintained by the basolateral membrane Na^+ pump (1) (see Chapter 11). *(Redrawn from Dantzler WH, Wright SH: Comprehensive toxicology. Vol 7. Renal toxicology, Oxford, 1997, Pergamon.)*

Indeed, this efficient transport system can clear (i.e., remove) 90% of some agents from plasma in a single passage through the kidneys, provided that the plasma concentration of the agents is not too high. This high clearance is the reason that PAH^- can be used to estimate renal blood flow.

Net secretion of organic cations (OCs) such as verapamil and morphine occurs by an analogous mechanism. In this case the facilitated diffusion step is at the basolateral membrane: independent carriers mediate monovalent cation and divalent cation transport into epithelial cells. The basolateral membrane voltage ($≈-50$ to -70 mV) provides the energy to concentrate these solutes approximately 10- or 100-fold inside the cells. In the apical membrane an OC^+/H^+

exchanger that transports both OC^+ and OC^{2+} serves as the secondary active transport system to effect net secretion of the OCs into the renal tubular fluid. The protons are recycled across the apical membrane by an NHE (H^+ ions are extruded into the tubule lumen and Na^+ ions are taken up). The Na^+ is conserved: it is transported into the interstitial space by the Na^+ pump in the basolateral membrane.

SUMMARY

1. Biological membranes are poorly permeable to polar solutes. Specialized carrier and pump proteins are therefore needed to mediate the transfer of specific solutes across the PM and to maintain the intracellular concentrations of critical solutes.

2. Solute carriers are integral membrane proteins that have kinetic properties similar to those of enzymes, but carriers mediate *vectorial* transport across biological membranes.

3. Some carriers simply speed the rate at which their substrate solutes (e.g., glucose) diffuse down a transmembrane concentration or electrochemical gradient. These *simple carriers* mediate *facilitated diffusion*. They cannot maintain a solute concentration or electrochemical gradient.

4. Some carriers couple the transport of two solutes. These cotransporters or countertransporters can generate a concentration or electrochemical gradient for one solute when the coupled solute, often Na^+, moves down its concentration or electrochemical gradient. This process is sometimes called secondary active transport.

5. The NHE is an example of electroneutral ion countertransport. It is used to extrude H^+ rapidly when pH_i falls to less than approximately 7.4. The entering Na^+ is then extruded against its electrochemical gradient by the PM Na^+ pump (see Chapter 11).

6. Na^+-glucose cotransporters, which are expressed in the apical membranes of intestinal and renal epithelia, play important roles in the absorption and reabsorption of glucose from the intestinal and renal tubule lumen. These cotransporters mediate transport with a net movement of charge; therefore Na^+-glucose cotransport is sensitive to V_m.

7. Among the numerous other cotransporters are the Na^+-I^- cotransporter, various Na^+–amino acid and Na^+-neurotransmitter cotransporters, and the Na^+-K^+-Cl^- cotransporter.

8. The NCX is expressed in many types of cells, including muscle and neurons. It mediates the voltage-sensitive exchange of 1 Ca^{2+} for 3 Na^+ ions and uses the energy stored in the Na^+ electrochemical gradient to help maintain a resting $[Ca^{2+}]_i$ of approximately 100 nM.

9. Among the many other types of countertransporters are the Cl^-/HCO_3^- exchanger and the multispecific OAT-1.

10. The properties (substrate affinities or turnover rates) of some carriers can be regulated by mechanisms such as phosphorylation and dephosphorylation or the membrane insertion and retrieval of carriers. Solute transport rates are thus modulated to accommodate physiological needs.

KEY WORDS AND CONCEPTS

- Polar versus nonpolar solutes
- Aquaporin
- Transporter turnover number
- Carrier-mediated transport
- Facilitated diffusion
- Passive transport
- Transition state
- Occluded solute
- Simple glucose carrier
- Secondary active transport
- Cotransport (symport)
- Sodium-glucose cotransport
- Sodium-iodide cotransport
- Countertransport (antiport or exchange)
- Sodium/calcium exchange
- Sodium/proton exchange
- Chloride/bicarbonate exchange
- Tertiary active transport

STUDY PROBLEMS

1. Why are Na^+-dependent cotransport systems generally limited to epithelial cells, whereas Na^+-dependent countertransport systems (and other exchangers such as the Cl^-/HCO_3^- exchanger) are expressed in all types of cells?

2. What $[Ca^{2+}]_i$ concentration would be expected in a cell with $V_m = -60$ mV and $[Ca^{2+}]_o = 1$ mM if intracellular Ca^{2+} is at electrochemical equilibrium? Why is this physiologically unrealistic?

3. Photoreceptor (retinal rods and cones) Na^+/Ca^{2+} exchangers mediate the exchange of 4 Na^+ for 1 Ca^{2+} + 1 K^+. This exchanger, like the cardiac/neuronal NCX (which exchanges 3 Na^+ for 1 Ca^{2+}), operates close to equilibrium (i.e., $4\ \mu_{Na^+,out} + \mu_{Ca^{2+},in} + \mu_{K^+,in} = 4\ \mu_{Na^+,in} + \mu_{Ca^{2+},out} + \mu_{K^+,out}$). Can you explain why an exchanger with this coupling ratio is needed in the photoreceptors to maintain "resting" $[Ca^{2+}]_i$ concentration at approximately 100 nM? (Hint: Under dark conditions, retinal photoreceptor $[Na^+]_i \approx 30\text{-}40$ mM, $[K^+]_i \approx 120$ mM, and $V_m \approx -40$ mV.)

4. What is the possible advantage of using NHE rather than another transport system to extrude protons if this exchanger does not operate close to equilibrium?

5. What is the significance of having Na^+-glucose cotransporters with different coupling ratios located at different levels of the kidney proximal tubule to minimize the spillover of glucose in the urine? Assume that the plasma glucose (molecular weight 180) concentration is 5 mM. This is also the concentration in the glomerular filtrate (180 liters/day) that enters the proximal tubules; 67% of the fluid is reabsorbed (i.e., 120 liters/day) in the proximal tubules, the only renal tubule segment in which glucose is absorbed.

BIBLIOGRAPHY

Abramson J, Wright EM: Structure and function of Na^+-symporters with inverted repeats, *Curr Opin Struct Biol* 19:425, 2009.

Blaustein MP, Lederer WJ: Sodium/calcium exchange: its physiological implications, *Physiol Rev* 79:763, 1999.

De La Vieja A, Dohan O, Levy O, et al: Molecular analysis of the sodium/iodide symporter: impact on thyroid and extrathyroid pathophysiology, *Physiol Rev* 80:1083, 2000.

Karmazyn M, Gan XT, Humphreys RA, et al: The myocardial Na^+-H^+ exchange: structure, regulation, and its role in heart disease, *Circ Res* 85:777, 1999.

Klepper J, Wang D, Fischbarg J, et al: Defective glucose transport across brain tissue barriers: a newly recognized neurological syndrome, *Neurochem Res* 24:587, 1999.

Lytton J: Na^+/Ca^{2+} exchangers: three mammalian gene families control Ca^{2+} transport, *Biochem J* 406:365, 2007.

Philipson KD, Nicoll DA: Sodium-calcium exchange: a molecular perspective, *Annu Rev Physiol* 62:111, 2000.

Riedel C, Dohan O, De La Vieja A, et al: Journey of the iodide transporter NIS: from its molecular identification to its clinical role in cancer, *Trends Biochem Sci* 26:490, 2001.

Sher AA, Noble PJ, Hinch R, et al: The role of the Na^+/Ca^{2+} exchanger in Ca^{2+} dynamics in ventricular myocytes, *Prog Biophys Mol Biol* 96:377, 2007.

Simpson IA, Carruthers A, Vannucci SJ: Supply and demand in cerebral energy metabolism: the role of nutrient transporters, *J Cereb Blood Flow Metab* 27:1766, 2007.

Thorens B, Mueckler M: Glucose transporters in the 21st century, *Am J Physiol* 298:E141, 2010.

Wright EM, Hirayama BA, Loo DF: Active sugar transport in health and disease, *J Intern Med* 261:32, 2007.

Zachos NC, Tse M, Donowitz M: Molecular physiology of intestinal Na^+/H^+ exchange, *Annu Rev Physiol* 67:411, 2005.

ACTIVE TRANSPORT

OBJECTIVES

1. Understand how the Na^+ pump uses energy from ATP to keep $[Na^+]_i$ low and $[K^+]_i$ high by transporting Na^+ and K^+ against their electrochemical gradients.

2. Understand how Ca^{2+} is sequestered in the sarcoplasmic and endoplasmic reticulum and transported across the plasma membrane by ATP-dependent active transport systems.

3. Understand how intracellular Ca^{2+} is controlled and Ca^{2+} signaling is regulated by the cooperative action of many transport systems.

4. Understand the roles of ATP-dependent transport systems in the transport of such ions as protons and copper, as well as a variety of other solutes.

5. Understand how different transport systems in the apical and basolateral membranes of epithelia, which separate two different extracellular compartments, act cooperatively to effect net transfer of solutes and water across epithelial cells.

PRIMARY ACTIVE TRANSPORT CONVERTS THE CHEMICAL ENERGY FROM ATP INTO ELECTROCHEMICAL POTENTIAL ENERGY STORED IN SOLUTE GRADIENTS

In Chapter 10, we learned how energy stored in the Na^+ electrochemical gradient can be used to generate concentration (or electrochemical) gradients for other (coupled) solutes. This is called secondary active transport because a *preexisting* electrochemical energy gradient is dissipated in one part of the transport process (e.g., the downhill movement of Na^+) to generate the chemical or electrochemical gradients of other solutes (e.g., glucose or Ca^{2+}). There is no *net* expenditure of metabolic energy by these transporters.

The question we need to address here is: How does the Na^+ concentration gradient (typically, $[Na^+]_o/[Na^+]_i \approx 10$ to 15) become established in the first place? This brings us to the role of ATP in powering **primary active transport.** During active ion transport, adenosine triphosphatases (ATPases) interconvert chemical (phosphate bond) energy and electrochemical potential (ion gradient) energy. These straightforward chemical reactions can, depending on the concentrations of substrates and products, operate in either the forward or the reverse direction; that is, they can either use (hydrolyze) or synthesize ATP.

Three Broad Classes of ATPases Are Involved in Active Ion Transport

The three classes of ion transport ATPases are the F-, V-, and P-type ATPases. Mitochondria possess F-type (F_1F_0) ATPases that synthesize ATP with energy stored in the proton electrochemical gradient across the inner mitochondrial membrane; the proton gradient is generated by oxidative metabolism. Vacuolar (V-type) H^+–ATPases lower intraorganellar pH by concentrating protons in a variety of vesicular organelles, including lysosomes and secretory and storage vesicles. Neither the F-type nor the V-type ATPases form stable phosphorylated intermediates. **P-type ATPases,** which do form stable **phosphorylated intermediates** that can be isolated chemically, transport numerous ions and other solutes into and out of cells and organelles. Examples of P-type ATPases are the PM Na^+ pump (Na^+, K^+-ATPase), the PM and **sarcoplasmic reticulum/ endoplasmic reticulum (S/ER) Ca^{2+}-ATPases (PMCA and SERCA),** and the gastric mucosa proton pump (H^+,K^+-ATPase). These P-type ATPases are the focus of much of this chapter.

THE PLASMA MEMBRANE Na$^+$ PUMP (Na$^+$,K$^+$-ATPase) MAINTAINS THE LOW Na$^+$ AND HIGH K$^+$ CONCENTRATIONS IN THE CYTOSOL

Nearly All Animal Cells Normally Maintain a High Intracellular K$^+$ Concentration and a Low Intracellular Na$^+$ Concentration

In most cells in mammals, including humans, $[K^+]_i \approx$ 120–130 mM, and $[Na^+]_i \approx$ 5–15 mM. The extracellular fluid, however, has a high $[Na^+]_o$ ($\sim$145 mM) and a low K^+ concentration $[K^+]_o$ ($\sim$4–5 mM). Moreover, cells are not impermeable to Na^+ and K: Na^+ and K^+ channels and Na^+ gradient–dependent transport systems (see Chapters 7 and 10) permit Na^+ to enter cells and K^+ to exit as the ions move down their respective electrochemical gradients. Therefore, all cells expend energy in the form of ATP to generate and maintain their normal Na^+ and K^+ electrochemical gradients. The transporter that accomplishes this work is the **sodium pump** or **Na$^+$,K$^+$-ATPase.** In the nervous system and the kidneys, the Na^+ pump accounts for a very large fraction (75% to 85%) of total ATP hydrolysis. The transport of Na^+ and K^+ by the Na^+ pump compensates for the leak of these ions into and out of the cell, respectively. This is known as the **pump-leak**

model of Na^+ and K^+ homeostasis. The Na^+ pump not only maintains constant $[Na^+]_i$ and $[K^+]_i$, but also influences cell volume. How the Na^+ pump contributes to cell volume maintenance is addressed in the next section.

The Na$^+$ Pump Hydrolyzes ATP While Transporting Na$^+$ Out of the Cell and K$^+$ Into the Cell

The Na^+ pump is an integral PM protein whose major (α, or "catalytic") subunit has 10 membrane-spanning helices (Figure 11-1) and contains the ATP and ion binding sites. The α-subunit is closely associated with a smaller, highly glycosylated, β-subunit that has a single membrane-spanning domain. Complexes of α- and β-subunits, in a 1:1 ratio, are required for Na^+ pump activity, but how the β-subunit functions is unknown. The Na^+ pump is frequently called the Na^+,K^+-ATPase because the protein is an enzyme

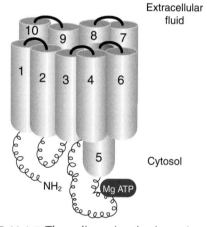

FIGURE 11-1 ■ Three-dimensional schematic model of the α- (catalytic) subunit of the Na^+ pump. This subunit consists of 10 membrane-spanning helical domains (cylinders in the figure). The large cytoplasmic loop between transmembrane helices *4* and *5* contains the ATP binding domain (shown) and the aspartate phosphorylation site. The ion binding sites are located in transmembrane helices *4, 5, 6,* and *8.* Residues that bind ouabain are located on the external surfaces of helices *1, 2, 5, 6,* and *7;* thus, bound ouabain may block access to the cation binding sites. *(Modified from Lingrel JB, Croyle ML, Woo AL, et al: Acta Physiol Scand Suppl 643:69, 1998.)*

(specifically, an ATPase) that requires both Na^+ and K^+ for its catalytic activity (ATP hydrolysis).*

The Na^+ pump hydrolyzes 1 ATP molecule to ADP and inorganic phosphate (Pi) while transporting 3 Na^+ ions out of the cell and 2 K^+ ions into the cell. The transport cycle begins with the binding of ATP (as the Mg^{2+}-ATP complex) at the hydrolytic site on the α-subunit (Figure 11-1). When 3 Na^+ ions bind to the pump on the cytoplasmic side, the ATP is cleaved and its terminal, high-energy phosphate is transferred to the α-subunit. This phosphorylation enables the protein to undergo a conformational change so that the bound Na^+ becomes transiently inaccessible ("occluded") to both the intracellular and extracellular fluids. The Na^+ binding site then opens to the extracellular fluid. This conformational change also markedly reduces the Na^+ affinity, while greatly increasing K^+ affinity. Thus, the 3 Na^+ ions are able to dissociate even though $[Na^+]_o \approx 145$ mM. Then, when 2 K^+ ions bind, the protein undergoes another conformational change. As the α-subunit–phosphate bond is cleaved, Pi is released into the cytoplasm, and the K^+ binding sites close to the external surface (i.e., the 2 K^+ ions are transiently occluded) and then open to the internal surface. The 2 K^+ ions are released into the cytoplasm because the affinity for K^+ decreases markedly during this conformational change. This sequence of steps in the Na^+ pump cycle is illustrated in Figure 11-2A.

The net reaction for the Na^+ pump can be written as Equation [1]:

$$3\ Na^+_{cyt} + 2\ K^+_{ECF} + 1\ ATP_{cyt} \rightleftharpoons$$
$$3\ Na^+_{ECF} + 2\ K^+_{cyt} + 1\ ADP_{cyt} + 1\ Pi_{cyt} \quad [1]$$

This net reaction can be diagrammed as shown in Figure 11-2B. Note that this is a straightforward chemical reaction; it can be reversed and can generate ATP if the product concentrations are greatly increased and the substrate concentrations are greatly reduced.

As a result of the 3 Na^+:2 K^+ coupling ratio, *inhibition* of the Na^+ pump will lead to a net gain of solute

(as Na^+ salts, to maintain electroneutrality) and a rise in osmotic pressure. The cells will therefore gain water and swell (see discussion of the Donnan effect in Chapter 4). Thus *the Na^+ pump participates directly in cell volume maintenance.*

The Na^+ Pump Is "Electrogenic"

The reaction sequence (Equation [1]) reveals that during each Na^+ pump cycle, one more positive charge leaves the cell than enters. This net flow of charge (i.e., outward "pump current") across the membrane generates a small voltage (cytoplasm negative). The Na^+ pump is therefore said to be electrogenic. Indeed, this voltage adds to the V_m, so that the actual resting V_m is slightly more negative than the V_m calculated from the Goldman-Hodgkin-Katz (GHK) equation (see Chapter 4). The maximum voltage that can be generated by the Na^+ pump with a coupling ratio of 3 Na^+: 2 K^+, under steady-state conditions, is approximately 10 mV. In practice, however, the contribution of the electrogenic Na^+ pump to the resting V_m (i.e., in the steady state) in most cells is only a few millivolts (1 to 4 mV) and is usually ignored. When $[Na^+]_i$ rises significantly, as in neurons after a long burst of action potentials, the rate of Na^+ transport by the Na^+ pump can increase considerably. Under these *non*–steady-state conditions, the Na^+ pump may transiently hyperpolarize the cells by 20 mV or more, thereby temporarily reducing the ability of stimuli to excite the cells.

The Na^+ Pump Is the Receptor for Cardiotonic Steroids Such as Ouabain and Digoxin

The Na^+ pump α-subunit is uniquely sensitive to a class of drugs known as **cardiotonic steroids**. Two examples, **digoxin** and **ouabain**, were originally discovered in plants, but ouabain also is synthesized in humans and other mammals (Box 11-1). Cardiotonic steroids inhibit the Na^+ pump and, as described later, thereby induce a cardiotonic effect (increased force of contraction of the heart, or *positive inotropic effect*). This is the key feature of cardiotonic steroid therapeutic efficacy in heart failure.

There are four molecular **isoforms** of the **Na^+ pump α-subunit**, α1 to α4, which differ in their affinities for Na^+, K^+, and cardiotonic steroids. These isoforms have been conserved during vertebrate

---

*The Na^+,K^+-ATPase was identified in 1957 by Jens Skou. He was awarded the Nobel Prize for this work in 1997.

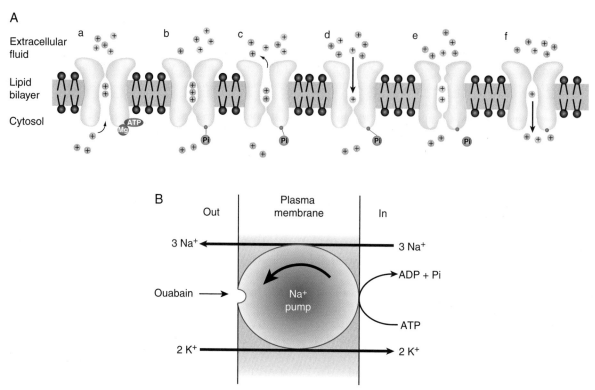

FIGURE 11-2 ■ **A,** Sequence of steps in the Na$^+$ pump cycle illustrates the mechanism of operation of the pump. The cycle begins with the binding of ATP to the large cytoplasmic loop *(a),* followed by the binding of three Na$^+$ ions *(gray circles)* from the cytosol *(a).* This enables the terminal phosphate of ATP to be transferred to the α-subunit, and the three Na$^+$ ions to be transiently occluded *(b).* The three Na$^+$ ions are then released to the extracellular fluid (ECF; *c).* Two K$^+$ ions *(blue circles)* from the ECF bind *(d)* and, following cleavage of the Pi, are transiently occluded *(e).* The cycle ends with the release of the two K$^+$ into the cytosol *(f).* **B,** Net reaction mediated by the Na$^+$ pump. Note that the ouabain binding site faces the ECF.

evolution. All cells express α1 and one other isoform; α1 is responsible for maintaining the low [Na$^+$]$_i$ in "bulk" cytoplasm.

Expression of specific α-subunit isoforms is upregulated or downregulated under various physiological and pathophysiological conditions. For example, in the heart, thyroid hormone increases, and heart failure decreases α2 expression. In kidney distal tubules, aldosterone upregulates α1, which then promotes Na$^+$ reabsorption and retention. In addition, several hormones, such as dopamine, vasopressin, and serotonin (5-hydroxytryptamine [5-HT]), modulate the activity of the Na$^+$ pump in a tissue- and isoform-specific manner. These hormones activate or inactivate the pump by promoting phosphorylation of the pump

at sites other than the site that is phosphorylated during ion transfer. New understanding about the significance of the isoforms is beginning to emerge (Box 11-2 and Figure 11-3).

INTRACELLULAR Ca^{2+} SIGNALING IS UNIVERSAL AND IS CLOSELY TIED TO Ca^{2+} HOMEOSTASIS

Intracellular Ca^{2+} signaling (a change in the concentration of free Ca^{2+} ions in the cytoplasm) is directly or indirectly involved in most cell processes, from sexual reproduction and cell division to cell death. Ca^{2+} ions are crucial in the fertilization of the ovum, in muscle

BOX 11-1

OUABAIN IS A HUMAN HORMONE IMPLICATED IN THE PATHOGENESIS OF HYPERTENSION (HIGH BLOOD PRESSURE)

The cardiotonic steroids (CTSs) derive their name from the fact that they improve the performance of the heart. Digoxin comes from the leaves of the foxglove plant, *Digitalis purpurea*, and ouabain comes from the bark of the ouabaio tree, *Acokanthera ouabaio*. Some pharmacologically related CTSs, the bufadienolides, are produced by poisonous toads of the genus *Bufo*. *Digitalis* steroids, such as digoxin, have been used clinically to treat heart failure and certain cardiac arrhythmias for more than 200 years, and they are still used frequently. Digoxin is lipid soluble; it can be administered orally and is readily absorbed. Ouabain is not used clinically because it is highly water soluble and, thus, poorly absorbed.

All cells have Na^+ pumps with a CTS binding site, but the physiological significance is unknown. Surprisingly, ouabain has been identified as a mammalian hormone that is secreted in the adrenal cortex and the hypothalamus.

Adrenocorticotropic hormone (ACTH) secreted by the pituitary gland, and catecholamines released by sympathetic neurons (see Chapter 13), stimulate secretion of ouabain by the adrenal gland. This "endogenous ouabain" apparently plays a role in the pathogenesis of some forms of hypertension. Excess circulating ACTH induces hypertension in humans and animals, but not in mice that have mutated ouabain-resistant $\alpha 2$ Na^+ pumps. This finding implies that the ouabain binding site is involved in the ACTH-induced-hypertension. Approximately 40% of patients with essential hypertension (i.e., hypertension of unknown cause) have significantly higher blood plasma levels of ouabain than are found in normotensive subjects (i.e., those with normal blood pressure). Moreover, chronic subcutaneous administration of ouabain, but not digoxin, induces hypertension in rodents; indeed, digoxin *counteracts* this effect of ouabain.

BOX 11-2

Na^+ PUMP ISOFORM LOCALIZATION, FUNCTION, AND PATHOPHYSIOLOGY

A clue to the isoform-specific functions of the Na^+ pump is that the high-affinity ouabain binding site has been conserved on the $\alpha 2$ and $\alpha 3$ isoforms during vertebrate evolution, whereas $\alpha 1$ ouabain binding affinity varies greatly. In addition, $\alpha 1$ is distributed relatively uniformly in the PM of many types of cells, whereas $\alpha 3$ (expressed in some neurons) and $\alpha 2$ are confined to PM microdomains that overlie sub-PM ("junctional") components of the S/ER. Interestingly, the NCX, but not PMCA, colocalizes with the $\alpha 2$ and $\alpha 3$ Na^+ pumps. Thus, the Na^+ and Ca^{2+} concentrations in these junctional cytosolic spaces and the adjacent S/ER may be governed by the $\alpha 2$ or $\alpha 3$ Na^+ pumps and the NCX. This organization of

transporters, diagrammed in Figure 11-3, may help explain how low doses of ouabain and other cardiotonic steroids exert large effects on $[Ca^{2+}]_i$ and Ca^{2+} signaling. This is exemplified by increased cardiac contractility (the *cardiotonic* effect; see Box 11-1).

Rare loss-of-function mutations in the $\alpha 2$-subunit give rise to *familial hemiplegic migraine* (FHM). Interestingly, certain gain-of-function mutations in voltage-gated Na^+ channels or Ca^{2+} channels also can cause FHM. A possible unifying feature is that FHM is the result of gain of Ca^{2+} which, in the case of mutant $\alpha 2$ Na^+ pumps or Na^+ channels, is mediated by NCX as a result of the elevated $[Na^+]_i$.

contraction, and in hormone and neurotransmitter secretion. Ca^{2+} ions are also involved in the control of electrical excitability (e.g., through Ca^{2+}-activated K^+ channels; see Chapter 8) and in the regulation of many protein kinases, protein phosphatases, and other

enzymes. Cell Ca^{2+} overload usually leads to cell death, and protection from Ca^{2+} overload may rescue damaged cells. Thus, an appreciation of cell Ca^{2+} homeostasis is essential for understanding many physiological and pathophysiological processes.

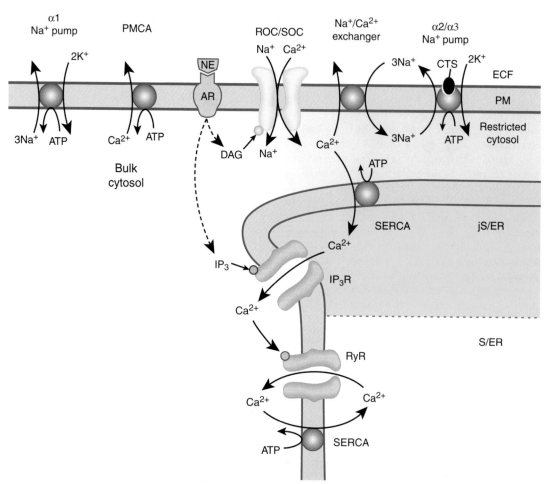

FIGURE 11-3 ■ Small portion of a cell showing the plasma membrane–junctional sarcoplasmic/endoplasmic reticulum (jS/ER) region. The PM that faces "bulk" cytosol contains $\alpha1$ Na^+ pumps, PMCA, and various ligand receptors such as the adrenergic receptor (AR) for norepinephrine (NE) shown here. The PM microdomain adjacent to the jS/ER contains $\alpha2$ (or, in neurons, $\alpha3$) Na^+ pumps, the NCX, and receptor-operated/store-operated cation channels (ROCs and SOCs, which are permeable to both Na^+ and Ca^{2+}; see Chapter 8). Ligand activation of PM receptors such as the AR promotes the synthesis of inositol trisphosphate and diacylglycerol (IP_3 and DAG, respectively; see Chapter 13). The DAG opens ROCs; S/ER Ca^{2+} store depletion opens SOCs. The S/ER membrane contains SERCA, as well as IP_3 receptors (IP_3Rs) and ryanodine receptors (RyRs), which are also Ca^{2+} release channels. Activation of RyRs (e.g., by elevating $[Ca^{2+}]_i$, as illustrated) or IP_3Rs (by IP_3 binding) opens the channels and releases Ca^{2+} from the S/ER into the bulk cytosol (see Chapters 13 and 15). Ouabain (or other cardiotonic steroids [CTS]) inhibits the $\alpha2/\alpha3$ Na^+ pumps and raises $[Na^+]$ primarily between the PM and jS/ER. The altered Na^+ gradient across the PM in this region reduces the driving force for Ca^{2+} extrusion by the NCX. This raises the $[Ca^{2+}]$ locally, enabling SERCA to store more Ca^{2+} in the jS/ER (*shaded area* of S/ER) so that more is released when the cells are activated. ECF, extracellular fluid. (*Modified from Blaustein MP, Wier WG: Circ Res 101:959, 2007.*)

The Ca^{2+} involved in cell signaling comes from the extracellular fluid (it may enter through a variety of Ca^{2+}-permeable channels; see Chapter 8) or from intracellular Ca^{2+} stores in the **endoplasmic reticulum (ER)** or, in muscle, the **sarcoplasmic reticulum (SR)**. This "signal Ca^{2+}" must then either be extruded across the PM or be resequestered in the S/ER. The PM NCX, which couples Ca^{2+} to Na^+ homeostasis, is described in Chapter 10. Here we consider other mechanisms involved in Ca^{2+} transport and their roles in Ca^{2+} homeostasis.

Ca^{2+} Storage in the Sarcoplasmic/Endoplasmic Reticulum is Mediated by a Ca^{2+}-ATPase

In Chapter 10 we noted that the cytosolic free (ionized) Ca^{2+} concentration ($[Ca^{2+}]_i$) in most cells at rest is approximately 100 nM (10^{-7} M or 0.0001 mM). The *total* intracellular Ca^{2+} concentration is generally approximately 1000 to 10,000 times higher than this, however, or approximately 0.1 to 1 mM. Thus more than 98% of the intracellular Ca^{2+} is sequestered in intracellular organelles, although a small amount is buffered (i.e., bound to cytoplasmic proteins, such as calmodulin, and to other molecules). The primary Ca^{2+} storage site is the ER or, in muscle, the SR, but a small amount is also normally concentrated in mitochondria. The S/ER is a system of interconnected tubules and sacs within the cytoplasm that plays a central role in Ca^{2+} signaling. Some elements of the S/ER lie just beneath the PM and are specialized for Ca^{2+} signal initiation or amplification. When cells are activated (e.g., by hormones, neurotransmitters, or depolarization), Ca^{2+} is often released from the S/ER stores. This released Ca^{2+} can trigger such processes as contraction and secretion (see Chapters 12 and 15). Subsequently, the Ca^{2+} is resequestered in the S/ER. How is this Ca^{2+} sequestration accomplished?

The S/ER Ca^{2+} pump, SERCA, uses 1 ATP to transport 2 Ca^{2+} ions from the cytosol to the S/ER lumen and 2 protons (H^+ ions) from the lumen to the cytosol by a transport mechanism analogous to that of the Na^+ pump.

$$2\ Ca^{2+}_{cyt} + 1\ ATP_{cyt} + 2\ H^+_{S/ER\ lumen} \rightleftharpoons$$
$$2\ Ca^{2+}_{S/ER\ lumen} + 1\ ADP_{cyt} + 1\ Pi_{cyt} + 2\ H^+_{cyt} \quad [2]$$

Details of the molecular conformations and transport mechanisms of both SERCA and the Na^+ pump have been elucidated by X-ray crystallography.

The 2 to 3 mM ATP in the cytosol provides enough energy to enable SERCA to concentrate Ca^{2+} in the S/ER lumen more than 1000-fold relative to the cytosol. The intra-S/ER *free* Ca^{2+} concentration is approximately 0.15 to 0.5 mM, but the S/ER lumen also contains Ca^{2+} binding proteins (e.g., calsequestrin and calreticulin) that bind and buffer the Ca^{2+}. Thus, if 80% to 90% of the intra-S/ER Ca^{2+} is bound, the *total* Ca^{2+} concentration in the lumen may be as high as several millimolar (Box 11-3).

SERCA Has Three Isoforms

The three isoforms of SERCA, SERCA1 to SERCA3, are the products of different genes whose expression is cell type specific. SERCA1 and SERCA2a are expressed in skeletal and cardiac muscles, respectively. Release of SR Ca^{2+} is essential for triggering contraction in both skeletal and cardiac muscles (see Chapter 15). Therefore, SERCA-mediated resequestration of the released Ca^{2+} plays a key role in muscle relaxation.

BOX 11-3

A LARGE QUANTITY OF Ca^{2+} IS STORED IN THE SARCOPLASMIC OR ENDOPLASMIC RETICULUM

A conservative estimate is that 80% to 90% of the Ca^{2+} in the SR or ER is bound to the proteins calsequestrin or calreticulin. The free (ionized) Ca^{2+} concentration in the S/ER is approximately 0.2 mM, and the total Ca^{2+} concentration (free + bound) in the S/ER is approximately 1 to 2 mM. If the S/ER encloses 2% to 5% of the cell volume, rapid release of *all* the stored Ca^{2+} should increase cytosolic $[Ca^{2+}]$ by 0.02 to 0.10 mM. These values are approximately 10 to 50 times larger than the largest Ca^{2+} signals evoked by physiological stimuli. The S/ER therefore contains more than sufficient Ca^{2+} to account for the observed increases in $[Ca^{2+}]_i$ even in the absence of Ca^{2+} entry from the extracellular fluid. Indeed, as discussed in Chapter 15, *all* the Ca^{2+} required to activate skeletal muscle contraction is derived from the SR.

In Brody's disease, a mutation in the SERCA1 gene impairs Ca^{2+} uptake into the SR and slows skeletal muscle relaxation (Box 11-4). Interestingly, genetic defects in certain Ca^{2+} pumps can also underlie some skin diseases, although the underlying mechanisms are unknown. Mutations in SERCA2 cause Darier's disease, which presents with wart-like blemishes over large areas of the skin and mucous membranes, and sometimes with neurological problems (impaired intellectual ability and epilepsy). Mutations in the gene that encodes a V-type Ca^{2+}-ATPase expressed in Golgi apparatus membranes, are associated with Hailey-Hailey disease (familial benign pemphigus). This presents with frequent outbreaks of painful rashes and blisters.

The Plasma Membrane of Most Cells also Has an ATP–Driven Ca^{2+} Pump

The PM contains, in addition to the NCX, an ATP-driven Ca^{2+} pump, PMCA, which is distinct from SERCA. The PMCA and NCX function in parallel to regulate $[Ca^{2+}]_i$. The NCX, with its 10-fold higher *rate* of Ca^{2+} transport (see Table 10-1) than PMCA, plays the dominant role in Ca^{2+} extrusion during recovery from activation, especially in cells with a large activity-induced Ca^{2+} influx such as cardiac muscle. Conversely, the PMCA has a 10-fold higher *affinity* for intracellular Ca^{2+} than the NCX, so the PMCA appears to be particularly important for keeping the $[Ca^{2+}]_i$ concentration very low under resting conditions.

BOX 11-4
A MUTATION IN THE SERCA1 GENE IMPAIRS SKELETAL MUSCLE RELAXATION

Brody's disease is a rare, nonlethal, inherited disorder of Ca^{2+} sequestration in skeletal muscle sarcoplasmic reticulum (SR). It is the result of a mutation in the SERCA1 gene that markedly slows Ca^{2+} transport into the SR. The disease is manifested as defective skeletal muscle relaxation that worsens rapidly during exercise. This impairment of function can readily be explained by the markedly decreased rate of Ca^{2+} sequestration into the SR that prolongs the contractile state (see Chapter 15).

The Roles of the Several Ca^{2+} Transporters Differ in Different Cell Types

The PMCA and SERCA, along with the NCX, govern Ca^{2+} homeostasis, but their functional interrelationships are complex and cell-type specific. In skeletal muscle, all the Ca^{2+} for contraction comes from the SR and is resequestered in the SR by SERCA1 during relaxation. In contrast, a large fraction of the Ca^{2+} for cardiac muscle contraction comes from the extracellular fluid and enters through voltage-gated Ca^{2+} channels. This Ca^{2+} must be extruded across the cardiac muscle PM (sarcolemma), and here the NCX plays a major role in removing Ca^{2+} from the cytosol during the relaxation phase ("diastole") of each cardiac cycle. Thus, in the heart, the Na^+ pump plays an important role in Ca^{2+} homeostasis because the Na^+ electrochemical gradient drives the NCX. In many smooth muscles, Ca^{2+} entry through voltage-gated and receptor-operated channels (see Chapter 8) and Ca^{2+} release from the SR (see Chapter 15) contribute to the rise of $[Ca^{2+}]_i$ that activates contraction. The NCX is involved not only in normal Ca^{2+} homeostasis, but also in pathophysiology: for example, NCX expression is greatly increased in arterial smooth muscle in several forms of hypertension.

Mitochondrial Ca^{2+} homeostasis also is crucial for cell function. Several mitochondrial Ca^{2+} transport systems are involved in controlling intramitochondrial $[Ca^{2+}]$. Indeed, rises in cytosolic $[Ca^{2+}]$ during cell activity induce increases in intramitochondrial $[Ca^{2+}]$. This stimulates the Krebs cycle enzymes and, thus, spurs oxidative metabolism and ATP production. Furthermore, mitochondrial Ca^{2+} overload, which may occur when cytosolic $[Ca^{2+}]$ cannot be adequately controlled by the PM and S/ER transporters, often plays a role in cell death.

Different Distributions of the NCX and PMCA in the Plasma Membrane Underlie Their Different Functions

Why do cells express both the NCX and PMCA, both of which can extrude Ca^{2+}? The specific localization of the Ca^{2+} transporters provides clues to transporter function. In many cell types, $\alpha2$ or $\alpha3$ Na^+ pumps, NCX, and receptor-operated/store-operated channels

BOX 11-5

DO CARDIOTONIC STEROIDS EXERT THEIR CARDIOTONIC EFFECT WITHOUT ELEVATING $[Na^+]_i$?

Cardiotonic steroids (CTSs) inhibit the Na^+ pump selectively and can be expected to elevate $[Na^+]_i$. Nanomolar concentrations of CTSs such as digoxin or ouabain, however, apparently exert their cardiotonic effects without measurably elevating $[Na^+]_i$ in the cell as a whole ("bulk" $[Na^+]_i$). How can this be explained? The high ouabain affinity of the $\alpha 2$ and $\alpha 3$ isoforms and the localization of the various Na^+ and Ca^{2+} transporters (see Box 11-2 and Figure 11-3) are consistent with the sequence of events shown at the right.

The main point is that negligible change in *total* cell Na^+ is needed to account for the augmented Ca^{2+} signaling induced by low-dose CTSs. Increase in $[Na^+]$ in the tiny space between the PM and junctional S/ER is sufficient to explain the positive inotropic effect induced by CTSs. Based on similar reasoning, a *reduction* of local $[Na^+]$ apparently underlies the relaxation of intestinal smooth muscle that is induced by β-adrenergic agonists such as isoproterenol, which stimulates the Na^+ pump (see study problems). Thus, control of

this local, sub-PM $[Na^+]$ apparently plays a critical role in regulating Ca^{2+} signaling in a large variety of cell types.

Inhibition of Na^+ pump $\alpha 2/\alpha 3$ isoforms
by nanomolar CTSs
⇓
$\uparrow[Na^+]$ in the tiny space between
the PM and junctional S/ER (jS/ER)
⇓
$\downarrow Ca^{2+}$ exit and/or $\uparrow Ca^{2+}$ entry via NCX
in PM microdomains adjacent to jS/ER
⇓
$\uparrow[Ca^{2+}]$ in the tiny space between the PM and jS/ER
⇓
$\uparrow[Ca^{2+}]$ in the lumen of the jS/ER
(mediated by SERCA; see Box 11-3)
⇓
$\uparrow Ca^{2+}$ release from the S/ER
whenever the cells are activated

(see Chapter 8) colocalize in PM microdomains that overlie junctional S/ER, or jS/ER (Figure 11-3). The PMCA and $\alpha 1$ Na^+ pumps are apparently excluded from these PM microdomains but are widely distributed elsewhere in the PM. Thus, the $\alpha 1$ Na^+ pumps and PMCA have housekeeping roles: they maintain low $[Na^+]_i$ and $[Ca^{2+}]_i$ in bulk cytosol. In contrast, the transporters in the junctional PM microdomains work together to regulate Ca^{2+} signaling; by delivering Ca^{2+} directly to SERCA, they modulate the pool of Ca^{2+} stored in the S/ER. The expression of these PM microdomain proteins is apparently coordinated: for example, NCX and certain cannonical transient receptor potential channel proteins (TRPCs; see Chapter 8) and, in some cases, $\alpha 2$ Na^+ pumps, are upregulated in arterial smooth muscle in several types of hypertension. The coordinated activity of $\alpha 2$ and NCX in the regulation of Ca^{2+} signaling also is illustrated by the cardiotonic and vasotonic effects of low

concentrations of cardiotonic steroids (Box 11-5 and Figure 11-3).

SEVERAL OTHER PLASMA MEMBRANE TRANSPORT ATPases ARE PHYSIOLOGICALLY IMPORTANT

H^+,K^+-ATPase Mediates Gastric Acid Secretion

The **gastric H^+,K^+-ATPase,** is a P-type ATPase that mediates acid secretion into the lumen of the stomach. Pepsin, the gastric peptidase, has optimum enzymatic activity at pH $\approx$ 3. The gastric glands secrete nearly isotonic hydrochloric acid (HCl) (145 mM; pH = 0.084); this is diluted in the gastric lumen to yield a final pH $\approx$ 3.

The gastric H^+,K^+-ATPase is a proton (H^+) pump in the apical membrane of the parietal cells in the gastric epithelium. This pump, which moves H^+ from the cytoplasm to the gastric lumen in exchange for K^+, is structurally homologous and functionally similar to the Na^+ pump. The regulation of the H^+, K^+-ATPase is, however, markedly different. Few copies of this transporter are present in the parietal cell apical membrane between meals; the H^+ pump molecules are, instead, located in the membranes of *tubulovesicles* that lie just beneath the apical membrane. This prevents digestion of the gastric epithelium. Ingestion of food activates neurons of the vagus nerve to promote secretion of gastrin (a local peptide hormone). Gastrin stimulates nearby enterochromaffin-like cells to release histamine. The histamine activates parietal cell histamine type-2 (H_2) receptors that, through a cAMP–mediated mechanism, induce the tubulovesicles containing the H^+,K^+-ATPase to fuse with the apical membrane. The cells can then pump H^+ into the gastric lumen and K^+ into the cells. At the same time, apical membrane Cl^- and K^+ permeabilities are both increased. The net effect is HCl secretion because Cl^- exits passively, through apical membrane Cl channels, whereas K^+ is recycled across the apical membrane (Figure 11-4). The Cl^- comes from the plasma and enters the cells across the basolateral membrane in exchange for

HCO_3^- (another role for Cl^-/HCO_3^- exchange; see Box 9-5). Knowledge of the mechanism of acid secretion has found widespread application in the treatment of gastric hyperacidity ("heartburn") and gastroesophageal reflux (Box 11-6).

Two Cu^{2+}-Transporting ATPases Play Essential Physiological Roles

Copper (Cu^{2+}) is an essential trace metal because several key metalloenzymes such as cytochrome c oxidase (involved in mitochondrial electron transport) and dopamine β-hydroxylase (required for catecholamine synthesis; see Chapter 13) require Cu^{2+}. Cu^{2+} is absorbed in the intestine by a two-step process: it most likely enters the cells passively across the apical membrane and is then actively transported out across the basolateral membrane. The Cu^{2+} is bound to albumin in the plasma and is carried to the liver, the critical organ for Cu^{2+} homeostasis. The liver, which synthesizes *ceruloplasmin*, a Cu^{2+}-binding protein, secretes free Cu^{2+} into the bile and secretes Cu^{2+}-ceruloplasmin complexes into the plasma. The ceruloplasmin then ferries the Cu^{2+} to all cells that must use small amounts of this cation.

Deficiency of the Cu^{2+}-requiring metalloenzyme activities has serious medical consequences. Genetic

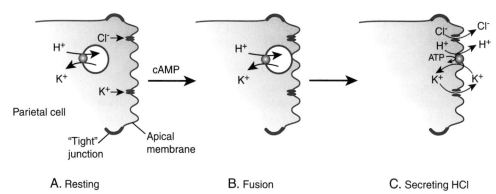

FIGURE 11-4 ■ Mechanism of hydrochloric acid (HCl) secretion by the gastric parietal cell. **A,** Before stimulation of the parietal cell by gastrin or histamine, most H^+,K^+-ATPase molecules are located in subapical vesicle membranes. Ingestion of a meal leads to stimulation of cAMP production in the parietal cell. **B,** The elevated [cAMP] promotes fusion of subapical vesicles with the apical membrane. **C,** At the same time, apical membrane Cl^- and K^+ channels are activated. The net effect is stimulation of HCl secretion and recycling of K^+ across the apical membrane.

BOX 11-6

TREATMENT OF GASTRIC HYPERACIDITY ("HEARTBURN") WITH AN H⁺,K⁺-ATPASE INHIBITOR

Postprandial (i.e., after a meal) gastric hyperacidity is a very frequent clinical problem. The most common treatment is acid neutralization with a mild alkali, such as Tums. A second, frequently used therapy involves block of parietal cell histamine (H_2)-receptors with drugs such as cimetidine (Tagamet), ranitidine (Zantac), and famotidine (Pepcid). If these treatments are inadequate, the H^+,K^+-ATPase can be blocked directly with omeprazole (Prilosec) or lansoprazole (Prevacid). The latter two agents are irreversible inhibitors of the ATPase and are long acting because the cells must synthesize new H^+, K^+-ATPase molecules to compensate for the loss of the original molecules.

analyses of two rare inherited diseases, Menkes' disease and Wilson's disease (Box 11-7), led to the discovery of two critical P-type **Cu^{2+}-transporting ATPases.** Serum Cu^{2+} and ceruloplasmin levels are low in both Wilson's disease and Menkes' disease. Menkes' disease is manifested as an apparent Cu^{2+} deficiency because Cu^{2+} is accumulated in the intestinal mucosa, as well as in the kidneys, lungs, pancreas, and spleen, but not in the liver or brain (where it is actually present in abnormally low amounts). Intestinal accumulation of Cu^{2+} results from a genetic defect in the ATPase that transports Cu^{2+} out of the intestinal mucosal cells across the basolateral membrane so that Cu^{2+} cannot be absorbed from the intestinal lumen.

In contrast, Wilson's disease is manifested as a toxic accumulation of Cu^{2+} primarily in the liver and brain, but also in the kidneys and cornea. The underlying problem is a genetic defect in a different Cu^{2+}-transporting ATPase, which exports Cu^{2+} across the hepatocyte apical (canalicular) membrane and into the bile.

BOX 11-7

MENKES' DISEASE AND WILSON'S DISEASE ARE CAUSED BY MUTATIONS IN DIFFERENT Cu²⁺ TRANSPORT ATPases

Menkes' disease is characterized by mental retardation, convulsions, progressive neurodegeneration, and multiple connective tissue disorders. The classic feature is kinky, steely hair (like steel wool). The disease is lethal, usually by 3 years of age. The disorder results from a defect in an X-linked recessive gene. The gene encodes the Cu^{2+}-transporting ATPase that exports Cu^{2+} from intestinal mucosal cells or renal tubule cells, across the basolateral membrane, to the interstitial space. Consequently, insufficient Cu^{2+} is absorbed from the intestinal lumen or reabsorbed by the kidneys. The manifestations of this disease are the result of greatly reduced activity of various Cu^{2+}-requiring metalloenzymes, such as cytochrome c oxidase and dopamine β-hydroxylase.

Wilson's disease is characterized by hepatitis or cirrhosis, neurological manifestations (e.g., tremors), and psychotic symptoms. A diagnostic feature is greenish yellow Kayser-Fleischer rings in the cornea, caused by copper deposits. The disease results from a defect in an autosomal recessive gene that encodes a Cu^{2+}-ATPase that exports Cu^{2+} from hepatocytes (liver cells) to the bile canaliculi. The inability to export Cu^{2+} from the liver accounts for the toxic Cu^{2+} accumulation in the liver and the consequent liver disease.

ATP-Binding Cassette Transporters Are a Superfamily of P-Type ATPases

Multidrug Resistance Transport ATPases Transport Many Different Types of Agents The human genome codes for three classes of ATPases that actively transport drugs (often as conjugates) across PMs. These classes are the P-glycoproteins, the breast cancer resistance proteins, and the **multidrug resistance proteins (MRPs)**. These proteins are all members of a superfamily of **ATP-binding cassette (ABC) membrane transporters**, 48 of which are encoded in the human genome. ABC transporters are, like the major facilitator superfamily (see Box 10-2), one of the largest superfamilies of proteins across all species. They all use energy from ATP hydrolysis to transport, actively, a large variety of chemically unrelated substances, including *xenobiotics* (foreign biologically active substances) such as chemotherapeutic agents. The various ABC transporters, which may be either exporters or importers, have different solute selectivities, but the precise mechanism of solute selectivity is not understood.

Two well-studied examples of the MRP class of transporters are MRP1 and MRP2. MRP1 is widely distributed, but its level of expression is normally low in the liver, where a homologous, functionally similar protein, MRP2, is highly expressed. ABC transporters are physiologically important (e.g., MRP1 transports leukotriene C_4, and hepatic MRP2 plays a key role in bilirubin glucuronide secretion into the bile). They also are involved in many medically important phenomena, including cystic fibrosis, resistance to anticancer agents, and bacterial resistance to antibiotics.

Figure 11-5 illustrates the novel mechanism of transport used by some MRPs. Some neutral solutes are cotransported with glutathione (GSH, the tripeptide γ-glutamyl-cysteinyl-glycine); some solutes are conjugated to GSH and then transported. In addition, MRPs transport some organic anions as free ions, and they also transport some solutes as glucuronate conjugates or sulfate conjugates.

In many instances, administration of cytotoxic agents (including anticancer drugs) can upregulate an MRP so that tumors that initially are sensitive to an agent such as doxorubicin (Adriamycin) can become resistant (i.e., the MRPs are cytoprotective). Because many MRPs have broad selectivity, this upregulation may cause the tumor to become resistant to multiple

FIGURE 11-5 ■ Two modes of transport mediated by multidrug resistance proteins (MRPs) such as MRP1 and MRP2. The transport may involve cotransport of glutathione (GSH) with a neutral organic ligand [e.g., the vinca alkaloid, vincristine (VNC), an anticancer agent] or extrusion of a glutathione (GS)-coupled solute [e.g., cisplatin (CSP), another anticancer agent]. Although not illustrated here, MRPs may also transport organic anions in an uncoupled manner or other solutes as glucuronate (e.g., bilirubin-glucuronide) or sulfate conjugates. Each MRP has its own, unique spectrum of substrates.

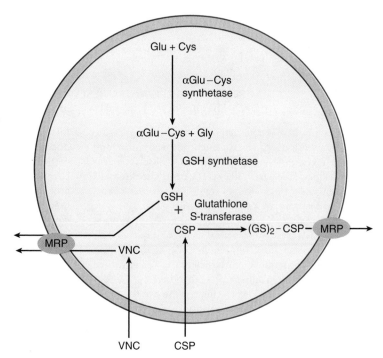

drugs, hence the name multidrug resistance proteins. Interestingly, during upregulation, mutant MRP genes may be preferentially expressed, so the substrate selectivity of the gene product may change with time.

Certain agents, including the Ca^{2+} channel blocker verapamil and the antiarrhythmic agent quinidine, block MRPs and thereby enhance the sensitivity to the antitumor agents. Use of these blockers has a serious drawback, however: normal cells are then also prevented from extruding these cytotoxic agents, and this may lead to intolerable side effects.

The Cystic Fibrosis Transmembrane Conductance Regulator Is a Cl⁻ Channel The **cystic fibrosis transmembrane conductance regulator (CFTR)** is another member of the ABC superfamily. CFTR is unusual in that it functions in part as a Cl⁻ channel and in part as a regulator of several other conductances. Various loss-of-function mutations in CFTR cause cystic fibrosis (see later). The most common mutations result in CFTR misfolding so that the Cl⁻ channel cannot be properly trafficked to and inserted into the PM.

NET TRANSPORT ACROSS EPITHELIAL CELLS DEPENDS ON THE COUPLING OF APICAL AND BASOLATERAL MEMBRANE TRANSPORT SYSTEMS

Epithelia Are Continuous Sheets of Cells

An epithelium is a sheet of cells that forms the lining of a surface or cavity in the body. Epithelial cells are joined by special **tight junctions** with variable permeability. These cells form a continuous sheet, usually one cell layer thick (Figure 11-6). A good structural analogy is a six-pack of beer cans joined by a plastic sheet (the tight junctions) with holes for the six cans (the cells). As we shall see, the "tightness" of the junctions (measured as "leakiness" or electrical conductance) varies considerably among epithelia and is thereby responsible for markedly different functional properties.

The cells in an epithelium are polarized so that each cell has an apical and a basolateral membrane. The apical surface faces the lumen of the cavity lined by the epithelium, whereas the basolateral membrane is in contact with the interstitial fluid (Figures 11-6 and 11-7). The apical surface is sometimes called lumenal or mucosal, and the basolateral surface is sometimes called serosal. Tight junctions separate the apical and basolateral membranes, which face solutions of different composition, express different sets of transport proteins, and can have different permeabilities to solutes and water. Solutes and water may move across the epithelium between cells through intercellular junctions (the **paracellular pathway**). Alternatively, these substances may move through the cells (the **transcellular pathway**). In the latter case, solutes are transported across the apical and basolateral membranes by different, selective transporters. In this two-step process, the transporters are arranged in series, as exemplified by the net uptake of glucose in the proximal small intestine and proximal renal tubules and by renal secretion of organic cations and anions (see Chapter 10).

Epithelia Exhibit Great Functional Diversity

Here we consider the general principles of transepithelial transport and the integration and coordination of multiple transport processes that contribute to the overall function of the intestinal, renal, and other epithelia. First, we explore the source of the Na^+ that is required for the numerous Na^+-coupled apical transport systems. Anion transport is then discussed, followed by water transport. Although epithelia are functionally diverse, one common feature is the presence of Na^+ pumps in the basolateral membranes. The identities of other transport proteins in apical and basolateral membranes of the epithelial cell, as well as the leakiness of the paracellular pathway (regulated by small proteins called *claudins*), determine the specific transport properties of the various epithelia. These other transport proteins then determine whether net transport of the various solutes is from lumen to interstitial fluid (absorption) or from interstitial fluid to lumen (secretion).

In Chapter 10 we learned how sequential expression of the Na^+-glucose cotransporters SGLT-2 and SGLT-1 along the nephron maximizes the reabsorption of glucose from the lumenal fluid in kidney proximal tubules. Similarly, other specific transport systems are expressed in the various epithelial cell types along the gastrointestinal (GI) tract and the renal tubules. In the more proximal segments this

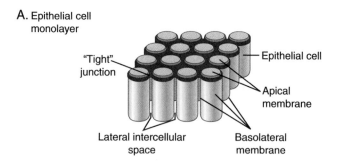

A. Epithelial cell monolayer

"Tight" junction

Epithelial cell

Apical membrane

Lateral intercellular space

Basolateral membrane

FIGURE 11-6 ■ Epithelial cell monolayer. **A,** Apical surface view. **B,** Cross-section through the epithelial cells shows the apical and basolateral surfaces, the lateral intercellular spaces, and the transcellular and paracellular pathways across the epithelium. **C,** Epithelial membrane potentials: V_a, potential across the apical membrane; V_{bl}, potential across the basolateral membrane; and V_{te}, the potential in the lumen relative to that in the interstitial space (i.e., the transepithelial potential, $V_{te} = V_{bl} - V_a$). *(Redrawn and modified from Friedman MH: Principles and models of biological transport, Berlin, 1986, Springer-Verlag.)*

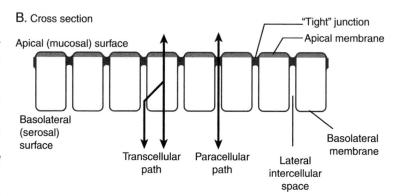

B. Cross section

Apical (mucosal) surface

"Tight" junction

Apical membrane

Basolateral (serosal) surface

Transcellular path   Paracellular path   Lateral intercellular space

Basolateral membrane

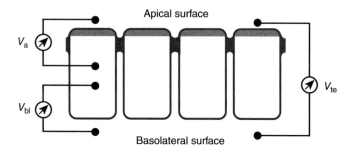

C. Epithelial potentials

Apical surface

V_a

V_{bl}

V_{te}

Basolateral surface

maximizes salt and water absorption, and in the more distal segments it refines the absorption of solutes and water and the secretion of solutes.

Ion gradients across the apical and basolateral membranes are established by the Na$^+$ pump and various secondary active transport processes. The membrane potentials across the apical and basolateral membranes, V_a and V_{bl}, are determined by the Na$^+$, K$^+$, and Cl$^-$ concentration gradients and by their relative permeabilities across the two membranes. The transepithelial potential, V_{te}, is thus defined as the electrical potential in the lumen relative to that in the interstitial space surrounding the basolateral surface of the epithelial cells (see Figure 11-6). V_{te} is equal to the difference, $V_{bl} - V_a$, where V_{te} may be either negative or positive. These electrical potentials are important for solute transport because the passive movement of an ion is driven not only by its concentration gradient, but also by the electrical potential gradient (see Chapter 9).

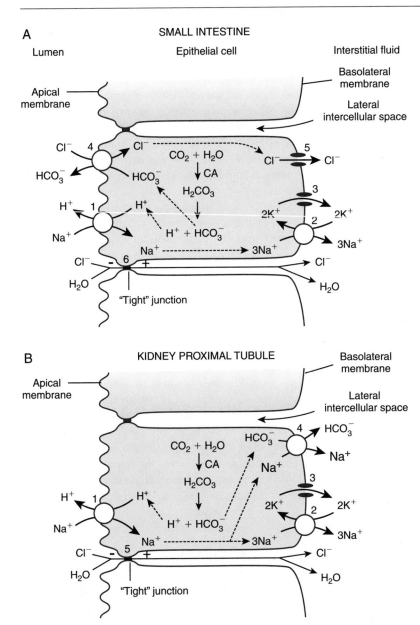

A SMALL INTESTINE

FIGURE 11-7 ■ Model epithelial cells such as an intestinal jejunal or ileal cell (**A**), or a renal proximal tubule cell (**B**), illustrate two slightly different mechanisms of net (re)absorption of NaCl and H_2O. The models show the uptake of Na^+ across the apical membrane through Na^+/H^+ exchangers (1), the extrusion of Na^+ by basolateral membrane α1 Na^+ pumps (2), and the recycling of K^+ through basolateral K^+ channels (3). H^+ and HCO_3^- are generated in the cells by carbonic anhydrase (CA). In the intestine (**A**), HCO_3^- is extruded into the lumen by the apical Cl^-/HCO_3^- exchanger (4), and the entering Cl^- is then extruded into the interstitium through basolateral Cl^- channels (5). Some Cl^- absorption also occurs through the paracellular pathway (6). In the kidney proximal tubules (**B**), the HCO_3^- is extruded into the interstitium by Na^+-HCO_3^- cotransporters (4), and transepithelial Cl^- movement occurs principally through the paracellular pathway (5). In both tissues, water absorption is a consequence of the osmotic gradient established by the solute movement. The ATP needed to drive the Na^+ pump is not shown. H_2CO_3, carbonic acid.

What Are the Sources of Na^+ for Apical Membrane Na^+-Coupled Solute Transport?

Humans normally ingest a modest amount of Na^+ (on average ~100 to 150 mmol/day), although dietary Na^+ may be extremely low (<15 mmol/day) in some nonindustrialized societies, such as the Yanomamo Indians of Northern Brazil. Nevertheless, extracellular Na^+ salts play an essential role in maintaining plasma volume. This implies that the body's Na^+ must be carefully conserved and that the Na^+ required for Na^+-solute cotransport must be recycled in the body. This principle is fundamental to transport processes in the GI tract and kidney tubules.

As noted earlier in this chapter, HCl is secreted into the lumen of the stomach. This acid must be neutralized in the small intestine because most digestion in the intestine occurs at neutral or alkaline pH. Consequently, gastric mucous glands, the biliary system, the exocrine pancreas, and Brunner's glands in the duodenal wall all secrete alkaline solutions rich in sodium bicarbonate ($NaHCO_3$). The $NaHCO_3$ neutralizes the HCl from the stomach and leaves NaCl in the intestinal lumen. Additional HCO_3^- is provided by an apical membrane Cl^-/HCO_3^- exchanger in the intestinal epithelial cells, driven by the Cl^- in the lumen. The Cl^- that enters the epithelial cell through the exchanger then escapes into the interstitial fluid through Cl^- channels in the basolateral membrane (Figure 11-7A). The luminal Na^+ promotes Na^+-solute cotransport across the apical membrane of the small intestine columnar epithelial cells (see Chapter 10). Reclamation of Na^+ occurs across the apical membrane principally through Na^+/H^+ exchange, which also introduces protons to neutralize excess luminal HCO_3^- (Figure 11-7A). The secreted protons originate in the epithelial cell cytoplasm by the action of the enzyme carbonic anhydrase on CO_2, a product of oxidative metabolism. The Na^+ that enters the epithelial cells is extruded into the interstitium by $\alpha 1$ Na^+ pumps, which are expressed only in the basolateral membrane. Thus, much of the Na^+ that was secreted higher up in the GI tract is reclaimed in the jejunum and ileum. The K^+ that enters the epithelial cell through the Na^+,K^+-ATPase is recycled through basolateral K^+ channels (Figure 11-7A).

An analogous mechanism for Na^+ recycling occurs in the kidneys (Figure 11-7B). Here, the glomeruli filter the blood and produce a nearly protein-free **ultrafiltrate** of plasma that, in normal adults, amounts to approximately 180 liters per day of a solution isoosmotic to plasma ($\sim$290 mOsm/kg), in which the major electrolytes are Na^+, Cl^-, and HCO_3^-. This fluid then enters the proximal tubules, but 99% of the Na^+, Cl^-, and H_2O is reabsorbed before the final urine is formed ($\sim$1.5 liters/day). Approximately 67% of the Na^+ is reabsorbed in the proximal tubules. Some of this Na^+ is cotransported with sugars and amino acids, but much of it is reabsorbed by Na^+/H^+ exchange (Figure 11-7B) as in the intestine. Some of the H^+ transported into the lumen is recycled through the organic cation/H^+ exchanger (see Chapter 10), but much of the H^+ reacts with HCO_3^- in the tubular fluid to form H_2O and CO_2. The CO_2 can then reenter the

cells to start another hydration cycle (i.e., to form more H^+). The HCO_3^- is extruded by a basolateral Na^+-HCO_3^- cotransporter and, thus, is conserved. Primary active transport of Na^+ across the basolateral membrane of the epithelial cells maintains the Na^+ and K^+ electrochemical gradients. These examples demonstrate that the Na^+ pump, directly or indirectly, drives *all* the aforementioned transport processes (Figure 11-7). Thus, it is not surprising that the *Na^+ pumps account for up to 85% of all the ATP hydrolysis in the kidneys.*

Another important aspect of the Na^+ pump activity in epithelia is the very large amount of K^+ transported into the cells. Most of this K^+ is recycled across the basolateral membranes and into the plasma through K^+ channels (Figure 11-7). In addition, in some cells K^+ is recycled by K^+-Cl^- cotransport (Figure 11-8).

Absorption of Cl^- Occurs by Several Different Mechanisms

Na^+ cannot be (re)absorbed without an accompanying anion. The main anion in the intestinal lumen and renal tubular lumen is Cl^-, which also must be recycled by a variety of mechanisms. Intestinal and renal cell cytoplasm is electrically negative relative to the intestinal or kidney tubule lumen. Thus, the Cl^- electrochemical gradient across the apical membrane may favor Cl^- movement from cell to lumen. Nevertheless, the lumen-negative transepithelial potential ($V_{te} \approx -3$ to -5 mV) in the small intestine and kidney proximal tubule provides an electrical driving force that favors the net movement of Cl^- across the epithelium from lumen to interstitial space. Moreover, the small intestine and renal proximal tubule "tight junctions" are actually somewhat leaky (i.e., they have relatively low electrical resistance). Cl^- therefore can move through the junctions between cells (the paracellular pathway) from lumen to plasma (Figure 11-7).

Two important Cl^- transporters in some intestinal and renal epithelial cell apical membranes are a Cl^-/HCO_3^- exchanger (see Figure 11-7A) and a 1 Na^+–1 K^+–2 Cl^- cotransporter (Figure 11-9). The Cl^- taken up at the apical membrane is extruded across the basolateral membrane by a K^+-Cl^- cotransporter (see Figure 11-8) or Cl^- channels (see Figures 11-7A, and 11-9), thereby averting a large rise in $[Cl^-]_i$. In the

Lumen   Interstitial fluid

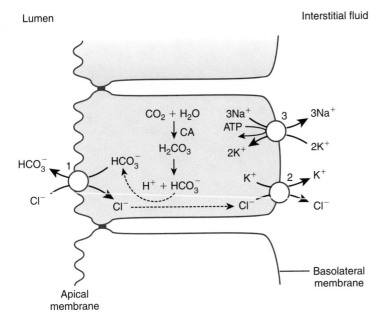

FIGURE 11-8 ■ Model epithelial cell shows another mechanism for Cl^- absorption by the transcellular route (also see Figure 11-7A). Cl^- is taken up across the apical membrane by a Cl^-/HCO_3^- exchanger (1). The Cl^- is then extruded across the basolateral membrane by a K^+-Cl^- cotransporter (2) using energy from the K^+ electrochemical gradient that is maintained by the Na^+ pump (3).

case of the basolateral Cl^- channels, Cl^- moves down its electrochemical gradient across the basolateral membranes when $[Cl^-]_i$ rises sufficiently to cause E_{Cl} to become more positive than V_{bl}.

Substances Can Also Be Secreted by Epithelia

Epithelia effect not only net solute (and fluid) transport from the lumen to the plasma (absorption), but also secretion of some substances into the lumen. Two examples, the net secretion of organic cations and organic anions, are discussed in Chapter 10.

Another example is K^+ secretion in certain renal epithelia. The K^+ is transported into the cells by the basolateral Na^+ pump; K^+ secretion into the lumen is then mediated by K^+ channels in the apical membrane (Figure 11-9). Depending on the body's needs, this K^+ can be either excreted in the urine or recycled. A genetic defect in the apical K^+ channels in the thick ascending limb of Henle's loop (TALH) in the kidney results in reduced K^+ permeability and reduced K^+ secretion. The inability of K^+ to recycle back into the kidney tubule lumen limits Na^+ absorption through Na^+–K^+–2 Cl^- cotransport and, thus, causes salt (NaCl) wasting and low blood pressure, or hypotension (Bartter's syndrome; Box 11-8).

Now, consider an epithelium (e.g., the colon) in which the cells possess a Cl^- entry mechanism (1 Na^+-1 K^+-2 Cl^- cotransport), as well as Na^+ pumps and K^+ channels in their basolateral membranes and Cl^- channels in their apical membranes (Figure 11-10). Under these circumstances, Na^+ drives Cl^- (and K^+) into the cells, across the basolateral membranes, by secondary active transport. Then, while the Na^+ is pumped out (recycled) across these membranes, the Cl^- is driven into the lumen, across the apical membrane, by its electrochemical gradient. At the same time, Na^+ moves from interstitial fluid to lumen through the paracellular pathway, driven by the lumen-negative V_{te} that is set up by the secretion of Cl^- (Figure 11-10). As noted earlier, genetically defective apical Cl^- channels in certain epithelia cause cystic fibrosis (Box 11-9).

The apical Cl^- channels and, thus, Cl^- secretion in intestinal epithelial cells are regulated by cyclic nucleotide–dependent protein phosphorylation. When cAMP or cyclic guanosine monophosphate (cGMP) is pathologically increased by enterotoxins, however, the result may be a massive secretion of Cl^- with loss of NaCl and water in the stool. This *secretory diarrhea* (Box 11-10) may be contrasted with the *osmotic diarrhea* described in Chapter 10, Box 10-8.

Lumen   Epithelial cell   Interstitial fluid

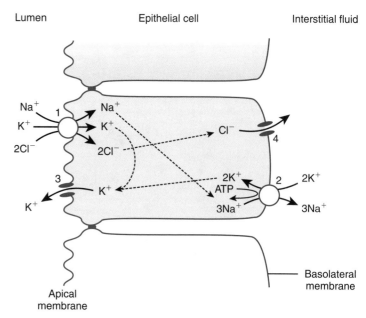

FIGURE 11-9 ■ Mechanism of net K^+ secretion across an epithelium (the thick ascending limb of Henle's loop). Na^+, K^+, and Cl^- enter the cell through an *apical* Na^+-K^+-2 Cl^- cotransporter *(1)*, driven by the Na^+ electrochemical gradient. K^+ also is pumped into the cell across the basolateral membrane in exchange for Na^+, by the Na^+ pump *(2)*. K^+ leaves the cell, down its electrochemical gradient, via K^+ channels in the apical membrane *(3)*. Cl^- moves down its electrochemical gradient, from cell to interstitial space, through Cl^- channels in the basolateral membrane *(4)*. Note that these cells also have K^+ channels and a K^+-Cl^- cotransporter (see Figure 11-8) in their basolateral membranes (not shown).

BOX 11-8

SALT WASTING, SALT RETENTION, AND BLOOD PRESSURE

Renal salt transport and net salt balance play critical roles in the regulation of plasma volume and blood pressure. Approximately 30% of the Na^+ filtered in the kidney glomerulus is reabsorbed in the thick ascending limb of Henle's loop. The mechanisms involved are illustrated in Figure 11-9. Genetic loss-of-function defects in the Na^+-K^+-2 Cl^- cotransporter, the apical K^+ channels (which enable K^+ recycling), or the basolateral Cl^- channels (which permit Cl^- to accompany Na^+ to maintain electroneutrality) result in severe salt (NaCl) wasting. All these defects cause low blood pressure (hypotension), which may be life-threatening in newborns. This salt wasting and hypotension, combined with excessive urinary Ca^{2+} loss, are known as Bartter's syndrome.

In contrast, salt retention leads to hypertension, a very prevalent disease and a problem that increases with age and with salt intake. All monogenic defects that enhance renal Na^+ reabsorption, such as mutations in the epithelial Na^+ channel, ENaC, or the proximal tubule Na^+/H^+ exchanger, NHE3, induce hypertension. Excessive aldosterone secretion, as may occur with certain tumors or hyperplasia (increased cell number) of the adrenal cortical glomerulosa cells (Conn's syndrome, or primary aldosteronism), also cause hypertension. Aldosterone increases expression of ENaC in the apical membrane, and the number of Na^+ pumps in the basolateral membrane of kidney cortical collecting tubule cells and thus enhances Na^+ reabsorption.

Net Water Flow Is Coupled to Net Solute Flow across Epithelia

The preceding examples show how solutes can be either absorbed or secreted across epithelia. We also need to consider the transepithelial movement of water. Epithelia do not actively transport water; water moves only passively, driven by the small osmotic gradients that are set up by the net solute transport. Water may move through the cells (i.e., the transcellular pathway) or, if the tight junctions are sufficiently leaky, through the paracellular pathway.

Water Transport Across Leaky Epithelia Is Osmotic and Obligatory The small intestine and the renal proximal tubule are examples of **leaky epithelia**. In

Lumen   Epithelial cell   Interstitial fluid

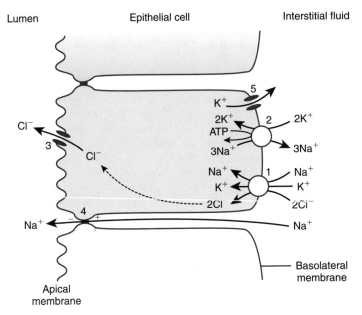

FIGURE 11-10 ■ Mechanism of net NaCl secretion across an epithelium. In this case Cl⁻ enters the cell through a *basolateral* Na⁺-K⁺-2 Cl⁻ cotransporter *(1)*, driven by the Na⁺ electrochemical gradient generated by the Na⁺ pump *(2)*. As the Cl⁻ concentration rises within the epithelial cell, the Cl⁻ electrochemical gradient across the apical membrane drives Cl⁻ out through Cl⁻ channels *(3)* that are regulated by cyclic nucleotides. The resulting transepithelial potential (V_{te}) (lumen negative) drives Na⁺ from the interstitial space to the lumen through the paracellular pathway *(4)*. The K⁺ that enters the cell across the basolateral membrane is recycled through K⁺ channels *(5)*. Secretion of NaCl provides the osmotic driving force for H_2O movement into the lumen.

BOX 11-9

CYSTIC FIBROSIS IS CAUSED BY MUTATIONS IN THE GENE THAT ENCODES THE CFTR Cl⁻ CHANNEL

Cystic fibrosis is an inherited autosomal recessive disease characterized by thick, viscous secretions from the mucous gland and airway epithelium, pancreatic insufficiency (greatly reduced exocrine secretions), and unusually high concentrations of Na⁺ and Cl⁻ in sweat. The cause of the disease is mutation of the gene that encodes an epithelial Cl⁻ channel and transport regulatory protein, the cystic fibrosis transmembrane conductance regulator (CFTR). The disease is most prevalent in people of European descent, with a disease incidence of ~1 per 1600 births and a mutated gene frequency is ~1 in 20.

The most common CFTR mutation in cystic fibrosis greatly reduces trafficking of CFTR Cl⁻ channels to the apical membranes of epithelia. The Cl⁻ conductance of the membrane is, therefore, decreased in patients with cystic fibrosis. In addition, regulation of certain other epithelial Cl⁻ channels and Na⁺ channels by CFTR may be altered in these patients. The reduction in Cl⁻ (and Na⁺) secretion reduces water secretion, so that the residual secretions are viscous. These thick, viscous secretions plug small pancreatic ducts and pulmonary airways and cause pancreatic insufficiency and a high rate of severe respiratory infections.

---

these tissues, which have a high rate of net solute transfer, the apical membrane permeability to water is high, in part because the membranes contain constitutive water channels (aquaporin-1). Thus, most of the net (osmotic) water flow occurs through the transcellular pathway. In addition, however, the net solute transport, from lumen to interstitial space, establishes a very small osmotic gradient across the epithelium.

This drives water flow through the tight junctions and into the narrow lateral intercellular spaces (the paracellular pathway; see Figure 11-7). Local hydrostatic pressure then propels the fluid (solvent and solutes) out of these lateral intercellular spaces and eventually into the blood. Despite the large amount of solute transfer, there is never a large osmotic gradient because constant osmotic water flow prevents build-up

---

BOX 11-10

ENTEROTOXINS THAT ACTIVATE Cl⁻ CHANNELS INDUCE SECRETORY DIARRHEA

Heat-stable enterotoxins from *Escherichia coli* activate guanylyl cyclase and increase the production of cGMP, whereas enterotoxins from *Vibrio cholerae* augment the production of cAMP. These cyclic nucleotides activate cGMP- or cAMP-dependent protein kinases, which phosphorylate and activate the cystic fibrosis transmembrane conductance regulator (CFTR) Cl⁻ channels in the apical membrane of certain intestinal epithelial cells. The consequent increase in Cl⁻ conductance enhances secretion of Cl⁻ and Na⁺ (Figure 11-10) and thereby provides an osmotic driving force for water flow from interstitium to intestinal lumen. The resulting excretion of watery stool is called *secretory diarrhea*. This loss of NaCl and water often causes severe dehydration

and may be fatal if not treated aggressively. This is a critical problem in developing regions of the world where unsanitary conditions prevail and where enterotoxigenic bacteria are endemic. These diarrheas can often be treated with oral rehydration using a solution containing NaCl and glucose (see Chapter 10). The cotransport of glucose and Na⁺ and, consequently, Cl⁻ into the body provides a source of nutrient and replenishes the salt and water lost through diarrhea. The action of the enterotoxins is blunted in individuals with a loss-of-function mutation in the CFTR gene, as would be expected from the aforementioned role of the CFTR Cl⁻ channels (see Box 11-9).

---

of a large osmotic pressure difference between the lumen and the interstitial space.

Another feature of water flow through leaky epithelia is that some solutes may move through the paracellular pathway with the water. This phenomenon, known as **solvent drag,** is an important mechanism for K⁺ and Ca²⁺ reabsorption in renal proximal tubules. The explanation is that the complete separation of water from solute takes a lot of energy. Therefore, if tight junctions are sufficiently leaky, dissolved solute will flow through these junctions along with the water (also called **bulk flow**).

Water Transport Across Tight Epithelia Is Regulated
The colon and renal late distal tubule and cortical and medullary collecting ducts are examples of **tight epithelia.** In these epithelia the transepithelial conductance is very low, and very little water normally flows across the tight junctions (i.e., through the paracellular pathway). Moreover, the apical membranes of these cells normally have very low water permeability unless water channels (aquaporin-2) are inserted into the membranes. When an increase in plasma osmolality signals a need to increase water reabsorption in the distal segments of the renal tubules, the posterior pituitary gland secretes antidiuretic hormone (ADH, or vasopressin). ADH acts on the cells in the distal nephron segments to promote the synthesis of cAMP.

The cAMP, in turn, stimulates the fusion of sub-PM vesicles, which contain aquaporin-2 in their membranes, with the apical membrane. In tight epithelia in the renal cortex the solute uptake systems generate a very small osmotic gradient across the apical membranes of the epithelial cells. Solute extrusion across the basolateral membrane then sets up a small osmotic gradient that drives water into the interstitial space. As a result, net water reabsorption is increased.

In contrast, if more water excretion is required to maintain water balance, the ADH level will remain low. In this case very little water is reabsorbed in the distal nephron and dilute urine (i.e., with a low osmotic pressure) is excreted. Defects in either the ADH secretory mechanism or the hormone receptors on the renal tubule cells, or mutations of the aquaporin-2 gene, result in pathological excretion of large amounts of dilute urine (diabetes insipidus; see Chapter 10).

The ultimate example of a tight epithelium is the urothelium that lines the urinary bladder. Once the urine is formed in the renal tubules, it is temporarily stored in the bladder. Virtually no transport of solute or water occurs either across the apical membranes of the urothelial cells or through the tight junctions in this epithelium. The urinary bladder is therefore simply a storage organ.

SUMMARY

1. Integral membrane proteins known as pumps or ATPases harness the energy from the hydrolysis of ATP to transport specific solutes such as Na^+, H^+, and Ca^{2+} against their electrochemical gradients. These transporters are said to mediate primary active transport.

2. The PM Na^+ pump mediates the export of 3 Na^+ ions and import of 2 K^+ ions while hydrolyzing 1 ATP to ADP and Pi. By exporting one net positive charge per cycle, this pump generates a small voltage and is, therefore, called an electrogenic pump. The Na^+ pump is uniquely sensitive to cardiotonic steroids such as ouabain and digoxin.

3. The Na^+ pump maintains the large Na^+ and K^+ electrochemical gradients across the PM of most cells. These gradients are critical for the electrical activity of excitable cells (see Chapters 7 and 8) and for powering secondary active transport (see Chapter 10). By maintaining a low $[Na^+]_i$, the Na^+ pump also plays a critical role in cell volume regulation: it enables cells to behave as if they are impermeable to Na^+ (see discussion of the Donnan effect in Chapter 4).

4. The Ca^{2+} pump in the S/ER membrane, SERCA, plays a key role in storing the Ca^{2+} in the S/ER that is required for Ca^{2+} signaling.

5. Certain Na^+ pump isoforms and the NCX act cooperatively to help regulate the Na^+ and Ca^{2+} concentrations in the tiny volume of cytosol between the PM and sub-PM ("junctional") S/ER in many cell types. This influences the storage of Ca^{2+} in the junctional S/ER and thus the Ca^{2+} signaling that depends on Ca^{2+} release from the S/ER.

6. Other transport ATPases such as the PM Ca^{2+} pump, two Cu^{2+} pumps, and proton pumps help to regulate ions in cells or their environment. For example, the gastric H^+,K^+-ATPase secretes protons into the lumen of the stomach to optimize the action of pepsin.

7. ABC proteins are involved in the ATP-dependent extrusion of some endogenous compounds and xenobiotics from cells. The CFTR, which behaves in part as a Cl^- channel, is also an ABC protein.

8. Transepithelial transport occurs in part through the paracellular pathway and in part through the transcellular pathway.

9. The Na^+ electrochemical gradient generated by the Na^+ pump provides the energy for net transport (either absorption or secretion) of solutes and water across epithelia.

10. Net transport of solutes across epithelia through the transcellular pathway requires two different transport mechanisms for each transported solute species, one in the apical membrane and one in the basolateral membrane.

11. Net solute transport through the paracellular pathway depends on the permeability of the tight junctions between cells and on the osmotic and electrical driving forces across the epithelium.

KEY WORDS AND CONCEPTS

- Primary active transport
- P-type ATPases
- Phosphorylated intermediate
- Sodium pump (Na^+,K^+-ATPase)
- Pump-leak model
- Cardiotonic steroids (e.g., digoxin and ouabain)
- Sodium pump catalytic (α) subunit isoforms
- Intracellular Ca^{2+} signaling
- Endoplasmic reticulum (ER)
- Sarcoplasmic reticulum (SR)
- SERCA (S/ER Ca^{2+}-ATPase)
- PMCA (plasma membrane Ca^{2+}-ATPase)
- Gastric H^+,K^+-ATPase
- Cu^{2+}-transporting ATPases
- ATP binding cassette (ABC) membrane transporters
- Multidrug resistance protein (MRP)
- Cystic fibrosis transmembrane conductance regulator (CFTR)
- Tight junction
- Transcellular pathway
- Paracellular pathway
- Ultrafiltrate

- Leaky epithelium
- Solvent drag
- Bulk flow
- Tight epithelium

STUDY PROBLEMS

1. Some intestinal smooth muscles relax when they are exposed to β-adrenergic agonists such as isoproterenol, which stimulate the Na^+ pump through a cAMP-mediated mechanism. The Na^+ pump stimulation is required for this relaxation. What is a likely mechanism for the relaxation?

2. Explain why so many secondary active transport systems are all coupled (indirectly) to the Na^+ pump.

3. Most transport systems, including the Na^+ pump, SERCA, PMCA, the Na^+/H^+ exchanger (NHE), the Na^+-glucose cotransporter (SGLT), and the simple glucose carrier (GLUT) are expressed in several different isoforms or splice variants. What are some possible reasons for the multiplicity of these transport systems?

BIBLIOGRAPHY

Anderson JM: Molecular structure of tight junctions and their role in epithelial transport, *News Physiol Sci* 16:126, 2001.

Blanco G, Mercer RW: Isozymes of the Na-K-ATPase: heterogeneity in structure, diversity in function, *Am J Physiol* 275:F633, 1998.

Blaustein MP, Wier WG: Local sodium, global reach. Filling the gap between salt and hypertension, *Circ Res* 101:959, 2007.

Borst P, Evers R, Kool M, Wijnholds J: A family of drug transporters: the multidrug resistance–associated proteins, *J Natl Cancer Inst* 92:1295, 2000.

Deen PM, Croes H, van Aubel RA, et al: Water channels encoded by mutant aquaporin-2 genes in nephrogenic diabetes insipidus are impaired in their cellular routing, *J Clin Invest* 95:2291, 1995.

Giachini FR, Tostes RC: Does Na^+ really play a role in Ca^{2+} homeostasis in hypertension? *Am J Physiol* 299:H602, 2010.

Green NM, MacLennan DH: Structural biology: calcium calisthenics, *Nature* 418:598, 2002.

Gutmann DAP, Ward A, Urbatsch IL, et al: Understanding polyspecificity of multidrug ABC transporters: closing in on the gaps in ABCB1, *Trends Biochem Sci* 35:36, 2009.

Koeppen BM, Stanton BA: *Renal physiology*, ed 4, New York, NY, 2006, Elsevier Health Sciences.

Kutchai HC: The gastrointestinal system. In Berne RM, Levy MN, editors: *Physiology*, ed 4, St Louis, 1998, Mosby.

Lifton RP, Gharavi AG, Geller DS: Molecular mechanisms of human hypertension, *Cell* 104:545, 2001.

Lingrel JB: The physiological significance of the cardiotonic steroid/ouabain binding site of the Na,K-ATPase, *Annu Rev Physiol* 72:395, 2010.

Lutsenko S, Barnes NL, Bartee MY, Dmitriev OY: Function and regulation of human copper-transporting ATPases, *Physiol Rev* 87:1011, 2007.

Poulsen H, Khandelia H, Morth JP, et al: Neurological disease mutations compromise a C-terminal ion pathway in the Na^+/K^+-ATPase, *Nature* 467:99, 2010.

Schwiebert EM, Benos DJ, Egan ME, et al: CFTR is a conductance regulator as well as a chloride channel, *Physiol Rev* 79(Suppl 1):S145, 1999.

Shin JM, Munson K, Vagin O, et al: The gastric HK-ATPase: structure, function, and inhibition, *Pflügers Arch* 457:609, 2008.

Welling PA, Cheng YP, Delpire E, et al: Multigene kinase network, kidney transport, and salt in essential hypertension, *Kidney Int* 77:1063, 2010.

Yatime L, Laursen M, Morth JP, et al: Structural insights into the high affinity binding of cardiotonic steroids to the Na^+, K^+-ATPase. *J Struct Biol* 174:296, 2011.

Zachos NC, Tse M, Donowitz M: Molecular physiology of intestinal Na^+/H^+ exchange. *Physiol Rev* 67:411, 2005.

Physiology of Synaptic Transmission

12

SYNAPTIC PHYSIOLOGY I

OBJECTIVES

1. Understand the structure and function of electrical synapses.

2. Describe the structure of a representative chemical synapse.

3. Understand the quantal nature of neurotransmitter release.

4. Understand the mechanism of transmitter release and the role of calcium.

5. Understand the synaptic vesicle cycle.

6. Understand the mechanisms that underlie short-term synaptic plasticity.

THE SYNAPSE IS A JUNCTION BETWEEN CELLS THAT IS SPECIALIZED FOR CELL-CELL SIGNALING

In Section II, we learned how the AP is generated and conducted in neurons and muscle cells. The critical issue in the nervous system is to get the right signal to the right place in the body at the right time. A key question then is, "How is the signal communicated from cell to cell, that is, from neuron to neuron, or from neuron to neuroeffector (muscle or gland) cell?" The intercellular junction through which the signals are transmitted is called the **synapse,**[*] and the communication across this junction is therefore called *synaptic transmission*. In this section (Chapters 12 and 13), we elucidate the cellular and molecular mechanisms that underlie synaptic transmission.

Approximately 100 billion neurons are present in the human brain. Moreover, neurons branch like trees, and the average neuron has approximately 1000 branches each ending in a small swelling, the *presynaptic* portion of the synapse, which is known as the **presynaptic terminal** or **synaptic bouton**. Thus the human nervous system has on the order of 100 trillion (10^{14}) synapses! Adding to the complexity is the fact that most neurons receive inputs from multiple neurons. The average neuron receives many more than 1000 synaptic inputs; indeed, a cerebellar Purkinje neuron may receive as many as 200,000! These neurons and synapses play essential roles in an enormous number of bodily activities from the control of respiration, blood circulation, and renal and gastrointestinal function, to sensory perception, body movements, and learning and memory. Our task here is to understand the mechanisms by which neurons communicate with one another.

---

[*]Charles Sherrington, the physiologist who coined the term *synapse* in the late nineteenth century, was a recipient of the 1932 Nobel Prize in Physiology or Medicine for his seminal work on spinal reflexes.

Synaptic Transmission Can Be Either Electrical or Chemical

In the nineteenth century, the classical morphological studies of Santiago Ramón y Cajal demonstrated that the nervous system, like other organs, is composed of cells (the **neuron doctrine**).* During the late nineteenth and early twentieth centuries there was fierce debate over two divergent views of synaptic transmission, dubbed the "war of soups and sparks."† As a result of the demonstration that nerves and muscle cells conduct electrical signals, one popular idea was that an electric "spark" at the end of a presynaptic neuron directly triggered the electrical signal in the *postsynaptic* neuron or muscle cell (i.e., synaptic transmission was thought to be purely electrical). Conversely, studies on the paralytic action of *curare*,†† and on the **autonomic nervous system**, hinted at the idea of chemical transmission.

The discovery of chemical synaptic transmission, and recognition that most synapses are chemical,

---

*Cajal and Camillo Golgi shared the 1906 Nobel Prize in Physiology or Medicine for their seminal work on neuronal structure. Ironically, Golgi, whose staining methods proved crucial for elucidating structure, favored the idea that the nervous system was a continuous reticulum rather than a network of discrete cells.

†Valenstein ES: *The war of the soups and the sparks: the discovery of neurotransmitters and the dispute over how nerves communicate,* New York, 2005, Columbia University Press.

††Curare, or D-tubocurarine, is an alkaloid toxin from the bark of a South American liana vine.

nearly led to the demise of the concept of electrical transmission. Nevertheless, some synapses in the mammalian central nervous system (CNS) are electrical. We will consider the mechanism of transmission at electrical synapses before turning to the more prevalent and diverse chemical synapses.

Electrical Synapses Are Designed for Rapid, Synchronous Transmission

Chemical and electrical synapses have distinct morphological features that are related to their differing functional properties. **Electrical synapses** are designed to allow current to flow directly from one neuron to another. At electrical synapses, the presynaptic and postsynaptic membranes are separated by only 3 to 4 nm (Figure 12-1A). At these narrow gaps, the two neurons are connected by **gap junction** channels. Each gap junction channel consists of two hemichannels: one in the presynaptic and one in the postsynaptic membrane. Each hemichannel, called a **connexon**, is an annular assembly of six peptide subunits, called **connexins**. The connexon forms a pore through the membrane (Figure 12-1B). The connexon in the presynaptic membrane docks face to face with a connexon in the postsynaptic membrane to form a conducting channel that connects the cytoplasm of the two neurons. Gap junction channels allow the passage of nutrients, metabolites, ions, and other small

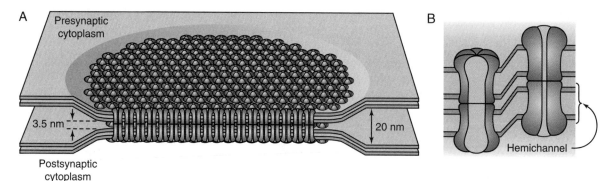

FIGURE 12-1 ■ Structure of an electrical synapse. **A,** The electrical synapse consists of a densely packed array of gap junction channels. The width of the synaptic cleft is 3 to 4 nm. **B,** Each hemichannel consists of an annular arrangement of six connexin subunits. Each gap junction channel consists of a hemichannel in the presynaptic membrane docked end to end with a hemichannel in the postsynaptic membrane. The cytoplasm of the presynaptic and postsynaptic cells is connected through the channel formed by each pair of hemichannels. *(Redrawn from Kandel ER, Schwartz JH, Jessell TM: Principles of neuroscience, ed 4, New York, 2000, McGraw-Hill.)*

molecules (≤1000 daltons). More than 20 connexin isoforms have been identified, and mutations in about half of the genes that encode these proteins are linked to human disease (Box 12-1).

The first description of electrical synaptic transmission was based on studies of the giant motor synapse of the crayfish. In this preparation, the presynaptic and postsynaptic axons are large enough to allow placement of intracellular stimulating and recording electrodes close to the synapse. These experiments demonstrated that an AP in the presynaptic neuron produces a depolarization in the postsynaptic neuron after a negligible synaptic delay (Figure 12-2), which is much shorter than the delay at chemical synapses (see later). Such nearly instantaneous transmission can be caused only by direct current flow between the cells. This current flows from the presynaptic cell through the gap junction channels and into the **postsynaptic cell**. Such direct flow of current from the presynaptic to the postsynaptic neuron does not occur at chemical synapses. Most electrical synapses are bidirectional: signals can be transmitted from either one of the connecting cells to the other. In contrast, chemical synapses are unidirectional. The conductance of gap junction channels is regulated by two distinct gating mechanisms (Box 12-2).

Electrical synapses between neurons have been identified in the mammalian CNS. They play a role in neuronal synchronization because they allow the direct, bidirectional flow of current from one cell to the other. For example, electrical synapses coordinate spiking among clusters of cells in the thalamic reticular nucleus. Similarly, electrical synapses in the suprachiasmatic nucleus help to synchronize spiking that may be necessary for normal circadian rhythm. Direct electrical communication between cells is also physiologically important outside the nervous system: For example, gap junction channels between heart cells enable the cells to depolarize and contract synchronously (see Chapter 14).

Most Synapses Are Chemical Synapses

At **chemical synapses**, the AP in the presynaptic nerve terminal releases molecules called **neurotransmitters** that generate electrical or biochemical signals in the postsynaptic cells. Early in the twentieth century, Henry Dale and Otto Loewi obtained critical evidence that dispelled doubts about chemical transmission.

BOX 12-1

CONNEXIN MUTATIONS LINKED TO DISEASE

Mutations in about half of the genes that encode the connexin family of proteins have been linked to several diseases. In some of these diseases, the connexin mutations result in dysfunctional gap junctions between glial cells. Mutations in the gene encoding connexin-32 (Cx32), for example, are associated with the X-linked form of Charcot-Marie-Tooth disease, one of the most common forms of hereditary neurological disorders. Charcot-Marie-Tooth disease is a motor and sensory neuropathy characterized by muscle weakness and various sensory defects. Many of the Cx32 mutants fail to form functional gap junctions between Schwann cells, and this leads to demyelination and axonal degeneration. Recessive mutations in the gene encoding connexin-47 (Cx47) are linked to Pelizaeus-Merzbacher–like disease, which is a rare disorder characterized by lack of CNS myelin development. The Cx47 mutants also fail to form functional gap junction channels.

Congenital cataracts are associated with mutations in Cx46 and Cx50. These connexins form gap junctions between lens fiber cells where they support normal lens function by helping to maintain cell transparency. Mutations in Cx26 are implicated in deafness. This connexin is normally expressed in the nonsensory epithelial cells in the cochlea, and not in the hair cells. The exact function of Cx26 in the cochlea is unknown, but it has been proposed to play a role in the recycling of K^+.

In the majority of connexin mutants that have been studied, the altered connexin subunits reach the cell surface and form gap junction-like structures. However, these structures either are nonfunctional or they form channels that function poorly compared with normal gap junction channels. In another class of mutants, the altered connexin subunits are retained in the endoplasmic reticulum and never reach the cell surface.

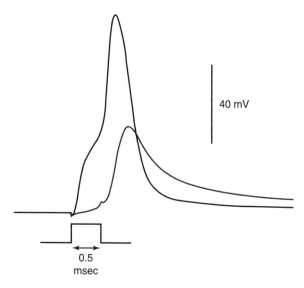

FIGURE 12-2 ■ Synaptic transmission at an electrical synapse proceeds without a synaptic delay. At the giant motor synapse in the crayfish, microelectrodes for passing current and recording potential are placed in both presynaptic and postsynaptic neurons. A 0.5-millisecond current pulse in the presynaptic cell *(at bottom)* evokes an AP in that cell *(black record)*, which begins at the time indicated by the *vertical dashed line*. At the same time, an AP is initiated in the postsynaptic cell *(blue record)*. *(Data from Furshpan EJ, Potter DD: J Physiol 145:289, 1959.)*

40 mV

0.5
msec

BOX 12-2

TWO DISTINCT GATING MECHANISMS IN GAP JUNCTION CHANNELS

The conductance of gap junction channels is physiologically regulated. This is accomplished through channel gating, and at least two distinct gating mechanisms operate within each gap junction hemichannel. The first is Vj-gating, which depends on the junctional voltage (V_j) across the gap junction. Vj-gating is responsible for rapid transitions between high and low conducting states of the channel. The low-conductance state that is entered as a result of Vj-gating does not completely close the channel. Hemichannels formed by some connexin isoforms close with depolarization; others close with hyperpolarization. The second type of gating mechanism involves slow transitions (10 to 30 msec) between the fully open and fully closed states. These slow transitions can be mediated by three distinct processes. First, slow transitions can occur in response to changes in voltage: this is called loop gating because it involves the extracellular loops that connect adjacent transmembrane domains in connexin. The loop gating voltage sensor and the Vj-gating voltage sensor are independent structures. Second, slow transitions can be caused by changes in pH or Ca^{2+}; this is called chemical gating. In cells that are normally coupled electrically and metabolically through gap junctions, an increase in $[Ca^{2+}]_i$ or a decrease in pH can close gap junction channels and uncouple the cells. This can serve as a protective mechanism, uncoupling damaged cells, which have elevated $[Ca^{2+}]_i$ or $[H^+]_i$, from healthy cells. Finally, slow transitions can be mediated by the docking or undocking of two hemichannels.

Dale showed that **acetylcholine (ACh)** was the most potent agent capable of mimicking **parasympathetic** nerve activation. His observation that the effects of ACh, injected into the bloodstream, were very rapid but short-lived led him to suggest that ACh was rapidly hydrolyzed. This presaged the discovery that the enzyme **acetylcholinesterase** terminates the action of ACh at synapses. Loewi subsequently demonstrated

that stimulation of the vagus nerve to a frog heart released a substance into the bathing solution. When a different frog heart was immersed in this bathing solution, its rate slowed.* This chemical neurotransmitter, released by the vagus nerve, was later shown to be

---

*Dale and Loewi shared the 1936 Nobel Prize in Physiology or Medicine for the discovery of chemical neurotransmission.

ACh. These discoveries laid the foundation for most of neuropharmacology and neurotherapeutics: agents that stimulate neurotransmitter release, mimic neurotransmitters, or interfere with their actions (see later) are among the most useful tools in the physician's arsenal.

The application of electron microscopy and ultracentrifugation methods in the 1950s and 1960s led to important advances in understanding the structure and chemistry of synapses. Key structural features of a representative chemical synapse are illustrated in Figure 12-3. The presynaptic terminal contains many small (~40 nm diameter) round structures, the **synaptic vesicles (SVs)**, which contain high concentrations of neurotransmitters (see later). The SVs tend to concentrate at or near the **active zone**, a specialized region of the presynaptic PM that is involved in transmitter release. This region is closely apposed to the postsynaptic cell, with its own *postsynaptic density* region that is enriched with **neurotransmitter receptors** (see Chapter 13). At the synapse, the two cells are separated by a **synaptic cleft** 20 to 40 nm wide.

NEURONS COMMUNICATE WITH OTHER NEURONS AND WITH MUSCLE BY RELEASING NEUROTRANSMITTERS

When a nerve AP is conducted down the axon to the presynaptic terminal, the resulting depolarization triggers the Ca^{2+}-dependent (see later) release of SV contents into the synaptic cleft. This process of **exocytosis** involves the fusion of the SV membrane with the PM and the consequent emptying of the vesicular contents into the synaptic cleft. The SV membrane is then recycled (see later).

The released neurotransmitter molecules diffuse across the synaptic cleft and interact with specific receptor molecules that are integral proteins in the PM of the postsynaptic neuron or neuroeffector cell. The interaction between the transmitter and its receptor can be characterized as a lock-and-key mechanism in which the transmitter (key) unlocks the receptor. This activates the receptor so that, depending on the receptor type (see Chapter 13), it either directly affects membrane conductance in the postsynaptic cell or,

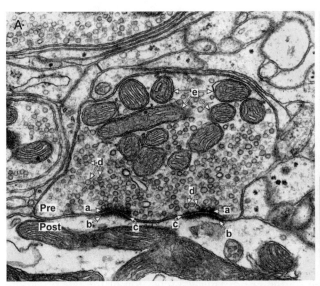
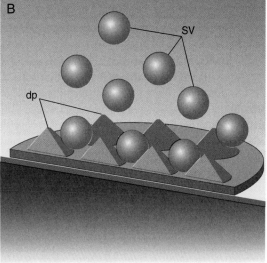

FIGURE 12-3 ■ Electron micrographic structure of a chemical synapse in the human hippocampus. **A,** The presynaptic terminal, or bouton, and the postsynaptic neuron are labeled (Pre and Post). Key structural features *(arrows)* are as follows: *a,* active zone; *b,* postsynaptic density; *c,* synaptic cleft (thin, pale region between the active zone and the postsynaptic density); *d,* synaptic vesicles (SVs); and *e,* mitochondria. **B,** Model of the presynaptic active zone, with dense projections (dp) and SVs, some of which are already docked at the active zone. (**A,** *Courtesy of R. Perkins and T. S. Reese, Laboratory of Neurocytology, National Institute of Neurological Disorders and Stroke, National Institutes of Health, Bethesda, Md.* **B,** *Redrawn from Zhai GR, Bellen HJ: Physiology, 19:262, 2004; used with permission from the American Physiological Society.*)

alternatively, initiates an intracellular signaling cascade that regulates a wide range of cellular processes including membrane conductance changes. These mechanisms modulate the postsynaptic neuron's excitability (i.e., the ability to fire an AP).

Most synapses in the mammalian nervous system are chemical synapses. The number of presynaptic neurons that synapse onto one postsynaptic cell varies widely. For example, one skeletal muscle fiber is usually innervated by only a single motor neuron, whereas each motor neuron usually innervates more than one muscle cell. In contrast, as already noted, many presynaptic neurons may synapse onto a single postsynaptic neuron. Synapses are not static structures: new synapses can form, synaptic connections can be strengthened or weakened, and some synapses can be eliminated. This flexibility contributes to the enormous complexity and rich diversity of synaptic transmission that underlies higher brain function.

The Neuromuscular Junction Is a Large Chemical Synapse

The large synapse formed between a spinal motor neuron and a skeletal muscle fiber is called the **neuromuscular junction (NMJ)**. Studies of neuromuscular transmission by Bernard Katz and his collaborators* greatly enriched our understanding of how chemical synapses work. The axon of the motor neuron contacts the muscle fiber at a region called the **end plate** (Figure 12-4). As the axon approaches the muscle it divides into several small branches, and each branch terminates in a knoblike swelling, the synaptic bouton.

Each synaptic bouton contains numerous SVs filled with the neurotransmitter ACh. The vesicles are clustered around active zones. At the NMJ, the synaptic boutons are separated from the postsynaptic membrane by a 100-nm synaptic cleft, which is wider than the synaptic clefts between neurons (typically, ~20 to 40 nm). Within the cleft of the NMJ is a basement membrane that anchors the enzyme acetylcholinesterase. This enzyme hydrolyzes ACh and thereby helps limit the duration of ACh action. Each active zone in each synaptic bouton lies directly over a *junctional*

fold, which is a deep invagination of the muscle cell membrane (Figure 12-4). A high density of ACh receptors (AChRs; ~20,000/μm^2) is localized near the top of each junctional fold. These NMJ AChRs, which are multimeric nonselective cation channels, are also activated by nicotine, the addictive drug from the tobacco plant, hence the name **nicotinic AChRs (nAChRs)**.

The NMJ preparation consists of a muscle and its attached nerve (e.g., the diaphragm and phrenic nerve), which can be easily removed and placed in an experimental chamber for recording. Stimulating electrodes are placed on the nerve trunk to initiate APs, and microelectrodes are placed in the muscle cell at the end-plate region to measure changes in V_m. Following an AP in the presynaptic neuron, a transient depolarization occurs in the muscle cell (Figure 12-5). This depolarization is called the **end-plate potential (EPP)**. The EPP is normally large enough to reach threshold for generating an AP in the skeletal muscle cell. To study the time course of the EPP, its size must be reduced to less than the AP threshold. This can be accomplished by lowering $[Ca^{2+}]_o$ and thus reducing the amount of transmitter released (see later) or by blocking some of the nAChRs (e.g., with curare). Under these conditions the EPP is revealed to have a rapid rising phase and a slower exponential decay (Figure 12-5). The rapid depolarization results from the sudden release of ACh from the presynaptic nerve terminal in response to the AP. The ACh diffuses rapidly across the synaptic cleft and binds to the postsynaptic receptors (nAChRs), which are ACh-gated ion channels. The binding of two molecules of ACh per receptor opens the channel gate to conduct inward current, thereby depolarizing the postsynaptic cell (i.e., the muscle fiber, in the case of the NMJ). As the ACh diffuses away and is hydrolyzed by acetylcholinesterase, the concentration of ACh in the synaptic cleft quickly declines to zero (even before the nAChRs close). The slow exponential decline in the EPP is largely a reflection of the rate of closure of the ACh-gated channels. Numerous diseases of neuromuscular transmission are recognized. Some are caused by defective ACh release, others involve impaired hydrolysis of ACh, and several result from defects in the nAChR channel (see Chapter 13).

---

*Katz shared the 1970 Nobel Prize in Physiology or Medicine for this work.

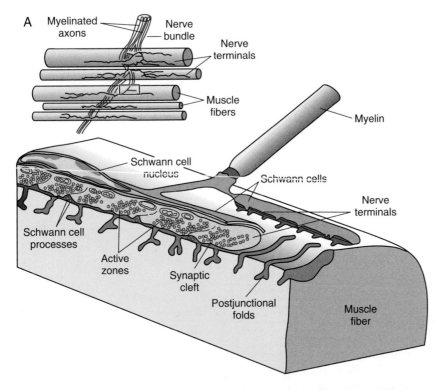

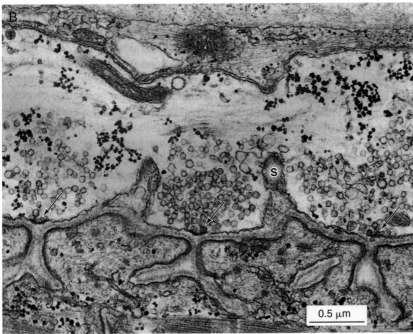

FIGURE 12-4 ■ The structure of the neuromuscular junction (NMJ). **A,** Schematic drawing of the innervation of several muscle fibers by motor neurons *(upper left inset)* and enlarged view of a portion of one NMJ (see *box in inset*). The nerve terminal contains numerous synaptic vesicles that cluster around active zones, which are the sites of transmitter release. The active zones are situated opposite the junctional folds in the muscle membrane. The nAChRs are clustered in the muscle membrane at the top of the junctional folds. **B,** An electron micrograph of an NMJ that illustrates many of the features shown in **A.** *Arrows,* active zones; S, Schwann cell process. *(From Kuffler SW, Nichols JG, Martin AR: From Neuron to brain, ed 2, Sunderland, MA, 1984, Sinauer.)*

FIGURE 12-5 ■ The end-plate potential (EPP) can be isolated by reducing its amplitude. The muscle V_m is recorded at the NMJ in response to motor nerve stimulation. Normally, nerve stimulation induces an EPP that is higher than the threshold for generating an AP. In the presence of curare, the amplitude of the EPP is reduced and it does not reach the muscle AP threshold. The isolated EPP has a rapid rising phase and a slower exponential decay. (Physostigmine was used to block hydrolysis of ACh by acetylcholinesterase; this increased the duration of the EPP and the muscle AP.) *(Data from Fatt P, Katz B: J Physiol 115:320, 1951.)*

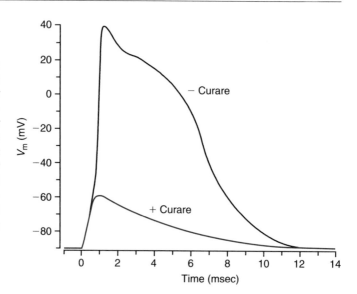

The significant **synaptic delay** between the arrival of the presynaptic AP at the nerve terminal and the beginning of the postsynaptic response (Figure 12-6) is a characteristic of all chemical synapses. The following events all contribute to the synaptic delay: (1) the presynaptic AP causes voltage-gated Ca^{2+} channels (VGCCs) to open; (2) Ca^{2+} enters the cell through the open Ca^{2+} channels and triggers neurotransmitter release; and (3) the neurotransmitter rapidly diffuses across the synaptic cleft, binds to postsynaptic receptors, and opens ion channels in the postsynaptic membrane. These processes all take time and thus contribute to the delay between the presynaptic AP and the postsynaptic response. Empirically, the opening of VGCCs is the slowest process and thus the major contributor to the synaptic delay. The synaptic delay at chemical synapses contrasts with transmission at electrical synapses where current flows directly, without delay, from the presynaptic cell to the postsynaptic cell through gap junction channels (see Figure 12-2).

Transmitter Release at Chemical Synapses Occurs in Multiples of a Unit Size

Neurotransmitters are released in discrete packets, called **quanta**. Initial evidence for this was obtained by Katz and his colleagues from electrical recordings at the NMJ. Small spontaneous depolarizations of the muscle cell can be observed, even in the absence of presynaptic APs (Figure 12-7). These spontaneous

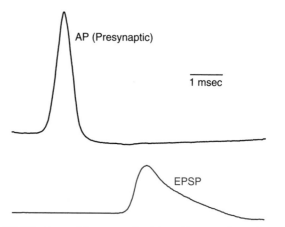

FIGURE 12-6 ■ The synaptic delay. V_m recordings from presynaptic and postsynaptic neurons are made simultaneously from the giant synapse in the squid stellate ganglion. There is a delay of approximately 2.5 milliseconds between the presynaptic AP *(top trace)* and the postsynaptic response *(bottom trace)*. EPSP, excitatory postsynaptic potential. *(Data from Bullock TH, Hagiwara S: J Gen Physiol 40:565, 1957.)*

depolarizations have many features in common with the EPPs. Although the spontaneous depolarizations are normally much smaller than the EPPs, they are identical in time course to the EPP, with a rapid rising phase and a slower exponential falling phase. Like the

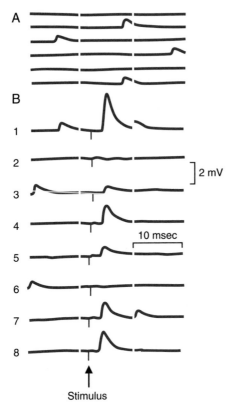

FIGURE 12-7 ■ At low $[Ca^{2+}]_o$, the EPP amplitude fluctuates randomly from one stimulus to the next. The muscle V_m is recorded at the end plate. **A,** Spontaneous depolarizations of the muscle at the end plate, MEPPs, have an amplitude of ~0.4 mV. **B,** Eight consecutive responses to motor nerve stimulation *(at arrow)* are shown, and each response (or "sweep") is numbered from 1 to 8. In sweeps 2 and 6 there was no response to nerve stimulation (synaptic failures). In sweeps 3 and 5 the EPP amplitude is the same size as the MEPP amplitude. In sweeps 4, 7, and 8 the EPP amplitude is approximately twice the MEPP amplitude, and in sweep 1 it is approximately four times larger, suggesting the release of ACh from, respectively, 2 and 4 synaptic vesicles. *(Data from Liley AW: J Physiol 133:571, 1956.)*

EPPs, the spontaneous depolarizations are largest when recorded at the end-plate region of the muscle cell. Both signals are reduced in amplitude by drugs (e.g., curare) that block nAChRs, and both are augmented by drugs that interfere with ACh hydrolysis.

Because they are similar to EPPs, but smaller, the spontaneous depolarizations are called **miniature end-plate potentials (MEPPs)**. The MEPPs have a uniform size of approximately 0.4 mV (Figure 12-7), which is approximately 2000 times larger than the depolarization resulting from the opening of a single nAChR. Because two molecules of ACh are required to open each channel, and not all the released ACh binds to postsynaptic receptors, each quantum must contain more than 4000 molecules of ACh. In fact, investigators have shown that approximately 5000 to 10,000 molecules of ACh are required to produce an MEPP. This implies that, in an SV with an outer diameter of 40 nm, the ACh concentration could be as high as 500 mM (see later).

The presynaptic AP triggers release of neurotransmitters in quantal packets that are identical in size to the spontaneously released quanta. This can be demonstrated by studying neuromuscular transmission after decreasing $[Ca^{2+}]_o$. In the presence of low $[Ca^{2+}]_o$ the EPP amplitude is greatly reduced from its normal size of approximately 70 mV to approximately 1 to 2 mV in amplitude. Furthermore, the EPP size fluctuates randomly from one stimulus to the next (Figure 12-7). Occasionally, nerve stimulation elicits no EPP; this is called a *failure*. After recording the responses to many stimuli, the number of EPPs of a given amplitude can be plotted in a histogram (Figure 12-8). Analysis of such an amplitude distribution shows that EPP amplitudes occur in integer multiples of the smallest EPP amplitude, and the smallest EPP is identical in size to the spontaneous MEPP amplitude (Box 12-3). Thus both spontaneous and nerve-evoked release of neurotransmitter at the neuromuscular junction are quantal.

Synaptic vesicles are the morphological correlates of the physiological quanta. Each vesicle stores one quantum of ACh, and the content of the vesicle is released by exocytosis when the vesicle fuses with the presynaptic membrane at the active zone. Quantal release also has been demonstrated at a variety of CNS chemical synapses. The most extensively studied is the calyx of Held, which is an unusually large glutamatergic synapse in the brainstem. Glutamate is stored in SVs in the presynaptic terminal, and release of the content of a single SV evokes a miniature **excitatory postsynaptic current (EPSC)** in the

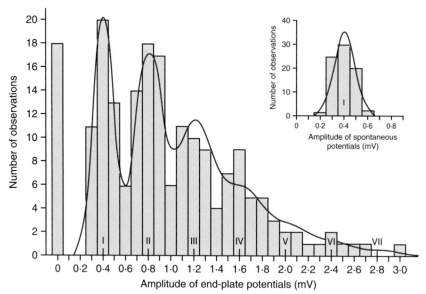

FIGURE 12-8 ■ Acetylcholine is released in fixed packets, or quanta, at the neuromuscular junction. Amplitude histograms were constructed after recording many EPPs and MEPPs like those shown in Figure 12-7. The numbers of EPPs at each amplitude are counted and plotted as an amplitude histogram (the MEPP amplitude distribution is plotted in the *inset*). Several peaks in the EPP amplitude distribution are present, all are at integer multiples of the MEPP amplitude. This finding implies quantal release of transmitter (see Box 12-3). *(From Boyd IA, Martin AR: J Physiol 132:74, 1956.)*

postsynaptic neuron. The type of quantal analysis used to describe transmitter release at the NMJ (Box 12-3) is also used to describe glutamate release at the calyx of Held. Spontaneous and AP-evoked release are both quantal and Ca^{2+} dependent. At the frog NMJ, a very specialized synapse, approximately 100 to 300 quanta are normally released in response to a presynaptic AP. Depending on stimulation frequency, from 10 to a few hundred quanta may be released at the calyx of Held. In stark contrast, at most CNS synapses, which are much smaller, presynaptic APs often fail to trigger neurotransmitter release, and when release is triggered successfully, only 1 or 2 quanta are released.

Ca^{2+} Plays an Essential Role in Transmitter Release

As described in the previous section, lowering $[Ca^{2+}]_o$ decreases the size of the EPP. The smaller size of the EPP results from the fact that fewer quanta are released in the presence of low $[Ca^{2+}]_o$. Because

transmitter release is an intracellular process, this result implies a pathway for Ca^{2+} entry into the presynaptic neuron. Additional insights into how Ca^{2+} regulates transmitter release were obtained from studies of another synaptic preparation, the squid giant synapse, whose large size has the distinct advantage of allowing both presynaptic and postsynaptic neurons to be voltage-clamped. Application of this method provided direct evidence for the existence of VGCCs in presynaptic membranes. These studies showed that the amount of transmitter that is released depends on the amount of Ca^{2+} that enters the cell. Furthermore, blocking VGCCs abolishes transmitter release. These and similar studies at the NMJ and other synapses have elucidated the essential role of VGCCs in quantal release of neurotransmitters. Subsequent studies, including some with neurotoxins from spiders and predatory snails (Box 12-4), revealed that two subtypes of VGCCs are involved in transmitter release in neurons: P/Q- and N-type Ca^{2+} channels (see Chapter 8).

BOX 12-3

THE PROBABILITY OF QUANTAL TRANSMITTER RELEASE

The EPP amplitude histogram (see Figure 12-8) reveals several peaks in the distribution. The first peak, at 0 mV, represents the number of failures. The next peak is centered at 0.4 mV, which is the same as the mean MEPP amplitude and thus reflects the nerve-evoked release of a single quantum of transmitter. The other peaks in the distribution occur at integer multiples of 0.4 mV. This suggests that the second peak results from the release of two quanta, the third peak results from the release of three quanta, and so on. The smooth curve drawn over the amplitude histogram is a theoretical distribution based on quantal release, and it clearly gives a good fit to the data.

The number of events in one of the peaks of the amplitude distribution divided by the total number of events is an estimate of the probability that the corresponding number of quanta are released. For example, the peak centered at 0.8 mV represents the release of two quanta. Thus the number of events in this peak divided by the total number of events is a measure of the probability that two quanta are released in response to an action potential. If the release of a quantum of transmitter is an independent, random event, then this probability should fit a binomial distribution. A binomial distribution describes a process in which an experimental trial results in two possible outcomes, success or failure. A binomial distribution has two parameters: p, the probability of success (i.e., the release of a quantum), and n, the number of "trials" (i.e., the number of sites that can release a quantum). Using a binomial distribution, the probability that x quanta will be released when n sites are available can be calculated as follows:

$$P(x:n) = \frac{n!}{x!(n-x)!} p^x (1-p)^{n-x}$$

By fitting a binomial distribution to the data in the EPP amplitude distribution we can estimate p and n. The product of p and n is the mean number of quanta released, and this is referred to as the *quantal content*. The quantal content can be as high as 300 at the NMJ or as low as 1 to 10 at some CNS synapses. The probability of release, p, is as high as 0.7 to 0.9 at the NMJ and as low as 0.1 at some central synapses, and n ranges from 1000 at the NMJ to 1 in the CNS.

BOX 12-4

CONE SNAIL AND SPIDER TOXINS INHIBIT VESICLE EXOCYTOSIS BY BLOCKING VOLTAGE-GATED Ca^{2+} CHANNELS

The venoms from certain animals contain from one to more than a hundred toxins. Many of these toxins potently reduce neuronal excitability, often by inhibiting AP generation or by blocking synaptic transmission. Synaptic transmission can be subdivided into presynaptic processes (transmitter synthesis, storage, and release) and postsynaptic processes (binding of transmitter to receptors and activation of ion channels; see Chapter 13). Specific toxins target most of these steps. For example, cone snail and spider venoms contain toxins that interfere with Ca^{2+}-dependent transmitter release. Cone snails are predatory marine snails that shoot a venomous "dart" into their prey. The venom contains hundreds of peptides, called *conotoxins*, that immobilize or kill the victim. One of these toxins, ω-conotoxin, selectively blocks N-type ($Ca_v2.2$) voltage-gated Ca^{2+} channels. The block of N-type Ca^{2+} channels in presynaptic nerve terminals prevents the Ca^{2+} influx required for transmitter release and thereby inhibits synaptic transmission. The funnel web spider, *Agelenopsis aperta*, kills its prey using venom containing several *agatoxins*. ω-Agatoxin type IVA inhibits P/Q-type ($Ca_v2.1$) voltage-gated Ca^{2+} channels with high affinity. Block of P/Q-type Ca^{2+} channels inhibits Ca^{2+}-dependent neurotransmitter release in the hippocampus and Ca^{2+}-dependent hormone secretion from both pancreatic β-cells and adrenal chromaffin cells. Because of their selectivity, these neurotoxins have been useful in identifying subtypes of voltage-gated Ca^{2+} channels and in studying the functional roles of these channels.

THE SYNAPTIC VESICLE CYCLE IS A PRECISELY CHOREOGRAPHED PROCESS FOR DELIVERING NEUROTRANSMITTER INTO THE SYNAPTIC CLEFT

In the presynaptic nerve terminal, SVs accumulate neurotransmitter and "dock" at the active zone. When an AP reaches and depolarizes the nerve terminal, voltage-gated Ca^{2+} influx triggers exocytosis—the fusion of docked synaptic vesicles to the PM. Consequently, neurotransmitter molecules are disgorged from the vesicle lumen into the synaptic cleft to activate the post-synaptic cell. Thereafter, through endocytosis, the empty synaptic vesicles are retrieved back into the nerve terminal, to be refilled with neurotransmitter for a new round of neurotransmission. This cyclical process, the **SV cycle**, is temporally and spatially precise and requires the intricate and coordinated interplay of dozens of proteins, which reside on both the synaptic vesicle and the PM at the active zone (Figure 12-9). Important aspects of the cycle are discussed here.

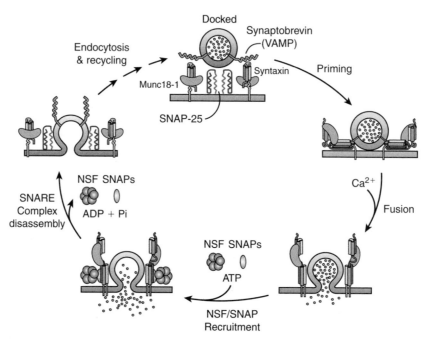

FIGURE 12-9 ■ Schematic representation of the synaptic vesicle cycle. Docked SVs *(at the top)* attach to the active zone (region just above the PM) potentially through the interaction of the proteins Rab and Rab3-interacting molecule (Rim), residing on the SV and in the active zone, respectively. For visual clarity, Rab and Rim are not shown in the diagram. Proteins on the SV and in the active zone that participate in the SV cycle are labeled. Munc18-1 is an active zone protein that is complexed with syntaxin. SVs become primed for exocytosis through formation of a complex between soluble NSF-attachment protein receptor (SNARE) proteins on the SV (synaptobrevin/vesicle-associated membrane protein [VAMP]) and in the PM at the active zone (SNAP-25 and syntaxin-1). Munc18-1 switches from interacting exclusively with syntaxin to interacting with the entire SNARE complex. After priming, Ca^{2+} can bind to synaptotagmin and trigger fusion of the SV to the PM, with formation of a fusion pore through which neurotransmitter molecules are released into the synaptic cleft (at the bottom). After fusion-pore opening, SNAPs (not related to SNAP-25) and *N*-ethylmaleimide-sensitive factor (NSF, an ATPase) bind to, and disassemble, the SNARE complexes, in a process that requires ATP hydrolysis. This frees the SVs to be endocytosed and to recycle (see Figure 12-10). *(Modified from Südhof TC: Neurotransmitter release. In Südhof TC, Starke K, editors:* Handbook of experimental pharmacology. Vol 184, Pharmacology of neurotransmitter release, *Berlin, Germany, 2008, Springer-Verlag.)*

The Synaptic Vesicle Is the Organelle that Concentrates, Stores, and Delivers Neurotransmitter at the Synapse

Like many other intracellular organelles, the SV consists of a bilayer membrane enclosing a lumenal space. Most SVs are tiny, with an outer diameter of only ~40 nm. Because the thickness of the lipid bilayer is ~5 nm, the diameter of the SV lumen measures a mere 30 nm. For comparison, two common proteins, hemoglobin and ferritin, are approximately 6 and 12 nm in diameter, respectively. As discussed later, the extraordinarily small size of the SV gives rise to some unusual biophysical properties.

The energy stored in the Na^+ electrochemical gradient established by the Na^+ pump drives most cellular transport processes (see Chapters 9-11). In contrast, transport of neurotransmitters into the SV depends on a proton (H^+) electrochemical gradient generated by a proton pump known as the vacuolar H^+-ATPase, or V-ATPase. At a mass of approximately 1 million daltons, the V-ATPase is a massive multi-subunit complex, with a cylindrical shape that is approximately 14 nm wide and 24 nm long. Therefore it is not surprising that each SV typically has only a single molecule of the V-ATPase. Under physiological conditions, the V-ATPase couples the hydrolysis of 1 ATP to the transport of up to 4 H^+ ions into the lumen of the SV, thus simultaneously increasing lumenal $[H^+]$ and generating a lumen-positive membrane potential. Secondary active transporters couple the downhill movement of H^+ (from the vesicle into the cytosol) to the uptake of neurotransmitter molecules into the SV lumen. Neurotransmitter transporters that have been cloned include those for glutamate, γ-aminobutyric acid (GABA), glycine, ACh, and biogenic amines such as dopamine, serotonin, and histamine. Relative to the cytosol, the SV lumen is more acidic by approximately 1.5 pH units and more electrically positive by 40 to 70 mV (Box 12-5).

When an SV fuses to the PM at the active zone, neurotransmitter molecules leave the vesicle lumen and enter the synaptic cleft. Because the neurotransmitter content of the SV is usually released in an all-or-none, or quantized, manner, the term quantum is used to denote the population of neurotransmitter molecules contained in a single SV. The neurotransmitter content of an SV can range from several thousand to a few tens of thousands of molecules, depending on the neurotransmitter. Thus the neurotransmitter concentration in the SV may range from approximately 0.5 to 1.0 M or more (Box 12-6)! When the content of an SV is released, the neurotransmitter concentration in the synaptic cleft can rise transiently to hundreds of micromolar or even millimolar levels.

Neurotransmitter-Filled Synaptic Vesicles Dock at the Active Zone and Become "Primed" for Exocytosis

Once an SV has been filled with neurotransmitter, it makes its way to the active zone, apparently by diffusion (Box 12-7). The SV then attaches to the active zone through a process known as **SV docking**; the molecular details of the docking process are still poorly understood. It is reasonable to suppose that docking requires the interaction of proteins at the active zone with proteins on the surface of the SV. At least one protein on the SV, Rab3, and one protein in the active zone, Rim (*Rab3-i*nteracting *m*olecule), are thought to participate in the docking interaction (Box 12-8).

After docking, formation of a *core complex* between SNARE* proteins in the active zone and on the SV surface causes the SV to become primed for fusion (Figure 12-9; Box 12-9). Three **SNARE proteins** are crucial for **SV priming**: synaptobrevin (also known as *v*esicle-*a*ssociated *m*embrane protein, or VAMP) resides on the SV, whereas syntaxin-1 and SNAP-25 are on the PM at the active zone. SNARE proteins exhibit characteristic SNARE motifs. Each SNARE motif is a specific amino acid sequence that adopts an α-helical conformation (Box 12-9). Synaptobrevin and syntaxin-1 each have a single SNARE motif, whereas SNAP-25 incorporates two such motifs. When the three SNAREs interact, their SNARE motifs align to form a four-helix bundle, which is essential to the stability of the core complex. With the formation of the core complex, the SV membrane and the PM are "riveted" into extremely close apposition and become primed for Ca^{2+}-**triggered membrane fusion** and exocytosis.

---

*SNARE stands for *s*oluble *N*SF-*a*ttachment protein *r*eceptor; NSF is *N*-ethylmaleimide-*s*ensitive factor.

BOX 12-5

HOW TO THINK ABOUT pH AND MEMBRANE POTENTIAL FOR A SYNAPTIC VESICLE

With an inner diameter of 30 nm, an SV has an inner membrane surface area of 0.00283 μm^2 and a lumenal volume of 1.41×10^{-20} L. We will consider what it means for the SV to have a V_m and an internal pH.

A V_m is established by separating positive and negative charges on opposite sides of a membrane (see Chapter 4). At any V_m, the amount of charge, q, that is separated by the membrane is given by $q = C_m V_m$, where C_m is the membrane capacitance. For biomembranes, C_m is 1 $\mu F/cm^2$ of membrane area, or 1×10^{-14} $F/\mu m^2$, which means that the capacitance of the SV membrane is $C_m = 2.83 \times 10^{-17}$ F. Taking the SV V_m to be +50 mV relative to the cytosol, we can calculate the amount of positive charge inside the SV to be $q = 1.41 \times 10^{-18}$ coulombs. Because 1 mol of charge is equivalent to 96,485 coulombs (Faraday's constant), the amount of excess positive charge in the SV is 1.47×10^{-23} mol, or just 9 positive charges! Therefore, at the minimum, the V-ATPase needs to transport only 9 H^+ ions into the SV to generate a membrane potential of +50 mV.

Being more acidic than the cytosol by ~1.5 pH units, the lumenal pH is 5.7, corresponding to $[H^+] = 10^{-pH} = 2.00 \times 10^{-6}$ M. This means that the SV contains 2.82×10^{-26} mol of H^+, or 0.017 H^+! A fraction of a proton would seem to be nonsensical. Moreover, this implies that random entry or escape of even a single H^+ would change the pH inside the SV wildly. We must remember, however, that in living organisms, the most heavily buffered biochemical parameter is pH. Recall that a pH buffer consists of a mixture of a weak acid (HA) and its salt (A$^-$), and $[H^+]$ is uniquely defined by the *ratio* of [HA] and [A$^-$]:

$$[H^+] = K_a \frac{[HA]}{[A^-]} \qquad [B1]$$

where K_a is the dissociation constant of HA. Thus, even though $[H^+]$ may be small, [HA] and [A$^-$] need not be.

The large reserve of HA molecules is ready to release H^+ as needed, whereas the reserve of A$^-$ molecules is ready to "annihilate" any excess H^+. As a result, the buffer resists changes in $[H^+]$. Suppose that inside the SV, there is a pH buffer with $pK_a = 5.7$ ($K_a = 2.00 \times 10^{-6}$), at a concentration of 100 mM, then at a pH of 5.7, there are 425 molecules each of HA and A$^-$ in the lumen. If 25 extra H^+ ions are pumped into the SV, the HA and A$^-$ populations will become 450 and 400, respectively. This changes [HA]/[A$^-$] from 1.00 to 1.125, with a corresponding pH change from 5.70 to 5.75—good evidence that a buffer keeps the pH relatively stable.

The presence of pH buffers implies that more H^+ ions must be transported to cause acidification of the SV lumen. When the SV fuses with the plasma membrane, its lumen becomes continuous with the extracellular fluid, which is at pH $\geq$ 7. Using the foregoing example, and assuming that the SV lumen attains pH = 7 on exocytosis, we can calculate how many protons need be pumped into the SV to bring the luminal pH back down to 5.7. At pH 7.0, [HA]/[A$^-$] = $[H^+]/K_a$ = 0.05, so in the SV lumen, HA and A$^-$ number 40 and 810, respectively. When the SV is retrieved back into the terminal by endocytosis, the lumenal pH will be restored to 5.7, at which point HA and A$^-$ will again each number 425. In the process, the V-ATPase needs to pump 385 H^+ ions—far more than the 9 that would be required to generate a 50 mV V_m (indeed, we can estimate that moving 385 H^+ ions into the SV would generate a V_m greater than 2000 mV!). To regulate pH and V_m independently, therefore other transporters on the SV dissipate the V_m even as H^+ transport by the V-ATPase acidifies the lumen. For example, ClC-3, a Cl$^-$ channel, allows Cl$^-$ ions to enter the SV to neutralize the buildup of positive charges as H^+ ions are transported into the SV.

Botulinum toxin (BoTox), the most potent toxin known, blocks neurotransmitter release by preventing the formation of the core complex. Surprisingly, BoTox is used frequently as therapy for a variety of movement disorders and in cosmetic surgery because of its ability to block transmitter release (Box 12-10). Two other neurotoxins, *tetanus toxin* and the black widow spider toxin, α-*latrotoxin*, also interfere with transmitter release (Box 12-10) and, like BoTox, have been used to help dissect steps in the neurotransmitter release process.

BOX 12-6

THE TRANSMITTER CONCENTRATION IN A SYNAPTIC VESICLE

It is instructive to estimate the neurotransmitter concentration in an SV. Assuming the SV holds 5000 neurotransmitter molecules (8.3×10^{-21} mol), and knowing that an inner diameter of 30 nm implies a lumenal volume of 1.4×10^{-20} L, we can calculate the neurotransmitter concentration to be ~0.6 M! Such a high solute concentration would make the SV lumen very hypertonic with respect to the cytosol (which has an osmolarity of ~0.3 M); therefore osmotically driven water flux into the SV would cause swelling and

rupture. A deterrent to osmotic catastrophe would be a gel-like matrix in the SV lumen that can bind, or complex, neurotransmitter molecules. On being transported into the SV, neurotransmitter molecules would bind or adsorb to the matrix and become effectively removed from solution, with a corresponding reduction in lumenal osmolarity. This may be the solution that nature has evolved: At least in some SVs, a proteoglycan matrix with the ability to adsorb neurotransmitters has been found.

BOX 12-7

SYNAPTIC VESICLES MOVE TO THE ACTIVE ZONE APPARENTLY BY DIFFUSION

As monitored by imaging microscopy, SVs in the nerve terminal appear to move by diffusion; that is, each SV seems to undergo a random walk (see Chapter 2). Although some variation exists, the measured diffusion coefficient (D) is typically of the order of 10^{-3} $\mu m^2 \cdot sec^{-1}$ (or in more commonly used units, 10^{-11} $cm^2 \cdot sec^{-1}$).

For a spherical particle such as the SV, D can be estimated using the Stokes-Einstein relation:

$$D = \frac{kT}{6\pi\eta r}$$

where k is Boltzmann's constant, T is the absolute temperature, η is the viscosity of the medium in which the particle is moving, and r is the particle radius. The formula is intuitively reasonable: mobility should increase with temperature but decrease as the particle size or the viscosity of the medium increases. Knowing $k = 1.381 \times 10^{-16}$ $g \cdot cm \cdot sec^{-2} \cdot K^{-1}$, and taking $T = 310K$ (37°C), $r = 20$ nm, and the measured cytosolic viscosity of ~1 centipoise (0.01 $g \cdot cm^{-1} \cdot sec^{-1}$), the calculated diffusion coefficient of an SV is $1.14 \times$

10^{-7} $cm^2 \cdot sec^{-1}$, or 11.4 $\mu m^2 \cdot sec^{-1}$. We see that the calculated diffusion coefficient is approximately 10,000 times larger than what is actually observed. One possible rationalization is that the cytosol in the nerve terminal is vastly more viscous than the cytosol elsewhere in the cell. This is not supported by measurement. Another interpretation is that the SV is "sticky"; that is, it interacts with many microscopic binding sites in the cytoskeleton of the nerve terminal. Constant binding and unbinding do not change the random nature of diffusive motion, but they do slow down the random walk. This view is supported by experimental measurement.

The small diffusion coefficient observed for SVs suggests that an SV moves to the active zone very slowly. We must remember, however, that the presynaptic terminal is a very small structure with diameter $d \leq 1$ μm. We can estimate that diffusion of SVs over this distance would occur on the time scale of $t = d^2/6D$ (see Chapter 2), that is, of the order of 10 seconds. This is sufficiently fast for neurophysiology.

Binding of Ca^{2+} to Synaptotagmin Triggers the Fusion and Exocytosis of the Synaptic Vesicle

At the active zone, Ca^{2+} influx through Ca^{2+} channels triggers the fusion of primed SVs to the PM. Synaptotagmin is the protein on the SV that acts as a Ca^{2+} sensor. An N-terminal transmembrane domain anchors synaptotagmin in the SV membrane, whereas two C-terminal C_2-domains* act as Ca^{2+}-binding

*The C_2-domain is a Ca^{2+}-binding motif comprising approximately 130 amino acid residues. Originally identified in protein kinase C (PKC), C_2 domains have been found in a variety of signaling proteins that interact with cellular membranes. Structural diversity enables different C_2 domains to bind a wide range of molecules, including lipids, inositol phosphates, and proteins, in addition to, or even instead of, Ca^{2+}.

BOX 12-8

Rab3 GTPase MAY CONFER DIRECTIONALITY AND FIDELITY ON THE SYNAPTIC VESICLE CYCLE

Rab3 is a low-molecular-weight GTPase (molecular weight 24,000 kDa). Only in its GTP–bound form can Rab3 attach to the SV surface. An integral component of the active zone is a large protein called Rim (*Rab3-interacting molecule*; 1553 amino acid residues). Rim has an N-terminal zinc-finger domain* that interacts with GTP-Rab3; Rim also binds to several other proteins in the active zone that are critical to SV exocytosis. Thus the interaction of GTP-Rab3 on the SV with Rim in the active zone potentially underlies the docking process. Interestingly, during, or at the end of,

the SV cycle, Rab3 hydrolyzes its bound GTP to GDP. GDP-Rab3 dissociates from the SV surface only after Ca^{2+}-triggered exocytosis of the SV. In the cytosol, exchange of GTP for GDP regenerates GTP-Rab3, which can once again attach to SVs that are ready to undergo docking. Thus GTP-Rab3 binding potentially enables the SV to dock, and after GTP hydrolysis to GDP, GDP-Rab3 is stripped from SVs that had fused at the active zone. It is therefore possible that Rab3 serves to confer directionality on the SV cycle and to ensure the fidelity of the fusion process.

*A zinc-finger domain is a motif consisting of an α-helix and at least one β-strand, each of which contributes amino acid side chains that coordinately bind a Zn^{2+}. Binding of Zn^{2+} is essential for maintaining the finger-like structure. Structural variations enable different zinc fingers to bind to nucleic acids or proteins.

BOX 12-9

SNARE PROTEINS PROMOTE EXOCYTOTIC MEMBRANE FUSION

Exocytosis of secretory vesicles in different cell types follows the same underlying mechanistic strategy. Exocytosis requires a secretory vesicle to fuse with the PM to release vesicular content into the extracellular space. An absolute requirement for this membrane fusion is the formation of a *core complex* between SNARE proteins. Historically, SNAREs were classified by their location: v-SNAREs reside on the vesicle membrane; t-SNAREs are found on the target membrane to which the vesicle will fuse. X-ray crystallography has revealed that central to the core complex is a four-helix bundle formed from four highly conserved helical SNARE motifs. Because some SNAREs contain two SNARE motifs, the core complex can comprise three or four SNARE proteins. The helical SNARE motifs contain either highly conserved arginine (single-letter code R) or glutamine (single-letter code Q) residues that are essential to

formation of the four-helix bundle. Therefore, in the structure-based nomenclature, SNAREs that bear an arginine-containing motif are called R-SNAREs, whereas Q-SNAREs incorporate glutamine-containing motifs. Although there is only one type of R-motif, three types of Q-motifs are known: Q_a, Q_b, and Q_c. To form a stable core complex, all four types of SNARE motifs must be present in the four-helix bundle.

For exocytosis of an SV, three SNARE proteins are required: synaptobrevin (also known as vesicle-associated membrane protein, or VAMP) on the SV, and syntaxin-1 and SNAP-25, which are on the PM at the active zone. To form the core complex at the active zone, synaptobrevin contributes an R-SNARE motif, and syntaxin-1 contributes a Q_a motif, whereas the Q_b and Q_c motifs are contributed by SNAP-25, which contains two SNARE motifs.

sites. At the active zone, clusters of VGCCs are located close to the primed SV. When an AP depolarizes the nerve terminal, the VGCCs at the active zone open, and the resulting influx of Ca^{2+} ions creates a microdomain of very high $[Ca^{2+}]$ (at least tens

of micromolar) that envelops the nearby, primed SV. The high $[Ca^{2+}]$ in this microdomain enables the synaptotagmin on the SV to bind multiple Ca^{2+} ions. On Ca^{2+} binding, the C_2-domains of synaptotagmin become activated and undergo a twofold

BOX 12-10

BOTULINUM NEUROTOXINS INHIBIT NEUROMUSCULAR TRANSMISSION AND ARE WIDELY USED TO TREAT MOVEMENT DISORDERS

Botulinum toxin (BoTox), from the bacterium *Clostridium botulinum*, is the most toxic substance known, with a median lethal dose (LD_{50}) of approximately 1 ng/kg body weight in humans. BoTox, which does not appear to cross the blood-brain barrier, blocks neuromuscular transmission; death is caused by respiratory failure resulting from paralysis of the respiratory muscles. The seven distinct subtypes of BoTox, designated A through G, are zinc proteases that act in cholinergic and other nerve terminals. The toxins inhibit neurotransmitter release by cleaving, and thereby inactivating, some of the proteins involved in synaptic vesicle exocytosis. BoTox-A and BoTox-E cleave SNAP-25, BoTox-C cleaves both SNAP-25 and syntaxin and all other forms of BoTox cleave synaptobrevin. Thus BoTox prevents the formation of the core complex that is essential for SV exocytosis. BoTox block of transmitter release is relieved only after removal of the toxin and resynthesis of the cleaved peptides.

The ability of BoTox to block neuromuscular transmission has led to its use in treating disorders caused by overactive muscles. The intramuscular injection of BoTox is used to treat blepharospasm (excessive blinking), cervical dystonia (a condition in which the head is tilted to one side and the chin is elevated), strabismus, and numerous other movement disorders. Sudden onset of focal dystonia prevented the concert pianist, Leon Fleisher, from using his right hand to play the piano, although he continued playing with his left hand only. After receiving BoTox injections in his hand, however, Fleisher was able to resume concert playing with both hands. Because BoTox action is reversible, treatment must be repeated every 3 to 4 months.

BoTox also has extensive cosmetic applications: it is used to eliminate facial wrinkles caused by facial muscle contraction. In fact, BoTox injection has become the most common cosmetic procedure, with approximately 5 million injections performed per year.

Tetanus toxin, from *Clostridium tetani*, causes spastic paralysis by blocking inhibitory synapses in the spinal cord. This toxin, which is a metalloprotease, specifically prevents the release of inhibitory transmitters by cleaving synaptobrevin.

α-Latrotoxin (α-LTX), from black widow spider venom, induces massive release of neurotransmitters and depletion of synaptic/secretory vesicles in nerve terminals and endocrine cells. α-LTX creates Ca^{2+}-permeable channels in the PM; the Ca^{2+} influx through these channels in presynaptic nerve terminals is primarily responsible for the effect.

---

interaction: with the already-formed SNARE complex and with the phospholipids in the PM. This interaction is thought to destabilize the junction between the SV and the PM; this triggers fusion of the two lipid bilayers. Fusion between the SV membrane and the PM, which can be measured electrically (Box 12-11), opens a pore that releases the contents of the SV into the synaptic cleft.

Retrieval of the Fused Synaptic Vesicle Back into the Nerve Terminal Can Occur through Clathrin-Independent and Clathrin-Dependent Mechanisms

On fusion, the SV membrane becomes continuous with the PM at the active zone. Because exocytosis of SVs would steadily increase the total amount of PM in the active zone, there must be a compensatory process that retrieves the SV membrane into the terminal after neurotransmitter release. Three modes of retrieval have been observed—two fast and one slow (Figure 12-10). During fast retrieval, the recently formed fusion pore closes and the now empty SV can (1) remain docked at the active zone and be refilled with transmitter in preparation for another episode of exocytosis or (2) undock from the active zone and rejoin the SV pool near the active zone. Little is known about the molecular details of these fast retrieval processes. The process whereby an SV releases neurotransmitter and is retrieved immediately into the local SV pool is whimsically described as **kiss-and-run** (Figure 12-10B), whereas retention and refilling of the SV at the active zone are known as **kiss-and-stay** (Figure 12-10A).

In the slower mode of SV retrieval (Figure 12-10C), monomers of the protein clathrin coat the cytoplasmic surface of the SV. The clathrin molecules assemble

BOX 12-11

FUSION OF A SYNAPTIC VESICLE TO THE PLASMA MEMBRANE AT THE ACTIVE ZONE CAN BE MONITORED ELECTROPHYSIOLOGICALLY

When an SV fuses at the active zone, the SV membrane becomes continuous with the PM. As a result, the PM surface area increases slightly. Because biological membranes have electrical capacitance ($\sim$1 microfarad/cm^2; see Chapter 4), the increased surface area should be detectable as an increase in the capacitance associated with the PM. This has indeed been measured in the calyx of Held synapse, which occurs in the auditory brainstem and is the largest mammalian synapse. Extending well over 10 μm, the presynaptic terminal is extremely large. Similar to the calyx of a flower, the terminal is a cup-shaped structure that envelops the cell body of the postsynaptic neuron. Synaptic contacts between the calyx-like terminal and the postsynaptic neuronal cell body number approximately 500 (i.e., $\sim$500 active zones are present at this giant nerve terminal). Owing to their large size, the calyx terminal and the postsynaptic neuron can both be monitored electrophysiologically with patch electrodes. Spontaneous release of neurotransmitter at any of the 500 active zones can evoke an EPSC (*excitatory postsynaptic current*) in the postsynaptic cell. Because every observed EPSC must have been caused by SV exocytosis, every EPSC must be concomitant with an increase in presynaptic membrane capacitance, which should be measurable electrically.

A practical difficulty in measuring the capacitance change ensuing from fusion of a single SV is that the capacitance change is exceedingly small. With a diameter of 40 nm, an SV has a surface area of approximately 5000 nm^2, with a corresponding capacitance of 5×10^{-17} farad. In stark contrast, even the presynaptic terminal by itself has an area close to 10^9 nm^2, corresponding to a capacitance of 10^{-11} farad. Therefore fusion of an SV would increase the capacitance of the presynaptic terminal membrane by 5 parts per million—essentially impossible to measure in the presence of noise associated with even the best electrical measurements. In one study,* to overcome the problem of poor signal-to-noise ratio, researchers made 2.66 million temporally correlated recordings of the postsynaptic current and the presynaptic capacitance. Using the onset of the EPSC as a time reference, the noisy capacitance recordings were aligned, summed, and averaged (Figure B-1). The result is a trace, now with excellent signal-to-noise ratio, of the average presynaptic capacitance change associated with fusion of a single SV at an active zone. The amplitude of the capacitance change was 6.1×10^{-17} farad, which implies an average SV surface area of 6100 nm^2, which, in turn, corresponds to an SV diameter of 44 nm—in remarkable agreement with the value of 45 nm determined by electron microscopy of SVs at the calyx of Held.

Incidentally, the success of using averaging to improve signal-to-noise ratio stems from the nature of the random walk (see Chapter 2). Noise fluctuates randomly from positive to negative and thus can increase

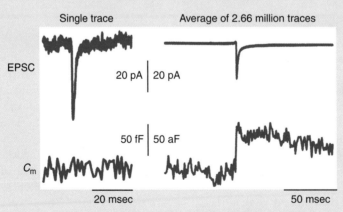

FIGURE B-1 ■ Simultaneously recorded traces of excitatory postsynaptic current (EPSC) and presynaptic membrane capacitance (C$_m$) at the calyx of Held synapse. Shown on the *left* is a pair of simultaneously recorded EPSC and C$_m$ traces for a single incidence of synaptic vesicle exocytosis. Shown on the *right* are corresponding traces resulting from averaging measurements from 2.66 million exocytotic events. Values marked on the *left* and *right* sides of each *vertical scale bar* apply to the traces on the corresponding sides; note difference in scales for the C$_m$ traces: fF = 10^{-15} farad; aF = 10^{-18} farad.

*Wu X-S, Xue L, Mohan R, et al: The origin of quantal size variation: vesicular glutamate concentration plays a significant role, *J Neurosci* 27:3046, 2007.

FUSION OF A SYNAPTIC VESICLE TO THE PLASMA MEMBRANE AT THE ACTIVE ZONE CAN BE MONITORED ELECTROPHYSIOLOGICALLY—cont'd

or decrease the measured signal randomly. When N traces containing random noise are added, the noise amplitude should increase with $\sqrt{N}$. In contrast, true signal occurs in a definite direction (either positive, like a capacitance increase, or negative, like an inward current). When N signal traces are summed, the signal amplitude grows in direct proportion to N. Therefore, when summing traces containing signal and random noise, the signal-to-noise ratio (S/N) is given by

$$\frac{S}{N} = \frac{N}{\sqrt{N}} = \sqrt{N}$$

That is, signal averaging causes the signal to overtake noise gradually in proportion to the square root of the number of data traces used in the average. For 2.66 million measurements at the calyx of Held, averaging is expected to improve the S/N by 1630-fold, as was verified experimentally (Figure B-1).

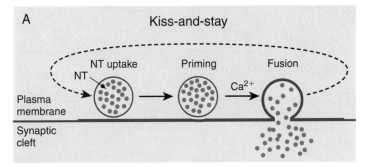

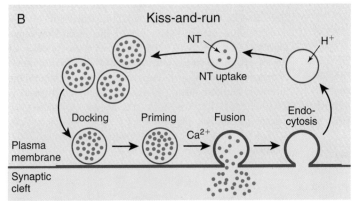

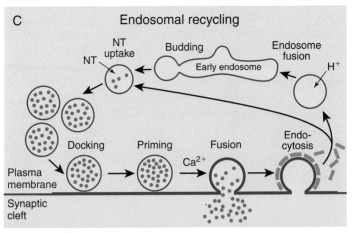

FIGURE 12-10 ■ Three modes of synaptic vesicle recycling. **A,** Kiss-and-stay: After neurotransmitter (NT) release, an SV is endocytosed by closing the fusion pore and, while remaining docked at the active zone, becomes refilled with neurotransmitters. The refilled SV is then ready to undergo neurotransmitter release. **B,** Kiss-and-run: After neurotransmitter release, an SV is endocytosed and undocked from the active zone to recycle by joining the reserve pool of SVs. Kiss-and-run does not require clathrin. **C,** Endosomal recycling: After exocytosis, an SV becomes clathrin coated *(lower right;* blue bars represent clathrin molecules) and is then endocytosed. After clathrin molecules have been stripped, the SV can recycle either directly by rejoining the reserve pool of SVs or through endosomes. *(Modified from Südhof TC: Annu Rev Neurosci 27:509, 2004.)*

into a mesh that envelops the SV to form a coated vesicle. Thereafter, just as in nonsynaptic endocytosis, the clathrin-coated SVs are detached from the PM; the clathrin coat is then disassembled in an ATP-requiring process. The uncoated SVs either rejoin the SV pool in the nerve terminal or are recycled through the endosomal system. During low-frequency electrical stimulation (e.g., ≤2 APs per second), SV exocytosis occurs at a low rate, and the clathrin-independent retrieval modes predominate. When the nerve terminal is stimulated at high frequency, **clathrin-dependent endocytosis** becomes the dominant mechanism for retrieval of SV membrane after neurotransmitter release.

SHORT-TERM SYNAPTIC PLASTICITY IS A TRANSIENT, USE-DEPENDENT CHANGE IN THE EFFICACY OF SYNAPTIC TRANSMISSION

At many synapses, repeated presynaptic stimulation leads to either enhancement or diminution of the postsynaptic response. The enhancement or diminution is short-lived, with a duration ranging from a fraction of a second to minutes. Such transient changes in the efficacy of neurotransmission in response to repeated stimulation are referred to as *short-term synaptic plasticity*. Two prominent forms of positive plasticity are **synaptic facilitation** and **post-tetanic potentiation** (PTP). A commonly observed form of negative plasticity is **synaptic depression**.

Synaptic depression is a progressive decrement in postsynaptic response on repeated stimulation. Figure 12-11 shows a typical experiment in which a synapse is given two successive stimuli (paired-pulse stimulation). The second stimulus evokes a significantly smaller postsynaptic response than the first; that is, the second response is *depressed* relative to the first (**paired-pulse depression**). The extent of synaptic depression depends on the interstimulus interval, Δt: the shorter the interval, the stronger the depression (Figure 12-11B).

The mechanism most commonly invoked to explain synaptic depression is *SV depletion*. Recall that only docked and primed SVs can undergo Ca^{2+}-triggered exocytosis, and the total number of such vesicles is limited at any given active zone. After the

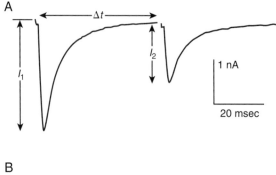

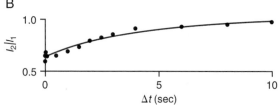

FIGURE 12-11 ■ Synaptic depression at the climbing fiber-Purkinje neuron synapse in the rat cerebellum. **A,** EPSCs evoked by a pair of stimuli separated by an interstimulus interval of $\Delta t = 50$ msec. The amplitude of the second response (I_2) is clearly depressed relative to the amplitude of the first (I_1); this is referred to as paired-pulse depression. **B,** The dependence of synaptic depression on the interstimulus interval. The ratio of the second response amplitude to the first (I_2/I_1) is plotted as a function of the interstimulus interval, Δt. The plot shows that synaptic depression is most pronounced at short interstimulus intervals but diminishes as the interstimulus interval is lengthened. *(Data from Regehr WG, Stevens CF: Physiology of synaptic transmission and short-term plasticity. In Cowan WM, Südhof TC, Stevens CF, editors: Synapses, Baltimore, 2001, Johns Hopkins University Press, p 135.)*

first stimulus has triggered SV exocytosis, the *readily releasable pool* of SVs is reduced. Therefore fewer SVs are available for exocytosis when the second stimulus arrives (Box 12-12). Although SV depletion is considered the "classic" mechanism, it cannot account for the large extent of depression observed at many synapses. Two additional mechanisms could contribute to synaptic depression. First, VGCCs in the presynaptic terminal may inactivate after they have been activated by the presynaptic AP; this can severely reduce the probability of transmitter release by subsequent APs. Second, synaptic depression could have a cytoarchitectural basis. SVs can dock and undergo exocytosis only

BOX 12-12
FUNCTIONAL CONSEQUENCE OF SYNAPTIC VESICLE DEPLETION

Depletion of the readily releasable pool of SVs at the active zone can affect the response of the synapse to subsequent stimulation. The situation can be made more quantitative by considering the simplest possible scenario. A synapse has a pool of R readily releasable SVs and a stimulus can trigger exocytosis of some fraction, f, of the SVs. This means that the first stimulus evokes a response that is proportional to fR and reduces the readily releasable pool to $(R - fR) = (1 - f)R$. The second stimulus releases the same fraction, f, of the readily releasable SVs, but because the readily releasable pool is now $(1 - f)R$, the postsynaptic response is proportional to $f(1 - f)R$, which is smaller than the first response. This simple model predicts that if the probability of neurotransmission (SV exocytosis) is low (i.e., f is small), then weak synaptic depression should be observed. Conversely, if the initial probability for release is high (i.e., f is large), then subsequent synaptic responses should show strong depression. This is indeed observed empirically.

at the active zone, which, at most synapses, spans a very small membrane area, typically 0.05 to 0.10 μm^2. This is consistent with active zones having a linear dimension of 100 to 300 nm. Because the SV diameter is approximately 40 nm, fusion of one or a few SVs could transiently disrupt the active zone structure and interfere with further docking and fusion of other SVs.

In response to repeated stimuli, synapses can also exhibit facilitation—a progressive increment in postsynaptic response to repetitive stimulation. In a typical experiment, a synapse receives two stimuli successively (Figure 12-12A, *inset*). The postsynaptic response to the second stimulus is significantly larger than the response to the first stimulus (**paired-pulse facilitation**). The process whereby stimulus-evoked neurotransmitter release is enhanced by prior stimuli is known as facilitation. Facilitation can increase synaptic strength by as much as 10-fold and persists for tens to hundreds of milliseconds. The degree of

facilitation depends on the interstimulus interval, Δt: the shorter the interval, the stronger the facilitation (Figure 12-12A).

Facilitation is the result of an increase in neurotransmitter release in response to stimulation; therefore the underlying mechanism must account for enhanced SV exocytosis. Because Ca^{2+} is the obligate signal that triggers SV exocytosis, any mechanism that enhances the presynaptic Ca^{2+} signal could lead to facilitation. Since the 1980s, much evidence has accumulated in favor of the **residual Ca^{2+} hypothesis**— that facilitation is caused by Ca^{2+} remaining in the presynaptic terminal after a prior stimulus. When a stimulus depolarizes the terminal to open clusters of VGCCs, the resulting Ca^{2+} influx creates microdomains at the active zones, with the local Ca^{2+} concentration, $[Ca^{2+}]_{loc}$, in the microdomains exceeding 10 μM, sufficient to trigger SV exocytosis. The microdomains of high $[Ca^{2+}]_{loc}$ are short-lived: after the stimulus, as soon as the VGCCs close, diffusion and buffering of Ca^{2+} rapidly eliminate spatial heterogeneities in less than a millisecond, leaving the terminal with a spatially uniform "residual" level of Ca^{2+} ($[Ca^{2+}]_{res} \leq 1 \mu M$). Thereafter, the normal homeostatic mechanisms— extrusion by the NCX and the PMCA, as well as uptake into the ER by SERCA—operate to lower $[Ca^{2+}]_{res}$. If another stimulus arrives before $[Ca^{2+}]_{res}$ decays completely and resting $[Ca^{2+}]$ is reestablished, the Ca^{2+} level in the newly created microdomains would be $[Ca^{2+}]_{loc} + [Ca^{2+}]_{res}$; that is, the local Ca^{2+} signal would be boosted by the residual $[Ca^{2+}]$ to yield a stronger trigger for SV exocytosis. In this model, facilitation persists for the duration required for residual Ca^{2+} to be cleared and for resting $[Ca^{2+}]$ to be reestablished in the terminal (Figure 12-12B). Two types of observations are consistent with the residual Ca^{2+} hypothesis: artificially elevating resting $[Ca^{2+}]_i$ in the terminal enhances stimulus-evoked neurotransmitter release, whereas buffering $[Ca^{2+}]_i$ reduces release. The minimal version of the residual Ca^{2+} hypothesis as presented here need not be the exact mechanism underlying facilitation; that is, residual Ca^{2+} need not promote SV exocytosis directly. Instead, residual Ca^{2+} could affect processes that control the stimulus-evoked Ca^{2+} signal and thus indirectly enhance neurotransmitter release. Irrespective of the microscopic details of the underlying mechanism,

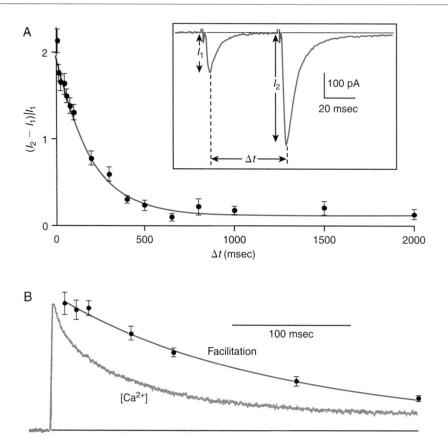

FIGURE 12-12 ■ Synaptic facilitation at the synapse formed between the granule cell and the Purkinje cell in the rat cerebellum. **A,** The graph shows facilitation as a function of interstimulus interval, Δt; facilitation is defined as $(I_2 - I_1)/I_1$, where I_1 and I_2 are, respectively, the amplitudes of the first and second EPSCs evoked by a pair of stimuli separated by a time interval, Δt. *Inset:* An example of EPSCs evoked by extracellular stimulation with a pair of stimulus pulses separated by Δt = 50 milliseconds; the second EPSC is clearly larger than the first (paired-pulse facilitation). **B,** Comparison of the time course of facilitation and residual Ca^{2+} in the presynaptic terminal. The continuous trace shows $[Ca^{2+}]$ in the presynaptic terminal, as measured with a fluorescent Ca^{2+} indicator. The *circles* represent facilitation at different times; a smooth curve was fit to the data points. *(Data from Regehr WG, Stevens CF: Physiology of synaptic transmission and short-term plasticity. In Cowan WM, Südhof TC, Stevens CF, editors:* Synapses, *Baltimore, 2001, Johns Hopkins University Press, p 135.)*

however, facilitation depends on $[Ca^{2+}]$ in the presynaptic terminal, and facilitation is linked to the persistence of the residual Ca^{2+} signal (Figure 12-12B).

Post-tetanic potentiation, or PTP, is an enhancement of neurotransmitter release at a synapse following rapid repetitive stimulation (a *tetanus*). The persistence of PTP ranges from tens of seconds to a few minutes. An example of PTP at a hippocampal synapse is illustrated in Figure 12-13. Although the

rise in $[Ca^{2+}]_i$ caused by tetanic stimulation is essential for PTP induction, PTP persists long after dissipation of the residual Ca^{2+} signal (Figure 12-13B). The disparate time courses of the Ca^{2+} signal and PTP indicates that $[Ca^{2+}]_i$ in the nerve terminal does not directly support and maintain PTP. Rather, the Ca^{2+} signal must trigger a signal transduction pathway that then supports PTP through a more enduring biochemical change, such as phosphorylation of key

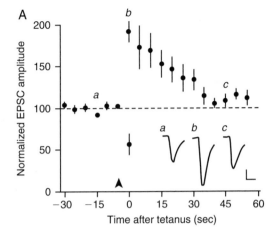

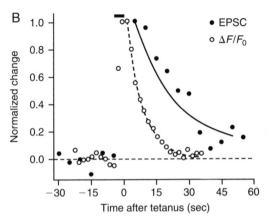

FIGURE 12-13 ■ Post-tetanic potentiation (PTP) at the synapse formed between a CA3 pyramidal neuron and a CA1 pyramidal neuron in the hippocampus. **A,** Tetanic stimulation resulted in an 83% increase in the amplitude of the EPSC evoked immediately after the tetanus. The potentiation of the synaptic response then diminishes over approximately 40 seconds after the tetanus. Average results from 10 synapses are plotted. *Inset:* EPSCs from one CA1 neuron at the three time points indicated, showing a pretetanus baseline response *(a)*, the maximally potentiated response immediately after the tetanus *(b)*, and a response after potentiation had decayed nearly back to baseline *(c); scale bars for inset,* 100 pA, 20 msec). The tetanic stimulation consisted of stimuli delivered at 100 Hz for 4.5 sec; the *arrowhead* marks the start of the tetanus. **B,** Comparison of the time course of PTP of EPSCs *(filled circles)* and the elevation of $[Ca^{2+}]$ *(open circles)* in the presynaptic nerve terminal as reported by the fluorescent Ca^{2+} indicator, fluo-4 (the data are expressed as change in fluorescence intensity relative to baseline fluorescence intensity, $\Delta F/F_0$). Each data set was normalized to its respective maximum value, and a single-exponential decay was fitted to the post-tetanic change in EPSC amplitude *(solid curve)* and $[Ca^{2+}]$ *(dashed curve). Solid bar* indicates the duration of the tetanus. Note that PTP persists significantly longer than the presynaptic Ca^{2+} signal. *(Data from Brager DH, Cai X, Thompson SM:* Nat Neurosci *6:551, 2003; copyright by Macmillan Publishers Ltd.)*

proteins in SV exocytosis. At the synapse illustrated in Figure 12-13, Ca^{2+}-activated protein kinase (protein kinase C or PKC) is implicated in the phosphorylation events that underlie PTP induction, although precisely which synaptic proteins are phosphorylated remains unknown.

At any particular synapse, multiple forms of short-term plasticity (both positive and negative) can occur. Therefore, on the time scale of milliseconds to many tens of seconds, the moment-by-moment response of the synapse to repeated stimuli is sculpted by the interplay and balance of the different processes of short-term plasticity.

SUMMARY

1. Signals are communicated from cell to cell in the nervous system by synaptic transmission, which can be either electrical or chemical.
2. At electrical synapses, current flows directly from one cell to the other through gap junction channels.

3. Most synapses are chemical synapses, in which the presynaptic terminal contains many SVs that are filled with high concentrations of neurotransmitters.
4. An AP in the presynaptic nerve terminal releases neurotransmitter molecules that diffuse across

the synaptic cleft and bind to receptors in the postsynaptic membrane. Receptor activation can either directly affect membrane conductance and/or initiate an intracellular signaling cascade.

5. The NMJ is an unusually large chemical synapse. Numerous SVs, clustered around active zones in the synaptic boutons, are filled with the neurotransmitter ACh. Each active zone lies directly over a junctional fold in the postsynaptic membrane that contains a high density of nAChRs. The enzyme acetylcholinesterase is anchored in the synaptic cleft.

6. Following a presynaptic AP, ACh is released, diffuses rapidly across the synaptic cleft, and binds to nAChRs. This opens the nAChR channels, which conduct inward current, thereby depolarizing the muscle cell and generating the EPP. The EPP is normally large enough to reach AP threshold in the muscle cell. ACh action is terminated as the transmitter diffuses away and is hydrolyzed by acetylcholinesterase.

7. The synaptic delay between the arrival of the presynaptic AP at the nerve terminal and the beginning of the postsynaptic response is characteristic of all chemical synapses. The time it takes to open presynaptic VGCCs accounts for most of the delay.

8. Neurotransmitters are released in discrete packets, called quanta; a quantum corresponds to the neurotransmitter content of a single SV. At the NMJ, spontaneous quantal release of 5000 to 10,000 molecules of ACh, stored in an SV, generates a MEPP. Nerve-evoked release of ACh is also quantal.

9. Neurotransmitter release requires Ca^{2+}. The presynaptic AP opens VGCCs, thus allowing Ca^{2+} to enter the nerve terminal. The number of quanta released depends on the amount of Ca^{2+} that enters the cell.

10. At most CNS synapses, APs often fail to trigger neurotransmitter release. When release is triggered successfully, only one or two quanta are released.

11. The SV concentrates, stores, and delivers neurotransmitter at the synapse. Transport of neurotransmitters into the SV depends on a V-ATPase, which couples ATP hydrolysis to the transport of H^+ into the SV. Secondary active transporters then couple the downhill movement of H^+ to the uptake of neurotransmitter molecules into the SV.

12. An SV attaches to the active zone through a process known as docking. After docking, formation of a core complex between SNARE proteins on the SV and in the active zone primes the SV for Ca^{2+}-triggered membrane fusion and exocytosis. BoTox blocks neurotransmitter release by preventing the formation of the core complex.

13. Synaptotagmin is the Ca^{2+} sensor on the SV. Ca^{2+} binding to synaptotagmin triggers fusion of the SV to the PM. This opens a pore that releases the contents of the SV into the synaptic cleft.

14. Following fusion, three modes of retrieval of SV membrane have been observed. During fast retrieval the fusion pore closes and the SV can (1) remain docked at the active zone or (2) undock and rejoin the SV pool near the active zone. In the slower mode of retrieval, clathrin-coated SVs are detached from the PM, the coat is removed, and the SV can rejoin the SV pool or be recycled through the endosomal system.

15. Synapses can exhibit paired-pulse depression, in which the postsynaptic response to the second of a pair of stimuli is smaller than the response to the first. The mechanism most commonly invoked to explain synaptic depression is SV depletion.

16. Synapses can exhibit paired-pulse facilitation, in which the enhanced postsynaptic response to a second stimulus is the result of an increase in neurotransmitter release caused by Ca^{2+} remaining in the nerve terminal after a prior stimulus.

17. Following tetanic stimulation, evoked neurotransmitter release is enhanced. This post-tetanic potentiation requires a rise in $[Ca^{2+}]_i$, which triggers a signal transduction pathway that enhances SV exocytosis.

KEY WORDS AND CONCEPTS

- Neuron doctrine
- Autonomic nervous system
- Synapse
- Synaptic bouton
- Presynaptic terminal

- Postsynaptic cell
- Chemical synapse
- Electrical synapse
- Gap junction
- Connexon
- Connexin
- Neurotransmitters
- Neurotransmitter receptors
- Acetylcholine (ACh)
- Acetylcholinesterase
- Parasympathetic
- Neuromuscular junction (NMJ)
- End plate
- Synaptic vesicle (SV)
- Active zone
- Synaptic cleft
- Synaptic delay
- End-plate potential (EPP)
- Nicotinic ACh receptor (nAChR)
- Miniature end-plate potential (MEPP)
- Synaptic vesicle cycle
- Exocytosis
- SV docking
- SV priming
- Ca^{2+}-triggered membrane fusion
- Quantum of neurotransmitter
- SNARE proteins
- Botulinum toxin (BoTox)
- "Kiss-and-run" mode of SV recycling
- "Kiss-and-stay" mode of SV recycling
- Clathrin-dependent endocytosis
- Excitatory postsynaptic current (EPSC)
- Synaptic facilitation (paired-pulse facilitation)
- Synaptic depression (paired-pulse depression)
- Post-tetanic potentiation (PTP)
- Residual Ca^{2+} hypothesis

STUDY PROBLEMS

1. You are studying synaptic transmission at an NMJ in the presence of low $[Ca^{2+}]_o$. You record a large number of spontaneous MEPPs, which have a mean amplitude of 0.5 mV. You also record a large number of EPPs in response to nerve stimulation. After fitting a binomial distribution to the EPP amplitude histogram you estimate that 50 sites can release a quantum of transmitter and that the probability that a quantum of transmitter is released in response to nerve stimulation is 0.06.
   a. Based on these numbers, what is the mean amplitude of the EPP?
   b. What is the probability that, under these low $[Ca^{2+}]_o$ conditions, there will be no postsynaptic response (no EPP) to nerve stimulation?

2. A new genetic disease has been identified that exhibits muscle weakness as one of its many symptoms. You suspect that neuromuscular transmission is depressed and decide to investigate.
   a. How would you demonstrate that neuromuscular transmission is depressed?
   b. Give three changes (molecular defects) that could occur at the neuromuscular junction that would decrease neuromuscular transmission.

BIBLIOGRAPHY

Atluri PP, Regehr WG: Determinants of the time course of facilitation at the granule cell to Purkinje cell synapse, *J Neurosci* 16:5661, 1996.

Brager DH, Cai X, Thompson SM: Activity-dependent activation of presynaptic protein kinase C mediates post-tetanic potentiation, *Nat Neurosci* 6:551, 2003.

De Camilli P, Haucke V, Takei K, Mugnaini E: The structure of synapses. In Cowan WM, Südhof TC, Stevens CF, editors: *Synapses*, Baltimore, MD, 2001, Johns Hopkins University Press, p 89.

Grumelli C, Verderio C, Pozzi D, et al: Internalization and mechanism of action of clostridial toxins in neurons, *Neurotoxicology* 26:761, 2005.

Kandel ER, Schwartz JH, Jessell TM: *Principles of Neuroscience*, ed 4, New York, NY, 2000, McGraw-Hill, p 175.

Südhof TC: Neurotransmitter release. In Südhof TC, Starke K, editors: *Handbook of Experimental Pharmacology*. Vol 184, *Pharmacology of Neurotransmitter Release*, Berlin, Germany, 2008, Springer-Verlag, p 1.

Südhof TC: The synaptic vesicle cycle, *Annu Rev Neurosci* 27:509, 2004.

Takamori S, Holt M, Stenius K, et al: Molecular anatomy of a trafficking organelle, *Cell* 127:831, 2006.

Wu X-S, Xue L, Mohan R, et al: The origin of quantal size variation: vesicular glutamate concentration plays a significant role, *J Neurosci* 27:3046, 2007.

Zhai GR, Bellen HJ: The architecture of the active zone in the presynaptic nerve terminal, *Physiology* 19:262, 2004.

13

SYNAPTIC PHYSIOLOGY II

OBJECTIVES

1. Understand what a chemical neurotransmitter and its receptor are.

2. Understand neurotransmission at the neuromuscular junction, a well-studied chemical synapse.

3. Understand the major classes of neurotransmitters and their actions.

4. Distinguish between ionotropic and metabotropic receptors.

5. Understand how excitatory and inhibitory neurotransmitters influence postsynaptic activity, and how synaptic transmission is terminated.

6. Understand how synaptic responses are integrated.

7. Understand short-term and long-term plasticity in synaptic transmission.

CHEMICAL SYNAPSES AFFORD SPECIFICITY, VARIETY, AND FINE TUNING OF NEUROTRANSMISSION

Many of the physiological principles of chemical synaptic transmission were elucidated by studies of the NMJ and squid giant synapse (see Chapter 12). We also noted that by far most synapses in mammals are chemical synapses. Chemical synapses afford the variety, specificity, and fine tuning of neurotransmission that underlie the enormous complexity of nervous system function. More than 100 different neurotransmitter molecules have been identified. This variety contributes to specificity so that two adjacent synaptic inputs from different presynaptic neurons may release different transmitters that act on different receptors to produce different actions in a single postsynaptic cell. Additionally, many presynaptic neurons can release more than one neurotransmitter. This provides a broad range of opportunities for chemical intervention to alter synaptic transmission; this is the basis of neuropharmacology and

psychopharmacology. To characterize chemical neurotransmission better, we first need to define what we mean by a neurotransmitter.

What Is a Neurotransmitter?

A neurotransmitter is an intercellular messenger molecule. It is generated in and released by a presynaptic neuron, and it acts on an adjacent postsynaptic neuron or neuroeffector cell, or even back on the neuron that released the molecule. With this definition, we can distinguish two classes of chemical neurotransmitter substances: conventional and unconventional transmitters.

Conventional Neurotransmitters Comprise Three Groups All conventional neurotransmitters are stored in SVs and are released in quantal fashion by Ca^{2+}–dependent exocytosis. These conventional transmitters can be divided into three groups. One group includes ACh, **γ-aminobutyric acid (GABA)**, **glutamate**, **glycine** (Figure 13-1), and certain purines, which are stored in approximately 40-nm, electron-lucent SVs.

1. Acetylcholine (ACh)

2. Amino Acid Neurotransmitters

Glutamate (Glu)   γ-Aminobutyrate (GABA)   Glycine (Gly)

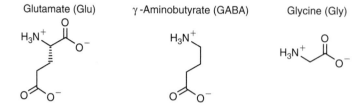

3. Biogenic Monoamines

Serotonin (5-HT)   Histamine (HA)

Dopamine (DA)   Epinephrine (Epi)   Norepinephrine (NE)

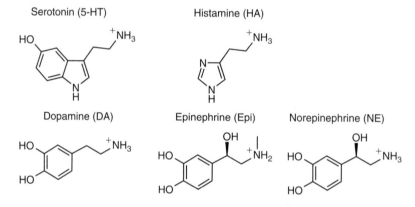

4. Endocannabinoids

2-Arachidonoylglycerol (2-AG)   N-arachidonoyl ethanolamide
(AEA, anandamide)

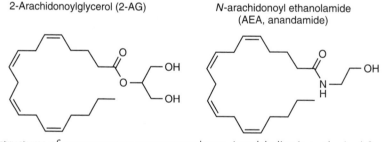

FIGURE 13-1 ■ The structures of some common neurotransmitters. Acetylcholine is synthesized from choline and acetyl-coenzyme A; γ-aminobutyric acid (GABA) is synthesized from glutamate; serotonin (5-HT) is derived from tryptophan; histamine is derived from histidine; and the catecholamines (dopamine, epinephrine, and norepinephrine) are sequentially derived from tyrosine. The following conventions have been used in these structural drawings: (1) implicit carbon: every unlabeled vertex, whether internal or terminal, represents a carbon atom; (2) implicit hydrogen: every carbon has a sufficient number of (undrawn) hydrogens to make the total number of bonds to that carbon equal to 4; and (3) explicit heteroatoms: noncarbon, nonhydrogen atoms (e.g., O, N) are labeled explicitly; hydrogens attached to the heteroatom are also explicitly drawn. For example, $\diagdown\diagup\diagdown\diagup$OH is equivalent to $CH_3-CH=CH-CH_2-OH$. In addition, a solid wedge bond means the attached group extends above the plane of the page; a hatched bond means the attached group extends below the plane of the page.

These transmitters are released and act at well-defined synapses with "classical" structures: the presynaptic cells, which have active zone densities in their boutons, are separated by narrow (20- to 40-nm) synaptic clefts from the postsynaptic cells, which have specialized postsynaptic densities (see Figure 12-3). This type of synaptic transmission is sometimes referred to as **wiring transmission** because direct connection of presynaptic and postsynaptic cells can be viewed as "hard wiring." Transmission at electrical synapses (see Chapter 12) also is a type of wiring transmission.

A second group of conventional neurotransmitters comprises **biogenic monoamines:** serotonin (5-hydroxytryptamine or 5-HT), **histamine**, and the **catecholamines**[*] (dopamine, norepinephrine or NE, and epinephrine or Epi, which are sequentially derived from tyrosine) (Figure 13-1). These transmitters are usually stored in small dense core (electron-opaque) SVs, which are approximately 40 to 70 nm in diameter. Biogenic amines can be released at classical synapses where the closely apposed postsynaptic cells have specialized postsynaptic densities. Biogenic amine transmitters are also released at **en passant synapses** where the presynaptic "terminals" are simply SV-containing *varicosities* (enlargements) that occur along axons as they pass postsynaptic cells. At en passant synapses, the postsynaptic cells often exhibit typical postsynaptic density regions to which the transmitter receptors are confined. In many instances, however, the transmitter receptors are more diffusely distributed on the postsynaptic cell surface. A good example of the latter is the synapse between a sympathetic neuron and a vascular smooth muscle cell.

At en passant synapses, the released transmitter may activate several postsynaptic cells. This type of neurotransmission is sometimes referred to as **volume transmission** because the transmitter activates all postsynaptic cells that express the appropriate receptor within the volume of tissue in which the transmitter concentration is sufficiently high. In contrast to the situation at classical synapses, neurotransmitters must diffuse across greater distances to reach postsynaptic cells; hence the synaptic delay is correspondingly longer for volume transmission.

The third group of conventional transmitters, the neuropeptides, range in length from 3 to approximately 100 amino acid residues, although most have fewer than 30 residues. These **peptide neurotransmitters** are often stored in **large dense core vesicles** ($\sim$100 to 150 nm diameter) in which monoamines are frequently cosequestered. These SVs are often concentrated in, and released from, boutons at classical synapses. Neuropeptide transmitter release is distinctive in requiring bursts of presynaptic APs, whereas most other conventional transmitters are usually released by a single AP. Ca^{2+}-evoked release from large dense core vesicles may also occur at various places along the presynaptic axon where no apparent synaptic structures are present. This type of transmitter release therefore more closely resembles the secretion of most hormones by endocrine cells, which do not have specialized active zones where the secretory vesicles cluster before release. Indeed, the storage and secretion of peptide hormones in the **magnocellular neurons** (large cell neurons) of the hypothalamus[†] were originally believed to be a peculiarity of these *neuroendocrine* cells. We now recognize that this phenomenon is much more widespread in the nervous system.

Unconventional Neurotransmitters Are Not Stored In Synaptic Vesicles In contrast to the aforementioned hydrophilic neurotransmitters, a few neurotransmitters are hydrophobic gases or lipids. These lipid-soluble transmitters cannot be stored in SVs, which have lipid bilayer membranes. The examples discussed here are the **gaseous neurotransmitters**, nitric oxide (NO) and carbon monoxide (CO), and several lipids known as **endocannabinoids**, such as **2-arachidonoylglycerol (2-AG)** (Figure 13-1). The *endogenous cannabinoids* are so named because many of their effects are mimicked by the major psychoactive agent in marijuana, Δ^9-tetrahydrocannabinol, an *exogenous* cannabinoid that is derived from the hemp plant, *Cannabis sativa*. Even though the unconventional transmitters are not stored in SVs, they

---

*Catecholamines all contain a catechol (1,2-dihydroxybenzene) moiety.

[†]Roger Guillemin and Andrew Schally shared (with Rosalyn Yalow) the 1977 Nobel Prize in Medicine or Physiology for their discoveries of peptide "releasing" hormones, which are stored in, and released by hypothalamic neurons. In 1931, Ulf von Euler first discovered a peptide neurotransmitter, substance P. von Euler and Julius Axelrod did pioneering research on norepinephrine as neurotransmitter, for which they shared the 1970 Nobel Prize in Medicine or Physiology with Bernard Katz (see Chapter 12).

fit the definition of synaptic neurotransmitters given earlier. They are generated in and released by a neuron, and they act on an adjacent neuron or effector cell or back on the neuron that released the molecule.

RECEPTORS MEDIATE THE ACTIONS OF NEUROTRANSMITTERS IN POSTSYNAPTIC CELLS

A neurotransmitter receptor is the target protein to which the transmitter molecule (the ligand) binds. This interaction is specific and is a mechanism for transducing the signal from the presynaptic neuron into a different signal in the postsynaptic cell. The receptors for conventional neurotransmitters are all integral membrane proteins.

Conventional Neurotransmitters Activate Two Classes of Receptors: Ionotropic Receptors and Metabotropic Receptors

Ionotropic Receptors Are Ligand-Gated Ion Channels (See Chapter 8) for Which the Ligands Are Neurotransmitters These receptors are composed of four or five subunits whose arrangement defines a central, gated pore. Binding of the transmitter to the ligand-binding site on the ionotropic receptor induces a conformational change that opens the channel. **Ionotropic receptors** generally mediate fast synaptic responses because the neurotransmitters directly gate ion channels. These receptors are important targets for pharmacotherapy.

Metabotropic Receptors Are Not Ion Channels but Are Monomeric Proteins with Seven Transmembrane Helices Binding of a neurotransmitter activates a metabotropic receptor, which in turn activates a *G-protein*. These receptors are therefore members of the **G-protein–coupled receptor (GPCR)** family.[*] This is the largest family of membrane proteins in the human genome; its members influence almost all biological responses and are the targets of approximately 60% of clinically useful drugs.

Most **metabotropic receptors** function as homodimers or heterodimers (or even oligomers). Binding of a neurotransmitter molecule to one monomer is usually sufficient to activate the dimer fully, and binding of a transmitter molecule to the second monomer may even reduce the activity of the dimer.

A G-protein is a heterotrimeric protein complex comprising α-, β-, and γ-subunits; the α-subunit has a binding site for *guanine nucleotide*. A wide range of different α-, β-, and γ-subunits is expressed, giving rise to many possible $\alpha\beta\gamma$ trimer combinations that can couple diverse receptors to downstream effectors (Figure 13-2). In the quiescent state, a *guanosine diphosphate* (GDP) is bound to the α-subunit of a G-protein ($\alpha_{GDP}\beta\gamma$). Some $\alpha_{GDP}\beta\gamma$ complexes are bound to appropriate GPCR dimers, thus forming inactive pentameric complexes. On activation by a neurotransmitter molecule, the metabotropic receptor catalyzes the exchange of *guanosine triphosphate* (GTP) for GDP to generate $\alpha_{GTP}\beta\gamma$. This enables the $\alpha_{GTP}\beta\gamma$ complexes to dissociate into α_{GTP} and $\beta\gamma$ moieties, each of which can interact with downstream effector such as phospholipase C, protein kinase C (PKC), or adenylyl cyclase, to initiate a cellular response (Figure 13-2). The signal is terminated when the α subunit hydrolyzes the bound GTP, causing the reversion to α_{GDP}, which then can recombine with $\beta\gamma$ to regenerate the inactive $\alpha_{GDP}\beta\gamma$ trimer. Because G-protein signaling depends on multiple enzymatic processes, synaptic responses mediated by metabotropic receptors are necessarily slower than those mediated by ionotropic receptors.

Metabotropic receptor-coupled G-proteins are molecular switches that modulate postsynaptic activity in a variety of ways. For example, the α-subunit of G-proteins activates enzymes that generate second messengers such as **inositol trisphosphate (IP$_3$), diacylglycerol (DAG)** (Figure 13-2), and the **cyclic mononucleotides**, cAMP and cGMP. The IP$_3$ releases Ca^{2+} from the S/ER, and DAG and the cyclic nucleotides can gate ion channels or stimulate protein kinases to phosphorylate other proteins. The $\beta\gamma$ subunits can directly modulate Ca^{2+} channels and K^+ channels (e.g., inward rectifier, Kir, channels; see Box 8-5). Phosphorylation at multiple sites on the cytoplasmic loop of the GPCR regulates its activity. This and the

---

[*]Alfred G. Gilman and Martin Rodbell were awarded the 1994 Nobel Prize for Physiology or Medicine for discovering G-proteins and characterizing their role in cell signaling.

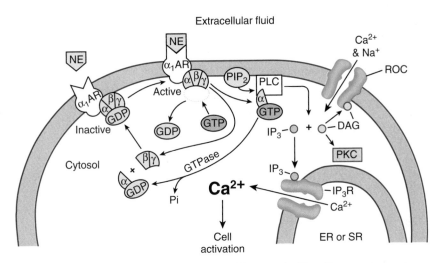

FIGURE 13-2 ■ The agonist receptor-G-protein–phosphoinositide cascade. The diagram shows the sequence of events that occurs following the binding of an agonist to its G-protein-coupled receptor in the plasma membrane. In the example illustrated here, NE activates a vascular smooth muscle α_1-adrenergic receptor, or α_1AR. The activated α_1AR promotes exchange of GTP for GDP on a heterotrimeric G-protein and its subsequent dissociation into α_{GTP} and $\beta\gamma$ complexes. The α_{GTP} activates phospholipase C (PLC), which catalyzes the cleavage of the membrane lipid, phosphatidylinositol 3,4-bisphosphate (PIP_2) into two second messengers. One of these is inositol 1,4,5-trisphosphate (IP_3), which triggers the release of Ca^{2+} from the ER or SR. The other is diacylglycerol (DAG), which can trigger the entry of Ca^{2+} from the extracellular fluid through receptor-operated channels (ROCs), as well as stimulate protein kinase C (PKC) to phosphorylate downstream signaling proteins. Although not shown here, the $\beta\gamma$ complex may also interact with downstream effector mechanisms.

variety of transmitter receptors and their subtypes, as well as the variety of G-protein subunit isoforms, result in a rich tapestry of responses.

GPCR Activation Ceases When the Neurotransmitter Dissociates from the Receptor or When the Receptor Is Desensitized When the transmitter molecule dissociates from the GPCR dimer, the receptor reverts to its inactive conformation and can no longer catalyze GTP-GDP exchange until it is reactivated by another transmitter molecule. Activated GPCRs can be phosphorylated by *G protein–coupled receptor kinases* (*GRKs*), members of the serine/threonine kinase family. This permits the binding of *β-arrestin* (a *scaffold protein*, an organizer of signaling molecules) to the GPCR and thereby sterically hinders coupling to G-protein complexes. This process is a

form of *desensitization*. The GPCR-β-arrestin complexes can be retrieved from the plasma membrane to enable recycling of the GPCRs. The GPCR-β-arrestin complexes also have another function: The bound β-arrestin can, in turn, bind and activate various kinases such as the *extracellular signal–regulated kinase* (*ERK*), whose downstream actions are G-protein independent.

In the ensuing discussion, it is not possible to provide a comprehensive description of all aspects of neurotransmitter action in the brain. We concentrate on a few representative examples in which the underlying mechanisms have been elucidated. Descriptions of more complex behaviors, in which the mechanisms are less well understood, are left to specialized neuroscience texts.

ACETYLCHOLINE RECEPTORS CAN BE IONOTROPIC OR METABOTROPIC

Nicotinic Acetylcholine Receptors are Ionotropic

The nAChR expressed at the NMJ is a representative ionotropic receptor (see Chapter 12). Five homologous nAChR subunits (α, β, γ, δ, and ϵ) have been identified. Each nAChR is a pentamer containing at least two α-subunits in combination with other subunits. At the NMJ, the subunit composition is 2α:β:δ:ϵ; in cholinergic neurons in the brain, the composition is often 3α:2β. Each subunit has five membrane-spanning domains and a large extracellular domain. Transmembrane segments from the five subunits form the channel pore, which opens when two of the α-subunits bind an ACh molecule simultaneously. The functional characteristics of nAChRs, which are nonselective cation channels, are discussed in a later section. As their name implies, nAChRs also can be activated by the tobacco plant alkaloid*, nicotine, which induces sensations of relaxation and euphoria but may become addictive. In **myasthenia gravis**, the most common disease affecting neuromuscular transmission, and in congenital myasthenic syndrome, nAChR function is disrupted (Box 13-1).

The enzyme acetylcholinesterase (AChE), which hydrolyzes ACh into acetate and choline, is highly concentrated in the synaptic cleft at cholinergic synapses, including the NMJ. Following the release of ACh, AChE rapidly reduces the extracellular ACh concentration and thereby contributes to the termination of its action. A Na^+–coupled cotransporter retrieves choline back into the presynaptic nerve terminals, where it is used to resynthesize ACh. Some paralytic drugs and toxins, including several clinically useful agents, act through interactions with nAChRs or AChE (Box 13-2).

Muscarinic Acetylcholine Receptors Are Metabotropic

Muscarinic AChRs (mAChRs) are GPCRs that mediate most of the actions of ACh in the **central nervous system (CNS)**. Five mAChR subtypes (M_1 to M_5) have been identified. The mAChRs are expressed by many CNS neurons as well as by various effector cells, including cardiac, smooth muscle, vascular endothelial, and exocrine gland cells. Many of these cells express two or more mAChR subtypes; this results in a diverse array of postsynaptic responses. Activation of mAChRs can lead to phosphoinositide hydrolysis and IP_3-activated Ca^{2+} release from the S/ER, to inhibition of adenylyl cyclase and reduction of cAMP levels, or to activation of Kir channels (see Box 8-5).

In the brain, activation of M_1-, M_2-, and M_5-mAChRs modulates region-specific release of some neurotransmitters, including ACh (through *autoreceptors*), dopamine, and GABA, and inhibition of certain nAChRs. The multiple underlying mechanisms include, among others, the phosphinositide-Ca^{2+}-PKC cascade (**phosphoinositide cascade**), opening of Kir channels, and protein kinase-mediated modulation of N- and P/Q type voltage-gated Ca^{2+} channels (see Table 8-2). Agonists and antagonists of specific mAChR subtypes have been developed to alleviate symptoms of Alzheimer's disease and schizophrenia by augmenting and antagonizing cholinergic transmission, respectively.

The mAChRs modulate sympathetic and parasympathetic ganglionic function by altering neuronal excitability. Classic examples of parasympathetic effects are slowing of the heart, contraction of bladder and airway smooth muscles, endothelium-mediated vasodilation, and secretion of fluid and electrolytes by the salivary glands. M_2-, M_3- and M_4-mAChRs mediate cardiac slowing by activating, respectively, a Kir channel and two different slowly opening voltage-gated K^+ channels. The mAChR-triggered, Ca^{2+}-dependent salivary secretions and bladder and airway smooth muscle contractions are activated, in part, by phospholipase C–mediated generation of IP_3 and consequent mobilization of Ca^{2+} from the S/ER (Figure 13-2).

Muscarinic AChRs are so named because they are activated by *muscarine*, a toxic alkaloid found in the poisonous mushroom, *Amanita muscaria*. Muscarine is called a *parasympathomimetic* agent because it mimics the natural parasympathetic agonist, ACh. Two other toxic alkaloids, *atropine*, from the deadly nightshade (*Atropa belladonna*), and *scopolamine*, from other members of the nightshade plant family, are mAChR antagonists that are employed clinically. Atropine is used routinely in ophthalmology to dilate the pupil of the eye. Atropine blocks mAChRs on the pupillary constrictor muscle; the unopposed dilator muscle, which is activated by NE, then dilates the pupil.

---

*Alkaloids are nitrogen-containing ringed compounds, usually of plant origin, that have physiological actions on animals and humans.

BOX 13-1

CONGENITAL AND ACQUIRED MYASTHENIC SYNDROMES RESULT FROM DEFECTIVE NEUROMUSCULAR TRANSMISSION

Myasthenia gravis is a neuromuscular disease characterized by muscle weakness and fatigability. It almost always affects the eyelids, eye muscles, and limb muscles. The most prevalent form of the disease is the autoimmune form, in which antibodies are produced against nAChRs. These antibodies bind to the α-subunit of the receptor and prevent activation of the receptor by ACh. In addition, the density of nAChRs is reduced in myasthenia gravis. As a result, the amplitude of the EPP is reduced to near threshold and many EPPs fail to activate the muscle, thus accounting for the weakness.

Congenital myasthenic syndromes (CMSs) are a heterogeneous group of disorders caused by presynaptic, synaptic, or postsynaptic defects at the neuromuscular junction. The clinical picture in all forms of these syndromes consists of respiratory and feeding difficulties at birth or weakness of the ocular and bulbar muscles (muscles of the tongue, lips, and pharynx) during the first 2 years of life. Postsynaptic CMS can be associated with changes in the gating kinetics of nAChR channels or with a reduction in the density of nAChR channels in the postsynaptic membrane. In the fast channel syndrome a mutation in the nAChR channel results in a greatly reduced postsynaptic response to ACh. Patch clamp recordings of single nAChR channels reveal that the mutation causes the channel open time to be much shorter than that of wild-type channels (Figure B-1). Because of the reduced inward current through ACh-gated channels, many skeletal muscle cells fail to reach AP threshold and thus fail to contract. This results in muscle weakness.

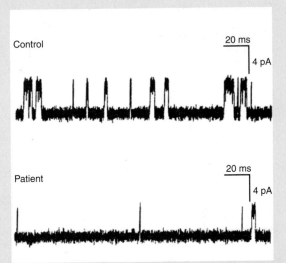

FIGURE B-1 ■ ACh-gated channel openings in a patient with CMS are briefer than normal. Single ACh-gated channel openings (shown as upward deflections) were recorded at an NMJ from a normal individual (control) and from a patient with CMS (patient). The single channel openings in the patient are briefer and less frequent than in the control subject. (Data from Ohno K, Wang H-L, Milone M, et al: *Neuron* 17:157, 1996.)

In addition, scopolamine, which can cross the blood-brain barrier (described in Box 3-5), is used to prevent nausea and motion sickness.

AMINO ACID NEUROTRANSMITTERS MEDIATE MANY EXCITATORY AND INHIBITORY RESPONSES IN THE BRAIN

Glutamate Is the Main Excitatory Neurotransmitter in the Brain

Glutamate (see Figure 13-1), which acts on both ionotropic and metabotropic glutamate receptors, does not cross the blood-brain barrier and must therefore be synthesized in the brain. The three types of ionotropic glutamate receptors are named for the pharmacological agents that activate them selectively: *N*-methyl-D-aspartate (NMDA) receptors (NMDARs), α-amino-3-hydroxy-5-methyl-4-isoxazole-propionate (AMPA) receptors (AMPARs), and *kainate* (kainic acid) receptors. The kainate receptors (KARs) and AMPARs are closely related structurally, whereas NMDARs are more distantly related. Ionotropic glutamate receptors are heterotetramers or homotetramers of protein subunits that are expressed as multiple isoforms.

Both NMDARs and AMPARs are expressed at most *glutamatergic* synapses, but some synapses may express only one receptor type. NMDARs are unusual because their activation requires, in addition to glutamate, a

BOX 13-2

SOME DRUGS AND TOXINS DISRUPT TRANSMISSION AT THE NEUROMUSCULAR JUNCTION

Certain drugs and toxins cause paralysis by binding to nAChRs and preventing ACh binding. One example, curare, is an arrow poison used in South America to paralyze prey. Curare is derived from various plants belonging to the genera *Chondrodendron* and *Strychnos*. The active component of curare is the toxin D-tubocurarine, an alkaloid. D-Tubocurarine is a competitive antagonist of ACh at nAChRs: it binds to the receptor but elicits no response, and it prevents the binding of ACh. D-Tubocurarine was once used as a muscle relaxant in anesthesia but is no longer used because of better alternatives. Another toxin with a similar mode of action is α-bungarotoxin, found in the venom of the banded krait *Bungarus multicinctus*. Both D-tubocurarine and α-bungarotoxin cause death by paralyzing respiratory muscles.

Another group of drugs that act at the NMJ are certain nerve gases, such as sarin, that have been used only as chemical weapons. Sarin gained notoriety in 1995 when a group of terrorists released the gas in the Tokyo subway system and it killed 12 people. Sarin is an organophosphate that binds covalently to the catalytic serine residue at the active site of AChE, and thus blocks ACh hydrolysis irreversibly. As a result, the concentration of ACh builds up in the synaptic cleft and causes continuous depolarization of some CNS neurons and skeletal muscle. Death results from loss of respiratory function.

In current practice, two classes of muscle relaxants are used as adjuncts to anesthetics. Depolarizing drugs such as succinylcholine, which cannot be hydrolyzed by AChE, activate nAChRs; these agents cause muscle depolarization and receptor desensitization. Nondepolarizing drugs such as pancuronium act like D-tubocurarine; they are competitive ACh antagonists that do not activate the receptors.

---

coagonist that is thought to be glycine or D-serine. Full activation requires simultaneous binding of two glutamate *and* two coagonist molecules. Activation of AMPARs and NMDARs also induce kinetically different synaptic responses. AMPARs are monovalent cation–selective channels that mediate most of the fast excitatory synaptic transmission in the brain. They gate rapidly, are poorly permeable to Ca^{2+}, desensitize strongly, and are blocked by intracellular polyamines such as spermine and spermidine. In contrast, NMDARs are slower-gating, nonselective cation channels that are permeable to Ca^{2+} but are blocked by extracellular Mg^{2+}. The physiological importance of the differences between AMPARs and NMDARs is discussed in a subsequent section.

KARs appear to be more complex. They have ionic conductances similar to those of AMPARs but, like metabotropic receptors, they may also activate intracellular signaling cascades at some synapses. Some KARs are located on presynaptic terminals, where they can facilitate release of neurotransmitters.

There are also three types of **metabotropic glutamate receptors** (mGluR groups I to III, with a total of eight subtypes). The mGluRs are widely distributed in the brain, on postsynaptic neurons, and on presynaptic terminals where they may modulate transmitter release. Many mGluRs are also expressed in glial cells, where they can be activated to increase $[Ca^{2+}]_i$ and open Ca^{2+}-activated K^+ (K_{Ca}) channels, and thus change $[K^+]_o$ in the immediate vicinity of the synapse. The consequent shift in the V_m of the presynaptic terminal may influence transmitter release.

Neuronal mGluRs exert their effects by regulating key intracellular signaling pathways. For example, activation of some group I mGluRs stimulates phospholipase C to augment NMDA-mediated excitation. Conversely, activation of some type II and III mGluRs can reduce excitation by inhibiting adenylyl cyclase. Some mGluRs also modulate synaptic transmission at GABAergic and monoaminergic synapses. Such a broad spectrum of actions results in mGluR involvement in diverse pathophysiological responses including mood disorders, epilepsy, and addictive behavior.

γ-Aminobutyric Acid and Glycine Are the Main Inhibitory Neurotransmitters in the Nervous System

Inhibitory synapses that use the amino acids GABA and glycine (see Figure 13-1) as neurotransmitters are the most abundant synapses in the CNS. Most of the neurons that release GABA are interneurons, which are neurons that make short-range connections with other neurons. Whereas GABA is the predominant inhibitory neurotransmitter in the brain, glycine is the main inhibitory neurotransmitter in the spinal cord. Many inhibitory synapses in the spinal cord, and even in the adjacent brainstem and in the cerebellum, however, release both GABA and glycine. The functional significance of this apparent redundancy is unknown.

Two types of GABA receptors, $GABA_AR$ and $GABA_CR$, and strychnine-sensitive glycine receptors (GlyR) (Box 13-3), are ionotropic receptors. These receptors are all pentameric transmitter-gated Cl^- channels. Reduced inhibitory synaptic transmission, in some cases the result of a genetic GlyR defect, is responsible for human startle disease (Box 13-3).

$GABA_ARs$ are activated by the cooperative binding of two GABA molecules. $GABA_ARs$ are also the target of benzodiazepines, agents such as diazepam (Valium) and chlordiazepoxide (Librium), which are widely used to treat anxiety disorders. These drugs bind to the $GABA_ARs$ and potentiate the action of GABA.

There also are metabotropic inhibitory transmitter receptors: the $GABA_BRs$. Activation of $GABA_BRs$ stimulates second messenger systems involving phospholipase C and adenylyl cyclase and leads to the opening of K^+ channels or inhibition of Ca^{2+} channels. Both mechanisms generate slow inhibitory signals in the postsynaptic neurons. When $GABA_BRs$ are located on presynaptic terminals (as they often are), their activation inhibits transmitter release.

NEUROTRANSMITTERS THAT BIND TO IONOTROPIC RECEPTORS CAUSE MEMBRANE CONDUCTANCE CHANGES

Chemical synapses can be broadly classified into two functional types: **excitatory synapses** and **inhibitory synapses**. The activation of an excitatory synapse increases the probability that the postsynaptic cell will generate an AP, whereas the activation of an inhibitory synapse decreases this probability. The neurotransmitters

---

BOX 13-3
GLYCINE RECEPTORS UNDERLIE STRYCHNINE TOXICITY AND HUMAN STARTLE DISEASE

The glycine receptors (GlyRs) are pentameric glycine-gated Cl^- channels. These receptors are present primarily in the spinal cord and brainstem, where they mediate inhibitory synaptic transmission. Five GlyR subunits have been identified: $\alpha 1$ to $\alpha 4$ and β. The α-subunits have 80% to 90% amino acid sequence identity, and the β-subunit shows approximately 47% similarity to the $\alpha 1$-subunit. Maximal activation of GlyRs requires a minimum of three bound glycines.

The plant alkaloid strychnine, isolated mainly from the seeds of *Strychnos nux vomica*, is a potent competitive antagonist of glycine with a K_d of 5 to 10 nM. Because of its high affinity for GlyRs, strychnine is an extremely toxic compound. It is still used in some rodent baits, and accidental human exposure can occur; ingestion of as little as 5-10 mg can cause death. By blocking GlyRs, strychnine decreases inhibitory neurotransmission in reflex circuits in the spinal cord and brainstem and thereby increases neuronal excitability. Increased motor neuron excitability enhances muscle contractions, leading to twitching and convulsions. Death is usually caused by respiratory failure

resulting from respiratory muscle spasms. Although no blocker of strychnine action exists, muscle relaxants that block nAChRs, such as pancuronium (see Box 13-2), are used to counteract the effects of strychnine poisoning.

Human startle disease, or hyperekplexia, is a rare neurological disease characterized by temporary muscle rigidity in response to unexpected stimuli. Muscle rigidity often causes an unprotected fall, producing chronic injuries that are also characteristic of the disease. Hyperekplexia is often misdiagnosed as epilepsy, but it can readily be distinguished because the patient never loses consciousness during a startle episode. The disease can be caused by mutations in $\alpha 1$ GlyR subunits. The mutations reduce the amplitude of glycine-activated Cl^- currents in at least three different ways. They can reduce the glycine affinity of GlyRs, reduce the conductance of single GlyR Cl^- channels, or reduce the surface expression of the receptors. These mutations all reduce inhibitory neurotransmission and enhance neuronal excitability in a manner similar to strychnine poisoning.

that cause such changes in excitability usually do so by opening or closing ion channels, thereby increasing or decreasing membrane conductance.

At Excitatory Synapses, the Reversal Potential Is More Positive Than the Action Potential Threshold

In Chapter 12, synaptic transmission at the NMJ, a well-studied excitatory synapse, was described by the following sequence of events: A presynaptic AP releases ACh molecules, which diffuse across the synaptic cleft and bind to nAChRs. This opens the nAChR channels and depolarizes the postsynaptic membrane.

When an nAChR channel opens near the resting V_m, inward current flows through the open channel. In response to a presynaptic AP at the NMJ, millions of molecules of ACh are released, and they cause a large number of nAChR channels to open nearly synchronously. The macroscopic (whole-cell) current through these open nAChRs is called the **end-plate current (EPC)**. The inward EPC causes the muscle membrane to depolarize, thereby producing the EPP described in Chapter 12 (see Figure 12-5).

The ionic species that flow through an open nAChR channel can be identified using a voltage clamp to characterize the effect of V_m on the amplitude of the EPC. When the postsynaptic V_m is made less negative, in the range of -100 mV to -20 mV, the amplitude of the inward EPC becomes smaller (Figure 13-3). When $V_m \approx 0$ mV, there is no detectable EPC (i.e., its amplitude is 0); at more positive membrane potentials the EPC is outward rather than inward (Figure 13-3). The V_m at which the EPC reverses direction from inward to outward is called the reversal potential (E_{rev}; see Chapter 6). If Na^+ movements were solely responsible for generating the EPC, E_{rev} would be equal to E_{Na}. In fact, at the NMJ E_{rev} is approximately 0 mV, which is much more negative than E_{Na} and far from the equilibrium potential for any ion. This implies that more than one ionic species moves through the open nAChR channel: indeed, these channels are almost equally permeable to Na^+ and K^+. When the channel opens near the normal resting potential, the current direction is inward because K^+ is close to equilibrium (i.e., $V_m \approx E_K$) so that there is little K^+ efflux. However,

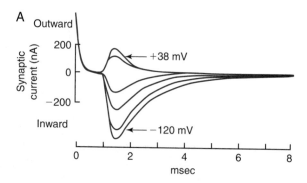

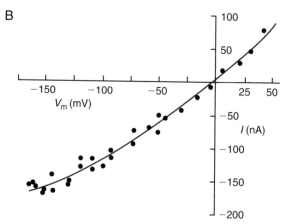

FIGURE 13-3 ■ The reversal potential of the EPC at the NMJ. The muscle V_m was voltage-clamped and an EPC was recorded in response to stimulation of the motor nerve. **A,** EPCs recorded at membrane potentials between -120 mV and $+38$ mV. At negative V_m, the EPC is inward; at positive V_m, the EPC is outward. **B,** The amplitude of the EPC is plotted as a function of V_m. The reversal potential (E_{rev}) of the EPC is close to 0 mV. *(Data from Nichols JG, Martin AR, Wallace BG: From neuron to brain, ed 3, Sunderland, Mass, 1992, Sinauer.)*

a large Na^+ influx results from the large inward driving force on Na^+ (i.e., $V_m - E_{Na} << 0$). The resulting inward EPC depolarizes the membrane to the AP threshold (~ -45 mV). Indeed, a general characteristic of *all* excitatory synapses is that E_{rev} is more positive than the AP threshold. Thus activation of ionotropic receptors at excitatory synapses causes inward current to flow. This current drives V_m towards the AP threshold and is therefore depolarizing and excitatory.

NMDAR and AMPAR are Channels with Different Ion Selectivities and Kinetics

All ionotropic glutamate receptors are nonselective cation channels permeable to both Na^+ and K^+, with $E_{rev} \approx 0$ mV. Thus activation of ionotropic glutamate receptors induces an inward EPSC. In turn, the EPSC generates an **excitatory postsynaptic potential (EPSP)**, which depolarizes the cell towards the AP threshold.

NMDARs have additional properties that are functionally important. First, extracellular Mg^{2+} blocks the NMDAR channel in a voltage-dependent manner (Figure 13-4 and Box 13-4). At negative V_m, Mg^{2+} is driven into the channel and acts like a plug—it lodges in the pore and blocks current flow through the channel. At positive V_m, Mg^{2+} is driven out of the channel and

relieves the block (Figure 13-4). Thus the channel exhibits outward *rectification*: it conducts outward current better than inward current. A second important property of NMDAR channels is that they are also permeable to Ca^{2+}. Ca^{2+} entry through these channels can increase $[Ca^{2+}]_i$ and activate Ca^{2+}-dependent signaling cascades in the postsynaptic cell. Because **Mg^{2+} block of NMDAR channels** is relieved by depolarization, current flow and thus Ca^{2+} entry through NMDAR channels occur at depolarized V_m (Figure 13-4).

Most neurons express both NMDARs and AMPARs at glutamatergic synapses. This feature facilitates a comparison of kinetics of the signals generated by these two populations of glutamate-activated channels. When presynaptic glutamatergic neurons are stimulated, the EPSC recorded at a positive V_m (e.g., +20 mV) has both a fast

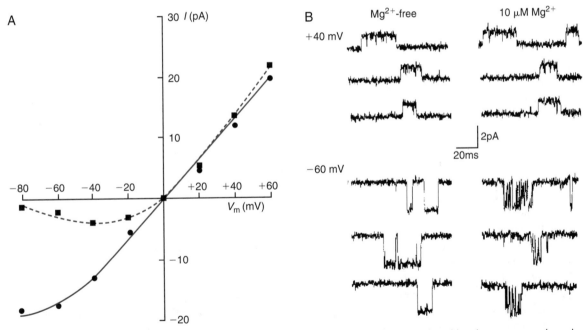

FIGURE 13-4 ■ Mg^{2+} blocks NMDAR channels. **A,** Macroscopic (whole-cell) currents induced by glutamate are plotted as a function of V_m. In the absence of Mg^{2+} *(circles)* the current-voltage relationship is approximately linear between +60 and −60 mV. After addition of 500 μM extracellular Mg^{2+} (squares), the currents are greatly reduced at negative potentials. At negative V_m, Mg^{2+} is driven into the channel where it blocks current flow. **B,** In the absence of Mg^{2+}, glutamate induces relatively long-lasting single-channel openings at both +40 and −60 mV. In the presence of 10 μM Mg^{2+}, the currents at +40 mV are unchanged, whereas at −60 mV, the single channel openings appear as rapid bursts. This is because Mg^{2+} ions rapidly enter and exit the channel, which causes the channel to switch rapidly between blocked and unblocked states. *(Data from Nowak L, Bregestovski P, Ascher P, et al: Nature 307:462, 1984.)*

ENERGY OF DEHYDRATION DETERMINES PERMEATION AND BLOCK OF NMDAR CHANNELS BY Ca^{2+} AND Mg^{2+}

Blockade of NMDAR channels by Mg^{2+} is strongly voltage-dependent, and the physiological importance of NMDARs is dominated by this property. Mg^{2+} blocks NMDAR channels by binding to a site in the permeation pathway and thereby preventing permeant ions from entering and moving through the channel. Extracellular, but not intracellular, Mg^{2+} is an effective blocker because Mg^{2+} can readily access the blocking site from the extracellular solution, but not from the cytoplasm. The block by external Mg^{2+} is voltage-dependent because Mg^{2+} binding is enhanced as the V_m becomes more negative (extracellular Mg^{2+} is attracted into the channel pore by the negative V_m). However, at very negative membrane potentials, Mg^{2+} can be driven through the channel, a finding indicating that it is a "permeant"

blocker. Ca^{2+} can also be considered a permeant blocker, but it is a very weak blocker with relatively high permeability in NMDAR channels. The ability of Mg^{2+} or Ca^{2+} to permeate or block NMDAR channels can be understood by considering their energies of dehydration. Ions in solution are surrounded by a shell of tightly bound water molecules, and some of these water molecules must be removed before an ion can move through a channel. In NMDAR channels, the permeability of Mg^{2+} or Ca^{2+} is inversely correlated with their energies of dehydration. Thus an ion with tightly bound water molecules (high energy of dehydration), such as Mg^{2+}, has difficulty moving through the channel and is an effective blocker. Ca^{2+} has a relatively low energy of dehydration; as a result it can move through the channel with ease.

transient component and a long-lasting component (Figure 13-5). The slow component is inhibited by DL-2-amino-5-phosphonovalerate (APV), which selectively blocks NMDAR channels. EPSCs recorded near the resting potential (e.g., -80 mV) have only a fast component that is not blocked by APV. These results show that AMPAR channels generate fast EPSCs and that NMDAR channels generate long-lasting EPSCs. When a single presynaptic AP activates a glutamatergic synapse with the postsynaptic membrane near the resting potential, only AMPARs are activated. If the synapse is stimulated repetitively to cause sustained depolarization of the postsynaptic membrane, then Mg^{2+} block is relieved to enable the NMDARs to conduct Na^+ and Ca^{2+}.

As the main excitatory neurotransmitter in the brain, glutamate and its receptors participate in many different normal brain activities, including neural development, pain perception, synaptic plasticity and memory, and behavior. Therefore it is not surprising that glutamate receptors, especially NMDARs, are also important in brain *patho*physiology. They have been implicated in some forms of epilepsy; furthermore, excessive Ca^{2+} entry mediated by NMDARs apparently contributes to the pathophysiology associated with

Alzheimer's disease, **Parkinson's disease** (see below), Huntington's chorea, and amyotrophic lateral sclerosis (ALS or Lou Gehrig's disease). As a result of a blood clot causing local ischemia and anoxia (a stroke), or rapidly repeated seizures, glutamate released from injured presynaptic terminals will induce excessive excitation (excitotoxicity) in the surrounding cells. This results in overwhelming Ca^{2+} gain by the cells because the prolonged NMDAR activation causes excessive Na^+ and Ca^{2+} influx. The gain of Na^+ itself may also promote Ca^{2+} entry or reduced Ca^{2+} clearance through the NCX (see Chapter 10). The net effect is Ca^{2+} overload, which can be the precursor to cell death.

Sustained Application of Agonist Causes Desensitization of Ionotropic Receptors

Most neurotransmitter receptors undergo **receptor desensitization**, which is a reversible reduction in receptor response during prolonged exposure to neurotransmitter. To illustrate the process, we will describe the desensitization of nAChRs. The characteristics of desensitization can be studied by measuring the current that flows through open nAChRs as a result of agonist application. When a high

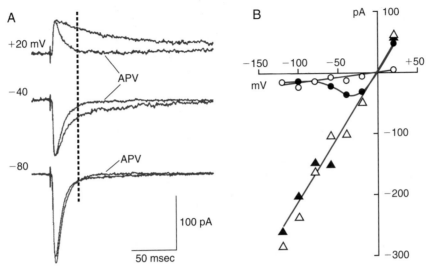

FIGURE 13-5 ■ EPSCs at glutamatergic synapses have both a fast transient component and a long-lasting component. **A,** EPSCs recorded at −80, −40, and +20 mV with and without DL-2-amino-5-phosphonovaleric acid (APV), which selectively blocks NMDAR channels. APV has no effect on the fast, transient EPSC at −80 mV, but APV blocks a long-lasting component at +20 mV. **B,** The current-voltage relationship at the peak of the EPSC was measured before *(solid triangles)* and after *(open triangles)* APV application. At the time indicated by the *dotted line* in **A,** when the fast component would have decayed, the current-voltage relationship was again measured in the absence *(solid circles)* and presence *(open circles)* of APV. The fast transient component, being mediated by AMPARs, is unaffected by APV. *(Data from Hestrin S, Nicoll RA, Perkel DJ, Sah P: J Physiol 422:203, 1990.)*

concentration of agonist is applied, nAChR channels are activated to conduct an inward current. In the continued presence of agonist, the inward current reaches a peak and then declines in magnitude as the nAChRs desensitize and the channels close (Figure 13-6). Following removal of the agonist, nAChRs recover from desensitization over several tens of seconds (Figure 13-6). The nAChR subunit composition determines the rate of desensitization. For the nAChRs at the NMJ, desensitization is relatively slow, developing over several seconds. In the CNS, nAChRs containing α7-subunits desensitize in milliseconds, whereas those that do not contain α7 desensitize in seconds. The rate and extent of desensitization and the rate of recovery from desensitization also depend on the nature of the applied agonist (Figure 13-6). Importantly, at low concentrations, agonists can desensitize receptors without activating them, a process called high-affinity desensitization. This is well-documented for nAChRs and AMPARs.

At Inhibitory Synapses, the Reversal Potential Is More Negative Than the Action Potential Threshold

Activation of certain synapses, called *inhibitory* synapses, decreases the probability that a postsynaptic AP will occur. GABA and glycine are the main inhibitory neurotransmitters at inhibitory synapses in the CNS. Both of these neurotransmitters bind to ionotropic receptors that conduct Cl^-. The synaptic current mediated by $GABA_A R$ or GlyR has an E_{rev} close to E_{Cl}. This indicates that the synaptic current is predominantly a Cl^- current.* Because E_{Cl} is often more negative than the resting V_m, activation of $GABA_A Rs$ or GlyRs usually induces an outward **inhibitory postsynaptic current**

---

*John C. Eccles discovered the role of the chloride conductance in synaptic inhibition. Eccles shared the 1963 Nobel Prize in Physiology or Medicine with Alan Hodgkin and Andrew Huxley (see Chapter 7) for his seminal work on the electrophysiology of synaptic transmission.

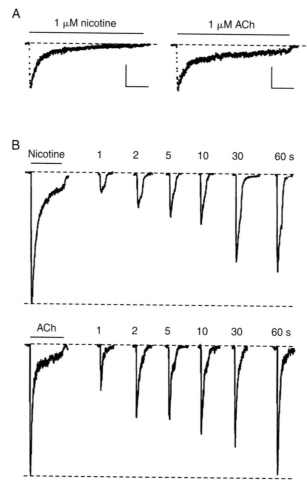

FIGURE 13-6 ■ Densensitization of nAChRs. **A,** Decline of inward currents through nAChRs in response to persistent application of nicotine or ACh is due to desensitization. Scale bars represent 110 pA and 1 sec. **B,** To characterize recovery from desensitization, the receptors are first desensitized with a 2-second exposure to agonist. This is followed, at various intervals (duration in seconds indicated above each response), by a second application of agonist. The response to the second application is a measure of the fraction of receptors that have recovered from desensitization. Recovery is faster after desensitization by ACh than after desensitization by nicotine. *(Data from Giniatullin R, Nistri A, Yakel JL: Trends Neurosci 28:371, 2005.)*

(IPSC). In turn, the IPSC generates a hyperpolarization in the postsynaptic cell—an **inhibitory postsynaptic potential (IPSP)** (Figure 13-7). This hyperpolarizing IPSP moves V_m away from the AP threshold (~ -55 mV) and is therefore inhibitory. Inhibitory synapses need not always produce a hyperpolarization. The open Cl^- channels tend to maintain V_m at E_{Cl}, which could be more negative than, the same as, or even more positive

than the resting potential. However, as long as E_{Cl} is more negative than the AP threshold, the open Cl^- channels exert an inhibitory effect.

Although GABA is an inhibitory neurotransmitter in the adult brain, it is one of the main excitatory neurotransmitters in the embryonic brain and early in postnatal life. For example, in the neonatal hippocampus, GABA, acting on $GABA_A Rs$, depolarizes

A

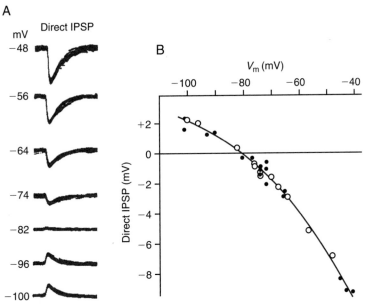

FIGURE 13-7 ■ The reversal potential of the IPSP. **A,** IPSPs recorded from a motor neuron in response to stimulation of inhibitory inputs. The initial V_m of the motor neuron (indicated next to each record) was changed by passing a steady current through an intracellular electrode. **B,** The amplitude of the IPSP is plotted as a function of V_m. The reversal potential of the IPSP is at approximately −80 mV, which is very close to E_{Cl}. Open and filled circles represent two separate sets of measurements. *(Data from Coombs JS, Eccles JC, Fatt P: J Physiol 130:326, 1955.)*

hippocampal pyramidal neurons and thereby activates voltage-gated Na^+ and Ca^{2+} channels, thus leading to further depolarization and Ca^{2+} entry. Early in development, excitatory glutamatergic synapses are poorly developed or absent, and GABA provides most of the excitatory drive required to open voltage-gated Ca^{2+} channels. Ca^{2+} entry through these channels is necessary for neuron outgrowth. The seemingly paradoxical ability of GABA to be either an excitatory or an inhibitory neurotransmitter acting on the same $GABA_A Rs$ is the result of a developmental change in $[Cl^-]_i$, and thus a change in E_{Cl}. $[Cl^-]_i$ is determined primarily by two cotransporters. The $Na^+-K^+-2Cl^-$ cotransporter, NKCC1 (see Figure 11-9), is the predominant cotransporter expressed early in development. Using energy from the Na^+ gradient, NKCC1 promotes Cl^- entry. This establishes a relatively high $[Cl^-]_i$ and consequently a relatively positive E_{Cl}, which underlies the excitatory GABA-induced depolarization in the neonate. The other cotransporter, the K^+-Cl^- cotransporter KCC2, becomes upregulated early

in postnatal life, whereas NKCC1 becomes downregulated. Using energy from the K^+ gradient, KCC2 promotes Cl^- extrusion, thereby decreasing $[Cl^-]_i$ and leading to a more negative E_{Cl}. The new E_{Cl} is lower than the AP threshold. This shift in E_{Cl} is responsible for the transformation of GABA from an excitatory to an inhibitory transmitter during development. Importantly, some sensory neurons do not undergo this shift in E_{Cl} during maturation, and GABA remains an excitatory neurotransmitter in these neurons.

Temporal and Spatial Summation of Postsynaptic Potentials Determine the Outcome of Synaptic Transmission

An average CNS neuron can receive as many as 200,000 synaptic inputs, some excitatory and some inhibitory. Activation of a single excitatory synapse initiates an EPSP that is insufficient to drive V_m to the AP threshold. For the postsynaptic neuron to reach threshold, several EPSPs must occur at about the same time so that their individual depolarizations can sum to produce a larger response. If, however, an IPSP coincides with an EPSP in the postsynaptic cell, then the EPSP will be reduced in amplitude. A neuron continuously sums its excitatory and inhibitory inputs; the result of this process—the net depolarization at the axon hillock (see Figure 5-1)—determines whether or not an AP is fired.

The neuronal membrane sums, or integrates, synaptic inputs occurring over space and time. This summation is affected by two passive membrane properties, the length constant (λ) and the membrane time constant (τ_m) (see Chapter 6). Recall that λ determines the distance over which a subthreshold change of V_m can spread passively (see Chapter 6). If two excitatory synapses at different locations on the postsynaptic neuron are active at the same time, the net effect of their depolarizations at any location (including the axon hillock) will be determined by a process called **spatial summation**. Spatial summation is affected by the length constant: a large λ means that inputs from distant synapses can sum effectively, whereas a small λ means that distant inputs will be attenuated and thus will sum poorly. If a single excitatory synapse is activated in rapid succession, the consecutive postsynaptic depolarizations sum in a process called **temporal summation**. Temporal summation is affected by τ_m, which determines the temporal persistence of a subthreshold change in V_m (Figure 13-8). A larger τ_m means that a synaptic response will last longer and thus be able to sum with a later synaptic input, to generate a larger resultant response.

Synaptic Transmission Is Terminated by Several Mechanisms

To serve as an effective means of conveying information, synaptic transmission must be punctuated (i.e., it must start and then stop). Activation of neurotransmitter receptors ceases when the transmitter dissociates from the receptor or desensitizes. The mechanism for initiating transmission—release of neurotransmitter into the synaptic cleft—is described in detail in Chapter 12. Here we discuss mechanisms that terminate neurotransmission by removing neurotransmitter molecules from the synaptic cleft. The three principal mechanisms for clearance of neurotransmitter are diffusion, enzymatic deactivation, and reuptake.

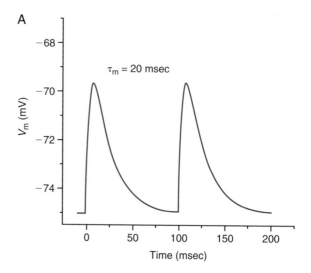

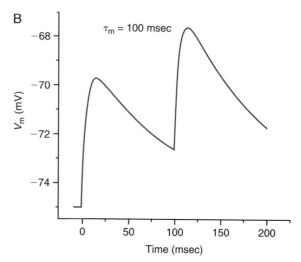

FIGURE 13-8 ■ Temporal summation of EPSPs. The traces simulate a pair of EPSPs generated in response to two presynaptic stimuli: one is delivered at 0 msec and the second at 100 msec. The duration of the EPSP is determined primarily by τ_m. **A,** When $\tau_m = 20$ msec, the EPSP is relatively brief and the first EPSP has declined to the resting potential when the second EPSP occurs at 100 msec. The second EPSP is the same size as the first (i.e., there is no summation). **B,** When $\tau_m = 100$ msec, the EPSP decays more slowly and V_m is still depolarized when the second EPSP occurs. As a result, the second EPSP sums with the first, generating a larger depolarization.

As soon as neurotransmitter molecules are released, they begin to diffuse away from the synaptic cleft; this causes the transmitter concentration in the cleft to decrease continuously with time. Diffusion therefore plays a role in terminating neurotransmission at all chemical synapses.

Neurotransmitter concentration at the synapse can be rapidly decreased by enzymes that efficiently convert certain neurotransmitter molecules into inactive species. The best-known example of such an enzyme is AChE, which was described in the context of neurotransmission at the NMJ (see Chapter 12). AChE destroys ACh by hydrolyzing it to acetate and choline. Transporters retrieve the choline back into the presynaptic cell for resynthesis of ACh. In **purinergic neurotransmission**, in which the neurotransmitter is ATP (see below), transmission is terminated by ectonucleotidases that rapidly degrade ATP to adenosine, which is transported back into cells. In contrast to choline, which is essentially inactive, adenosine itself can act as a transmitter. For example, it can modulate purinergic or cholinergic transmission by binding to, and activating, presynaptic adenosine receptors to inhibit ATP or ACh release, respectively. In peptidergic transmission, the peptides diffuse from the site of release and are cleaved by extracellular peptidases. This is the main mechanism for terminating neuropeptide action.

At many types of synapses, released transmitters are cleared by transporters expressed on neurons or glial cells in a process referred to as **neurotransmitter reuptake**. For example, at glutamatergic synapses, Na^+- and K^+-dependent excitatory amino acid transporters (EAATs) are responsible for reuptake of glutamate into glial cells. At GABAergic synapses, Na^+- and Cl^--dependent transporters mediate reuptake of GABA into the presynaptic neuron. Transporters also contribute to the clearance mechanism at synapses where the neurotransmitter is a biogenic amine (e.g., dopamine, NE, and 5-HT). Because biogenic amines in the CNS are important in pathways that govern mood and motivated behavior, substances that interfere with biogenic amine transporters, and thus raise the local transmitter concentration, are often antidepressant. For example, **selective serotonin reuptake inhibitors** (SSRIs; e.g., fluoxetine, one trade name of which is Prozac), are a class of **antidepressants**, and tricyclic antidepressants (e.g., amitriptyline and imipramine) inhibit uptake of NE and 5-HT. Cocaine, which alters mood, is a blocker of the dopamine transporter.

BIOGENIC AMINES, PURINES, AND NEUROPEPTIDES ARE IMPORTANT CLASSES OF TRANSMITTERS WITH A WIDE SPECTRUM OF ACTIONS

There are five biogenic monoamine neurotransmitters: 5-HT, histamine, and the three catecholamines (dopamine, Epi [or adrenaline], and NE [or noradrenaline]). These transmitters elicit a very wide range of physiological responses in the periphery as well as in the CNS. They are implicated, for example, in the regulation of appetite and feeding, in the central control of metabolism, cardiac function and blood pressure, in the coordination of movement, in sleep, in cognitive functions (e.g., awareness, reasoning, and memory), and in behavior. Although some *serotonergic receptors* are ionotropic, most of the effects of biogenic amines are mediated by transmitter-specific metabotropic receptors.

The biogenic amine neurotransmitters are all synthesized from amino acids. Histamine is generated by decarboxylation of histidine; 5-HT is synthesized by hydroxylation and then decarboxylation of tryptophan. Tyrosine is hydroxylated to form dihydroxyphenylalanine (DOPA), from which the three catecholamines are synthesized sequentially: first, DOPA is decarboxylated to form dopamine, which is then β-hydroxylated to form NE, which is, in turn, methylated to yield Epi.

Epinephrine and Norepinephrine Exert Central and Peripheral Effects by Activating Two Classes of Receptors

The two broad classes of adrenergic receptors (ARs), α (with six subclasses) and β (with three subclasses), are activated by both Epi and NE. The two classes, both of which are metabotropic, are distinguished primarily by their relative sensitivity to the synthetic adrenergic agonist, *isoproterenol* (N-isopropyl-noradrenaline). Isoproterenol is more effective than either Epi or NE in activating βARs, whereas the αARs are very poorly activated by isoproterenol.

Central ARs are involved in many complex behaviors. For example, noradrenergic neuronal projections from the brainstem (the region of the brain adjacent to the spinal cord) are important in inducing the alert, waking state known as arousal. Disturbances in this pathway contribute to a variety of stress-related disorders and insomnia. An example of NE action in the CNS is the attenuation of poststimulatory inhibition in hippocampal pyramidal neurons. βAR activation by NE stimulates adenylyl cyclase and increases cAMP synthesis. The cAMP activates protein kinase A (PKA), which phosphorylates and thus inhibits slow K_{Ca} channels, which normally generate a long-lasting hyperpolarization after an AP (often referred to as an **afterhyperpolarization**). Blocking this afterhyperpolarization enhances the ability of the pyramidal neurons to generate APs.

Peripheral ARs play a major role in regulating cardiovascular function. Sympathetic neurons in the heart release NE, which activates βARs (primarily $β_1$ARs) in cardiac myocytes, thereby turning on adenylyl cyclase. The resultant activation of PKA leads to phosphorylation of VGCCs to increase their open probability and thus promote Ca^{2+} influx. This inward Ca^{2+} current increases the rate of depolarization in cardiac pacemaker cells and consequently increases the heart rate. In the contractile myocytes, the enhanced Ca^{2+} influx causes a stronger contraction (see Chapter 15). Epi, the main hormone secreted by the adrenal medulla, a modified sympathetic ganglion, is an important activator of cardiac βARs. βAR blockers ("β-blockers")[*] are used to treat angina (pain resulting from cardiac hypoxia or anoxia), heart attacks, and hypertension, primarily because of their ability to reduce the heart rate.

Tonic sympathetic nerve activation is crucial in the maintenance of blood pressure. Peripheral sympathetic nerve activity releases NE, which activates $α_1$ARs on vascular smooth muscle cells. This triggers arterial contraction by promoting IP_3-induced release of Ca^{2+} from the SR (Figure 13-2). In light of this mechanism,

it is not surprising that pharmacological blockade of $α_1$ARs causes a marked drop in blood pressure. The sympathetic nerve terminals that innervate most arteries release, besides NE, ATP and neuropeptide Y (NPY); this is a good example of cotransmission. The roles of ATP and NPY are discussed later.

Dopaminergic Transmission Is Important for the Coordination of Movement and for Cognition

The largest concentration of dopaminergic neuron cell bodies in the brain is in the *substantia nigra*. These neurons, whose axons project to the *corpus striatum*, are crucial in coordinating movement, and their degeneration causes severe tremors and muscle rigidity (Parkinson's disease, Box 13-5). Other dopaminergic neurons in the CNS are involved in motivation and reward, and hyperactivity of these neurons is associated with addictive behaviors. This is exemplified by the fact that cocaine causes extracellular dopamine buildup in the *limbic system*[†] by inhibiting Na^+-dependent dopamine reuptake that normally terminates signaling. Still other dopaminergic neurons, particularly in the prefrontal cortex, participate in cognitive functions such as spatial learning and memory.

There are five types of dopamine receptors: D_1R to D_5R. These receptors, all of which are metabotropic, can be grouped into two classes: the D_1-like receptors (D_1R and D_5R) and the D_2-like receptors (D_2R, D_3R, and D_4R). Stimulation of D_1-like receptors *activates* PKA through adenylyl cyclase, and PKC though phospholipase C. In the hippocampus, for example, the resulting phosphorylation of Na^+ channels reduces their activity and thus decreases neuronal excitability. In contrast, activation of D_2-like receptors *inhibits* adenylyl cyclase and thus reduces PKA activity, which normally maintains VGCC activity through phosphorylation. One consequence of the reduced Ca^{2+} influx is inhibition of small conductance K_{Ca} channels; this is expected to increase neuronal excitability.

Much evidence points to a key role for overactive D_2Rs in the pathogenesis of schizophrenia, and many clinically useful antipsychotic agents such as the

---

[*]James W. Black, a pharmacologist who shared the 1988 Nobel Prize in Medicine or Physiology, developed the first βAR blocker (1964) and the first histamine H_2 receptor blocker (1972). Blocking H_2 receptors is an effective treatment for gastric hyperacidity ("heartburn"; see Chapter 11 and Epilogue).

[†]Structures at the inner border of the cerebral cortex involved in emotion, behavior and memory.

BOX 13-5
PARKINSON'S DISEASE RESULTS FROM REDUCED ACTIVITY OF DOPAMINERGIC NEURONS

Parkinson's disease (PD) is a movement disorder caused by progressive neurodegeneration in the CNS. Here we consider only primary parkinsonism, also known as idiopathic PD, which is distinct from related disorders called Parkinson-plus diseases. PD has four primary symptoms: a low-frequency *tremor* that is most prominent at rest, muscle rigidity resulting from increased muscle tone, slow movement *(bradykinesia)* or the inability to initiate movement *(akinesia)*, and *postural instability* resulting from the failure of postural reflexes that can cause falls in advanced PD. Other motor symptoms include shuffling gait, a stooped, forward-flexed posture, impaired swallowing and speech, and loss of fine motor control.

The primary symptoms of PD all result from reduced activity of dopaminergic neurons in the *substantia nigra;* these neurons project to the *corpus striatum* and are crucial in regulating movement. Loss of dopaminergic neurons leads to a movement disorder characterized by reduced motor activity.

Although only 15% to 20% of patients with PD have a family history of the disease, substantial evidence indicates that at least five genes contribute to the genetic origin of familial PD. One of these genes, *SNCA*, encodes the protein α-synuclein. The normal function of α-synuclein is unknown, although evidence indicates that the protein is involved in neurotransmitter release and vesicle turnover. PD patients have an accumulation of α-synuclein in cytoplasmic inclusions called *Lewy bodies* in dopaminergic neurons. The formation of α-synuclein fibrils in Lewy bodies is associated with cell death, but the mechanism is unknown.

---

phenothiazines (e.g., chlorpromazine) and other compounds (e.g., risperidone) are D_2R antagonists. As one would expect, these agents have the undesirable side effect of inducing Parkinson-like symptoms.

Serotonergic Transmission Is Important in Emotion and Behavior

The cell bodies of CNS neurons that employ 5-HT as a transmitter are clustered in several nuclei near the midline *raphe* (ridge) in the upper brainstem. This relatively small number of cells, through ascending and descending axon projections, innervates nearly all areas of the brain and thus helps regulate many different vital functions. These include (to name just a few) sleep and wakefulness, emotions, cognition, and cardiovascular, respiratory, and intestinal activities. Derangements of the serotonergic pathways contribute to various pathological states, including anxiety, depression, mania, and schizophrenia. Some of the most successful drugs used to treat anxiety and depression are SSRIs. These agents block the Na^+-5-HT cotransporter (see Chapter 10) that clears 5-HT from the synaptic cleft and thereby normally helps to terminate serotonergic signaling.

Two of the three main classes of 5-HT receptors, $5\text{-}HT_1R$ and $5HT_2R$, are metabotropic. Activation of $5\text{-}HT_1Rs$ inhibits adenylyl cyclase; activation of $5\text{-}HT_2Rs$ usually stimulates phospholipase C, which generates IP_3 and thus mobilizes Ca^{2+} from the S/ER. Members of the third class, $5\text{-}HT_3Rs$, are ionotropic receptors, which, like $5\text{-}HT_1Rs$ and $5HT_2Rs$, are also widely distributed in the CNS and in the peripheral nervous system. These 5-HT receptors are all involved in complex behavior patterns that are also influenced by other transmitters. For example, reduced $5\text{-}HT_1R$ activity plays a role in anxiety. Nevertheless, ionotropic $GABA_ARs$ and KARs also are involved in anxiety: activation of KARs, perhaps by promoting GABA release, and activation of $GABA_ARs$ (e.g., by benzodiazepines such as diazepam [Valium]) exert *anxiolytic* (anxiety-reducing) effects. Similarly, all three classes of 5-HT receptors have been implicated in addiction and withdrawal symptoms, but central dopaminergic and adrenergic transmission also play important roles.

The $5\text{-}HT_3Rs$, which are nonselective cation channels permeable to Na^+, K^+, and Ca^{2+}, are members of the superfamily of pentameric ligand-gated channels that includes the nAChRs, $GABA_ARs$, and GlyRs. Thus activation of $5\text{-}HT_3Rs$ leads to rapid membrane depolarization. $5\text{-}HT_3R$ antagonists such as ondansetron (Zofran) are widely used to treat postoperative or cancer chemotherapy–induced nausea and vomiting. These drugs exert their effect by blocking receptors in the *area postrema* (the vomiting center in the brainstem), as well as receptors in the vagus nerve, which normally activates the vomiting center.

Histamine Serves Diverse Central and Peripheral Functions

The cell bodies of histaminergic neurons in the brain are confined to the tuberomammillary nucleus in the posterior hypothalamus, and, like serotoninergic neurons, they project to most regions of the brain and spinal cord. Together with central ACh and NE projections, histaminergic neurons play a role in arousal and attention, as well as in memory, learning, and mood. Histamine's role in arousal is apparently reflected in the observation that drowsiness is a common side effect of antihistamines. The causal relationship is not straightforward, however, because central cholinergic transmission, which also plays a role in arousal, may also be inhibited by some antihistamines.

Three important sources of histamine are non-neuronal: basophils, mast cells, and the enterochromaffin-like cells of the gastric wall. Secretion of histamine by gastric enterochromaffin-like cells (through H_2R) triggers the secretion of protons by the H^+,K^+-ATPase in the parietal cells of the gastric mucosa (see Chapter 11). Secretion of histamine by mast cells and basophils is intimately involved in innate and acquired immunity, as well as in allergy and inflammation.

The four classes of histamine receptors (H_1R to H_4R) are all metabotropic. Stimulation of H_1Rs generally activates Ca^{2+} signaling cascades mediated by phospholipase C and DAG. In the hippocampus, the elevated $[Ca^{2+}]_i$ opens K_{Ca} channels and increases afterhyperpolarization and thereby reduces the firing rate of pyramidal neurons. In contrast, activation of H_2Rs usually stimulates adenylyl cyclase and thus activates PKA. In the hippocampus, this leads to phosphorylation and consequent inactivation of small conductance K_{Ca} channels. The result is reduced afterhyperpolarization of the pyramidal neurons, so that the firing frequency can increase. In other words, as this example shows, activation of H_1Rs and H_2Rs can have opposing effects. The net effect of histamine, however, tends to be enhancement of long-term potentiation (LTP) of cell firing (see later) in the hippocampus, a cellular

manifestation of memory. Activation of H_3Rs leads to inhibition of adenylyl cyclase and of N- and P-type Ca^{2+} channels—consistent with observations that activation of H_3Rs on presynaptic neurons can inhibit release of certain neurotransmitters.

The discovery of the histamine receptors and their functions led to the development of numerous therapeutic agents (see left footnote on page 198). For example, many allergic reactions can be successfully treated with H_1R antagonists such as loratadine (Claritin) and cetirizine hydrochloride (Zyrtec). "Heartburn" (gastric hyperacidity) can be treated effectively with H_2R antagonists such as famotidine (Pepcid) and ranitidine (Zantac) (see Chapter 11 and Epilogue).

ATP Is Frequently Coreleased with Other Neurotransmitters

Purinergic synapses are distributed throughout the central and peripheral nervous systems. ATP and adenosine (generated from ATP by ectonucleotidases on the cell surface) are the main purinergic neurotransmitters, although uridine triphosphate and uridine diphosphate also activate some purinergic receptors. ATP is frequently coreleased with other neurotransmitters. The effects of adenosine are mediated by P1 receptors, whereas ATP acts through P2X and P2Y receptors. All P1 and P2Y receptor subtypes are metabotropic receptors that serve diverse roles. For example, some P1 receptors (on presynaptic nerve terminals) modulate neurotransmitter release; others mediate smooth muscle relaxation, as do some P2YRs. Other P2YR-mediated actions include platelet aggregation and epithelial Cl^- secretion.

P2XRs are all homotrimeric or heterotrimeric ATP-gated cation channels. Each monomeric subunit consists of two transmembrane domains linked by an extracellular loop, so that both the N-terminus and the C-terminus are intracellular. These channels are permeable to small monovalent cations, and some are also Ca^{2+}-permeable. At sympathetic nerve endings in smooth muscles, ATP is coreleased with NE to help trigger Ca^{2+}-dependent smooth muscle contraction. The ATP rapidly opens Ca^{2+}-permeable P2XR channels to admit Ca^{2+} and initiate the contraction; the NE, by activating $\alpha 1ARs$, promotes IP_3-induced Ca^{2+} mobilization from the SR to prolong the contraction. P2XRs on sensory nerves in the viscera play a key role in detecting and signaling tissue damage and inflammation; ATP leakage from damaged cells activates these P2XRs.

Adenosine makes an important contribution to the sleep/wake state. Adenosine accumulates in certain areas of the hypothalamus, for example, as a result of neuronal activity associated with the release of ATP in the awake state; extracellular adenosine levels are elevated even further during sleep deprivation. One consequence of the activation of P1 receptors (of the A_1 subtype) at these high adenosine levels is an increase in an inwardly rectifying K^+ channel conductance in the cholinergic neurons that mediate arousal; this reduces excitability. Thus increasing adenosine levels promotes sleep while reducing adenosine promotes wakefulness.

Altered purinergic transmission has been implicated in many different pathological conditions. These include CNS ischemia and injury, neurodegenerative diseases, epilepsy, neuroimmune and inflammatory disorders, migraine and neuropathic pain, and mood disorders.

Neuropeptide Transmitters Are Structurally and Functionally Diverse

Numerous peptides, ranging from 3 to approximately 100 amino acid residues in length, serve as neurotransmitters. They are stored in approximately 100- to 150-nm dense core vesicles, and are released through Ca^{2+}-dependent exocytosis at sites with or without active zones (see Chapter 12). Many of these transmitters can be classified into a few groups based on such criteria as amino acid sequence and anatomical localization: (1) hypothalamic peptides, including several hormone-releasing transmitters such as growth hormone–releasing hormone (GHRH), and other hormones such as vasopressin, oxytocin, and somatostatin; (2) opioid peptides such as dynorphin and leu- and met-enkephalins; (3) neuropeptide Y (NPY) and related peptides; (4) tachykinins such as substance P and neurokinins A and B; and (5) members of the vasoactive intestinal peptide (VIP) –glucagon family. Numerous other peptide transmitters, however, do not fit into these classes. Irrespective of classification, all neuropeptide transmitters act on specific metabotropic receptors and often induce alterations in the excitability of postsynaptic cells.

The numerous peptide neurotransmitters have many different and important functions. For example, the endogenous opioids regulate responses to emotional and physical stress. Several hypothalamic neuropeptides, including orexin, ghrelin, pro-opiomelanocortin, melanin-concentrating hormone, and NPY, regulate feeding behavior and satiety. In most cases, the detailed mechanisms of action of these transmitters are not yet resolved. One example in which some of the mechanisms have been elucidated involves substance P, which, along with glutamate, NO, and other neuropeptides, plays a prominent role in chronic pain. When we injure an area of skin or a joint, we experience not only acute pain, but also long-lasting pain and even increased sensitivity to painful stimuli (hyperalgesia) in the same or adjacent areas. Acute pain is initiated at peripheral pain receptors and relayed through the spinal cord to the perception areas in the brain. This rapid signaling involves glutamatergic synapses. In addition, especially with high-frequency or prolonged stimulation, the sensory nerve terminals in the spinal cord release substance P, which activates neurokinin-1 receptors and, through phospholipase C, generates IP_3 and DAG. These second messengers elevate $[Ca^{2+}]_i$ and activate several protein kinases that phosphorylate inwardly rectifying K^+ channels, NMDARs, and AMPARs. The result is enhanced depolarization and signaling by the postsynaptic neurons. The net effect is central sensitization to pain stimuli and thus hyperalgesia and chronic pain.

Most neuropeptides are coreleased with other transmitters—either other peptides or low-molecular-weight transmitters. In the former case, the peptides are stored together in the same dense core vesicles; in the latter, the peptides and low-molecular-weight transmitters may be stored in separate vesicles that are differentially released. A case in point is the corelease of NPY (stored in large vesicles) with NE and ATP (in small vesicles) at the sympathetic nerve terminals that innervate arterial smooth muscle. NPY, by itself, does not activate contraction, but it modulates the vasoconstrictor effects of NE and ATP in two ways. Acting through postsynaptic NPY-Y1 receptors, NPY potentiates the contractile effects of ATP and NE, but acting through presynaptic NPY-Y2 receptors it reduces the release of ATP and NE. This is just one more example of the great variety and complexity of synaptic transmission.

UNCONVENTIONAL NEUROTRANSMITTERS MODULATE MANY COMPLEX PHYSIOLOGICAL RESPONSES

Unconventional Neurotransmitters Are Secreted in Nonquantal Fashion

Earlier in this chapter, we noted that NO, CO, and the lipidic endocannabinoids are unconventional neurotransmitters because they cannot be stored in SVs. Nevertheless, release of these transmitters may also be triggered by a rise in $[Ca^{2+}]_i$, even though their release is nonquantal. These unconventional transmitters are released immediately following synthesis (synthesis on demand), which may involve a Ca^{2+}-dependent step. Moreover, these transmitters may be synthesized and secreted by postsynaptic as well as presynaptic neurons. When a transmitter is released by a postsynaptic neuron but acts on the presynaptic neuron, the neurotransmitter is said to act in a retrograde fashion (**retrograde synaptic transmission**).

Many Effects of Nitric Oxide and Carbon Monoxide Are Mediated Locally by Soluble Guanylyl Cyclase

NO is synthesized from L-arginine by **NO synthase (NOS)**, which has two constitutively expressed isoforms, one in neurons (nNOS) and one in endothelial cells (eNOS).* nNOS, which is widely distributed in central and peripheral neurons, is activated by Ca^{2+}-calmodulin (Ca^{2+}/CaM) as a result of Ca^{2+} entry through NMDAR channels and consequent elevation of $[Ca^{2+}]_i$. Constitutive **hemoxygenase-2 (HO2)**, which produces CO by degrading *heme* (the other two products are iron and *biliverdin*), is found in brain neurons and astrocytes, as well as in myenteric plexus neurons and the testis. HO2 is turned on when it is phosphorylated by PKC, which is activated by a rise in $[Ca^{2+}]_i$. Because they are small, membrane-permeant molecules, NO and CO readily leave their sites of synthesis and diffuse rapidly to act on target neurons. An important distinction between the two gases is that whereas CO is relatively unreactive, NO is a free radical that is quite chemically reactive. One consequence is that once generated in the body, CO is long-lived, whereas NO is short-lived (at most a few seconds under normal physiological conditions).

Cytosolic **soluble guanylyl cyclase**, which contains a heme cofactor, is the major receptor for both gases. Binding of NO or CO to the heme activates the enzyme to convert GTP to cGMP. The cGMP can activate cGMP-dependent protein kinase to phosphorylate key proteins in neuronal function, or it can directly affect the electrophysiology of the neuron by binding to cGMP-gated or cGMP-modulated ion channels. Additionally, in the brain and in arterial smooth muscle, CO binds to and activates certain K_{Ca} channels that bear a heme cofactor; this leads to relaxation of cerebral and peripheral blood vessels. Finally, NO, which is chemically reactive, can covalently modify proteins. For example, NO can nitrosylate the thiol side chains of cysteine residues and thus modify protein function.

Endogenous NO is involved in a wide range of central and peripheral nervous system activities. For example, NO facilitates sleep, contributes to central sensitization to pain, influences synaptic plasticity and memory formation, and contributes to normal feeding behavior and gastrointestinal function. NO also is synthesized in and released by endothelial cells, and it plays an important role in the control of vascular tone (see Chapter 15). Although HO2, like eNOS, is prevalent in the brain, the function of CO as a neurotransmitter has been best studied in the intestine. NO and CO contribute about equally to nonadrenergic, noncholinergic synaptic transmission, from the myenteric neurons to the intestinal smooth muscle cells, that triggers the relaxation phase of peristalsis.

Endocannabinoids Can Mediate Retrograde Neurotransmission

The second major class of unconventional transmitters is the endocannabinoids. Two important members of this group are 2-AG and **N-arachidonoyl ethanolamide** (AEA or anandamide) (see Figure 13-1). Both of these highly lipophilic transmitters are synthesized from phospholipids that contain an arachidonic acid moiety. 2-AG appears to be the transmitter at CNS synapses, but both agents interact with cannabinoid type-1 receptors (CB_1Rs), the most abundant G-protein–coupled receptors in the brain. CB_1Rs are found in the presynaptic terminals at nearly all types of CNS synapses.

---

*Robert Furchgott, Louis Ignarro and Ferid Murad shared the 1998 Nobel Prize in Medicine or Physiology for discovering that the gas NO is an intercellular signaling molecule.

At CNS glutamatergic synapses, type I mGluRs play an important role in endocannabinoid signaling. These mGluRs are located in the perisynaptic membrane regions of postsynaptic cells, adjacent to the postsynaptic densities. During high-frequency activation of glutamatergic synapses, spillout of glutamate from the synaptic cleft may elevate the glutamate concentration in the perisynaptic region sufficiently to activate the mGluRs. The consequent G-protein activation triggers phospholipase C to cleave phosphatidylinositol bisphosphate to form DAG and IP_3 (see Figure 13-2). The DAG, in turn, is further cleaved by diacylglycerol lipase-α (DGL-α), which is tethered to the mGluRs by a scaffold protein. The product of this cleavage, 2-AG, crosses the synaptic cleft to bind to and activate presynaptic CB_1Rs, which then trigger activation of various K^+ channels and inhibition of voltage-gated Ca^{2+} channels. This acts as a *negative feedback* mechanism to reduce the release of glutamate by the presynaptic terminal (a synaptic circuit breaker). At other synapses, the 2-AG acts on presynaptic GABAergic neurons to suppress inhibitory neurotransmission.

As implied by their prevalence and broad distribution in the CNS, CB_1Rs have numerous, important physiological roles. They participate in the formation of short-term ("working") memories, including the fear response. Activation of CB_1Rs in the pain pathway reduces synaptic transmission (see preceding paragraph) and thus the perception of pain. CB_1Rs in some brain areas (e.g., the hypothalamus) help control appetite and feeding behavior. CB_1Rs are also involved in addiction behavior and in anxiety and depression. Finally, as these physiological roles would imply, CB_1Rs mediate the psychotropic and behavioral effects of Δ^9-tetrahydrocannabinol (Δ^9-THC), effects that include short-term memory impairment, altered sense of time, slowing of reaction time, and enhancement of appetite (the "munchies").

It should now be apparent that virtually all complex physiological processes are mediated by multiple neurotransmitters. Nevertheless, as we have noted, modulation of the concentration of a single transmitter, or modulation of receptor activation, is usually sufficient to induce profound physiological effects.

LONG-TERM SYNAPTIC POTENTIATION AND DEPRESSION ARE PERSISTENT CHANGES IN THE EFFICACY OF SYNAPTIC TRANSMISSION INDUCED BY NEURAL ACTIVITY

Acquisition of new knowledge and skills constitutes learning, and the newly acquired knowledge must be encoded and stored in memory. Knowing that facts and skills learned in youth (e.g., "1 + 1 = 2" or bicycle riding) remain accessible through decades of life, one may infer that learning and memory require enduring modifications in the nervous system. For this reason, long-term changes in synaptic transmission have been a major focus of neurophysiology research. In particular, two forms of plasticity at synapses have been studied extensively: long-term potentiation (LTP) and long-term depression (LTD).

Long-term Potentiation Is a Long-lasting Increase in the Efficacy of Transmission at Excitatory Synapses

Investigators have known since the early 1970s that "strong" stimulation (e.g., high-frequency burst of stimuli at 100 Hz) of neurotransmission at excitatory synapses in the CA1 region of the hippocampus can rapidly trigger a long-lasting increase in the efficacy of transmission at those synapses (Figure 13-9A). This activity-dependent increase in synaptic efficacy is known as **long-term potentiation (LTP)**. Once induced, LTP can last days to weeks, if not longer.

Because LTP is observed within seconds after stimulation, induction of LTP must involve activation of cellular signaling pathways that operate on a short time scale. Our knowledge of these signaling processes has grown rapidly (Figure 13-10). In response to electrical stimulation, the presynaptic terminal releases glutamate molecules, which activate the AMPARs and NMDARs on the postsynaptic membrane. The AMPARs conduct monovalent cation current to depolarize the postsynaptic cell. Persistent depolarization relieves Mg^{2+} blockade of the postsynaptic NMDARs, which then mediate Ca^{2+} influx into the postsynaptic cell. The consequent rise in $[Ca^{2+}]_i$ activates CaM, which in turn binds and activates Ca^{2+}/CaM-dependent protein kinase II (CaMKII). Activated

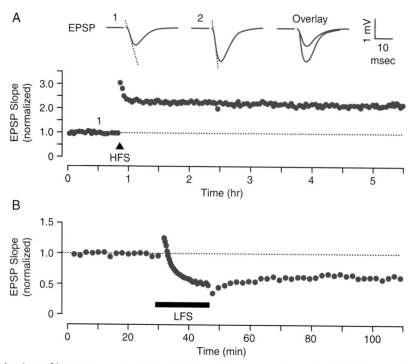

FIGURE 13-9 ■ Induction of long-term potentiation (LTP) and long-term depression (LTD) in the hippocampus. In an adult rat hippocampal slice, single extracellular electrodes delivered electric field pulses to stimulate nerve fibers that form excitatory synapses onto pyramidal neuron dendrites in the CA1 region. EPSPs from stimulated synapses were monitored through an extracellular recording electrode. Trains of pulses were delivered to induce long-term plasticity: High-frequency stimulation (HFS; 100 Hz for 1 second) was used to induce LTP; low-frequency stimulation (LFS; 1 Hz for 900 second) was used to induce LTD. **A,** Before LTP induction, single field pulses were used to evoke EPSPs to establish baseline behavior (labeled *1* on the graph). Shown above the graph is a representative baseline EPSP *(1)*, whose slope is indicated by the *dotted line*. EPSP slopes normalized to the baseline slope are plotted in the graph. HFS was delivered at the time marked by the *arrowhead*. There was a sharp augmentation of the EPSPs that was short-lived (post-tetanic potentiation, PTP; see Chapter 12), followed by a long-lasting enhancement (LTP; labeled *2* on the graph). A representative EPSP from the long-lasting phase is shown above the graph *(2)*; the *dotted line* indicates that the EPSP slope is steeper after LTP induction. An overlay of EPSPs 1 and 2 shows clearly that both the amplitude and slope were increased after LTP. **B,** Before LTD induction, single field pulses were used to evoke EPSPs to establish baseline behavior. As in **A,** EPSP slopes normalized to the baseline slope are plotted in the graph. The *black bar* below the trace marks the duration of LFS. The LFS induced attenuation of the EPSPs that was long lasting (LTD). *(Data from Bortolotto ZA, Anderson WW, Isaac JTR, Collingridge GL: Curr Protocols Neurosci Unit 6.13, 2001.)*

CaMKII can phosphorylate AMPARs to increase their conductance and thus contribute to LTP. More importantly, activation of CaMKII leads to the insertion of AMPARs into the postsynaptic density, so that the sizes of subsequent EPSCs are increased. Although activation of CaMKII is necessary for incorporation of new AMPARs at the synapse, the detailed molecular mechanism linking the two events is still unknown.

To maintain LTP over long periods of time, more enduring cellular changes must take place. Indeed, gene transcription and synthesis of new proteins are required for LTP maintenance. Beyond that, however, the processes that enable maintenance of LTP are poorly understood. Some observations suggest that persistence of LTP correlates with structural changes at the synapse—for example, an enlargement of the postsynaptic density

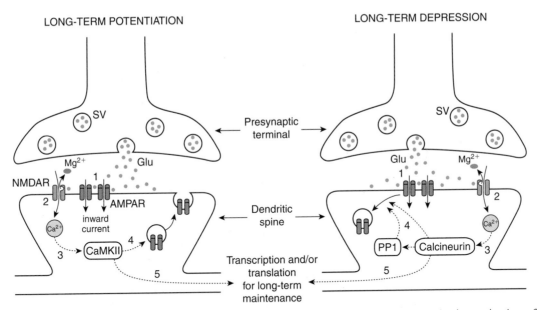

FIGURE 13-10 ■ NMDAR–dependent long-term potentiation (LTP) and long-term depression (LTD). The mechanism of NM-DAR-dependent LTP is shown on the *left*. The presynaptic terminal releases glutamate (Glu) stored in SVs. Thereafter: *(1)* the glutamate activates AMPARs on the postsynaptic cell; the activated AMPARs conduct inward current to cause depolarization; *(2)* the depolarization relieves blockade of NMDARs by Mg^{2+}; this enables NMDARs to conduct Ca^{2+} ions into the postsynaptic cell; *(3)* through activation of calmodulin (CaM), Ca^{2+} activates Ca^{2+}/CaM-dependent protein kinase II (CaMKII); *(4)* activated CaMKII promotes the insertion of more AMPARs into the synapse; and *(5)* activated CaMKII also initiates signals that trigger de novo transcription and translation, which are required for long-term maintenance of LTP. The mechanism of NMDAR-dependent LTD is shown on the *right*. The presynaptic terminal releases glutamate. Thereafter: *(1)* the glutamate activates AMPARs on the postsynaptic cell; the activated AMPARs conduct inward current to cause depolarization; *(2)* the depolarization relieves blockade of NMDARs by Mg^{2+}; this enables NMDARs to conduct Ca^{2+} ions into the postsynaptic cell; *(3)* through activation of CaM, Ca^{2+} activates the phosphatase, calcineurin, which in turn activates another phosphatase, PP1; *(4)* dephosphorylation of synaptic AMPARs by PP1 allows the AMPARs to be removed through an endocytic process activated by calcineurin; *(5)* activated calcineurin also initiates signals that trigger translation of previously transcribed mRNA, which is required for long-term maintenance of LTD.

(PSD) and the **dendritic spine,** as well as the presynaptic active zone. A reasonable scenario is that insertion of new AMPARs (as well as other synaptic proteins) into the PSD causes the PSD and the spine to increase in size. Subsequently, an unknown mechanism causes the active zone in the presynaptic terminal to grow to match the enlarged PSD. In this way, presynaptic neurotransmitter release and postsynaptic response can both be increased to enhance synaptic transmission.

In view of the complexity of the nervous system, LTP expression need not depend on a unique cellular signaling mechanism. For example, in the CA3 region of the hippocampus, LTP does not require NMDAR activation, but instead depends on cAMP signaling pathways and on enhanced neurotransmitter release rather than on augmented postsynaptic receptor function. Because neural plasticity is essential to nervous system function, we expect the phenomenon of LTP to be ubiquitous. The detailed cellular mechanisms that support LTP, however, may differ depending on the specific site of LTP expression.

Long-term Depression Is a Long-lasting Decrease in the Efficacy of Transmission at Excitatory Synapses

In contrast to induction of LTP, prolonged "weak" (low-frequency: 0.5 to 5 Hz) stimulation of neurotransmission at excitatory synapses in the CA1 region of the hippocampus can trigger a long-lasting *decrease* in the efficacy of transmission at those synapses (Figure 13-9B). This activity-dependent decrease in synaptic efficacy is known as **long-term depression (LTD)**. Like LTP, once induced, LTD has been observed to last days to weeks.

The phenomenology of LTD is well characterized, although many details of the underlying signaling mechanisms are still lacking (Figure 13-10). Blockade of NMDARs and increasing Ca^{2+} buffering capacity in the postsynaptic neuron both prevent LTD; therefore Ca^{2+} influx through NMDARs is crucial for LTD induction. An important role of the Ca^{2+} signal in LTD is to activate, through Ca^{2+}/CaM, the Ca^{2+}/CaM-dependent *phosphatase*, calcineurin. In turn, calcineurin can activate protein phosphatase 1 (PP1) by dephosphorylating and thereby inactivating the endogenous PP1 inhibitor, inhibitor-1. PP1 then dephosphorylates AMPARs at ser-831, a residue whose phosphorylation by PKA is required for retention of AMPARs at the synapse. Moreover, calcineurin promotes endocytosis by dephosphorylating key proteins in the endocytic pathway. The combined effect of the two phosphatases is to enable removal of AMPARs from the postsynaptic density. Thus, in LTP, the postsynaptic response is potentiated by insertion of AMPARs into the synapse, whereas in LTD, postsynaptic response is depressed through removal of AMPARs from the synapse. Long-term maintenance of LTD requires translation of existing mRNA into newly synthesized proteins but not new transcription. Also required is intact function of the *proteasome*, a fundamental protein-degrading machinery of the cell. Finally, as in the case of LTP, whereas LTD is expected to be a widespread phenomenon in the nervous system, its expression need not and indeed does not depend on a unique cellular mechanism.

Although LTP and LTD both require activation of CaM by an NMDAR-dependent rise in $[Ca^{2+}]_i$, Ca^{2+}/CaM ultimately activates different signaling mechanisms to trigger each process. Ca^{2+}/CaM activates CaMKII to establish LTP, whereas it activates calcineurin to establish LTD. That NMDAR-mediated Ca^{2+} influx can activate two forms of synaptic plasticity that are functionally very different implies that different spatial and temporal regulation of the Ca^{2+} signal is crucial for LTP and LTD. At least one aspect of such regulation is the temporal profile of the Ca^{2+} signal: LTP is induced by high-frequency stimulation, and the rapidly repeated Ca^{2+} influx through NMDARs accumulate in the postsynaptic neuron to cause a robust and sustained rise in $[Ca^{2+}]_i$. In contrast, LTD is induced by low-frequency stimulation that causes episodes of Ca^{2+} influx that are widely separated in time; this leads to a feeble, but longer-lasting rise in $[Ca^{2+}]_i$ in the postsynaptic neuron. These quantitative differences in the Ca^{2+} signals lead to activation of different downstream signaling processes.

SUMMARY

1. Chemical synapses enable neurons to communicate with each other or with their effector cells with great specificity and fine control. The variety and specificity are afforded by the more than 100 different neurotransmitters, which are intercellular molecular messengers.

2. The three groups of conventional neurotransmitters are (1) small molecules stored in small, electron-lucent SVs, (2) small molecules stored in small electron-opaque (dense core) SVs, and (3) the neuropeptides, which are stored in large dense core SVs.

3. Two broad classes of specific receptors, ionotropic and metabotropic receptors, mediate the actions of the neurotransmitters on postsynaptic cells. Iono-

tropic receptors are ligand-gated ion channels, for which the ligands are specific neurotransmitters. Metabotropic receptors are members of a family of GPCRs. Activation of GPCRs initiates G-protein cascades that generate a variety of second messengers including IP_3, DAG, cAMP, and cGMP. These messengers, as well as some G-proteins themselves, modulate synaptic transmission by (1) direct actions on ion channels, (2) regulation of protein kinase activity, and (3) triggering of Ca^{2+} entry into the cytosol from the extracellular fluid and internal stores.

4. ACh is the neurotransmitter at the NMJ and at many CNS synapses. Activation of ionotropic nAChRs opens these nonselective cation channels

that depolarize the postsynaptic cells. Metabotropic ACh receptors are also found at many synapses in the brain and spinal cord and in many organs such as the heart and urinary bladder.

5. Glutamate, the predominant excitatory neurotransmitter in the brain, and ionotropic glutamate receptors (NMDARs, AMPARs, and KARs) mediate most of this postsynaptic excitatory activity. The excitation results from the opening of these receptor-cation channels that depolarize the postsynaptic cells. In contrast, activation of mGluRs by glutamate stimulates phospholipase C, adenylyl cyclase or PKC, which modulate (either augment or inhibit) synaptic transmission.

6. GABA and glycine are the main inhibitory neurotransmitters in the CNS. Ionotropic receptors activated by these transmitters pass outward Cl^- current that hyperpolarizes the postsynaptic cells and makes it more difficult for an excitatory neurotransmitter to generate an AP.

7. In the CNS, activation of a single excitatory synaptic input generates an EPSP that is insufficient to reach AP threshold. The postsynaptic neuronal membrane integrates, in both space and time, input from multiple presynaptic neurons. This summation determines whether or not the postsynaptic neuron is depolarized to threshold at the axon hillock so that it can fire an AP.

8. The small biogenic amine neurotransmitters, Epi, NE, dopamine, 5-HT, and histamine, act primarily through metabotropic receptors. They influence, in one way or another, most of the activities of the brain, including cognitive function and memory, arousal, movement coordination, and mood and behavior. As a consequence, these transmitters are also involved in a large variety of pathophysiologic conditions including anxiety and depression, schizophrenia, Parkinson's disease, and drug addiction and withdrawal.

9. The biogenic amine transmitters also have important peripheral effects: for example, Epi and NE play key roles in cardiovascular regulation, 5-HT is involved in gastric acid secretion, and histamine triggers gastric acid secretion and airway smooth muscle contraction.

10. ATP, which acts through purinergic receptors, is coreleased with other neurotransmitters. Activation of these receptors modulates neurotransmitter release and smooth muscle relaxation and triggers platelet aggregation. Adenosine, which is generated from ATP by ectonucleotidases and acts on other purinergic receptors, helps modulate the sleep/wake state.

11. Peptide neurotransmitters include hormone-releasing transmitters, opioid peptides, neuropeptide Y, tachykinins, and the VIP-glucagon family. All these transmitters act on specific metabotropic receptors.

12. The gases NO and CO and certain lipids are unconventional neurotransmitters. They are not stored in vesicles, but are generated and released on demand. NO and CO are highly membrane-permeant; they rapidly diffuse to and enter postsynaptic cells. Both gases activate guanylyl cyclase, which converts GTP to cGMP, a second messenger. The endocannabinoids, 2-AG and anandamide, are highly lipophilic molecules that bind to CB_1Rs on presynaptic terminals at nearly all CNS synapses. At glutamatergic and GABAergic synapses, 2-AG suppresses transmitter release.

13. Neurotransmitter action is terminated by diffusion away from receptors, by enzymatic deactivation, and by reuptake. AChE inactivates ACh by hydrolyzing it to acetate and choline, and ectonucleotidases degrade ATP to adenosine. Reuptake of choline, glutamate, dopamine, NE, 5-HT, and adenosine helps to clear these substances from the synaptic cleft.

14. High-frequency stimulation can evoke LTP by activating AMPARs and NMDARs, which leads to activation of CaMKII and the eventual insertion of more AMPARs into the postsynaptic membrane. Gene transcription and synthesis of new proteins are required to maintain LTP over long periods of time.

15. Prolonged low-frequency stimulation can induce LTD. Ca^{2+} influx through NMDARs activates phosphatases that enable removal of AMPARs from the postsynaptic membrane. Maintenance of LTD over long periods of time requires translation of mRNAs but not gene transcription.

KEY WORDS AND CONCEPTS

- γ-Aminobutyric acid (GABA)
- Glutamate
- Glycine

- Histamine
- Catecholamines
- Biogenic monoamine neurotransmitters
- Dense core vesicles
- En passant synapses
- Wiring transmission
- Peptide neurotransmitters
- Magnocellular neurons
- Volume transmission
- Endocannabinoids
- Gaseous neurotransmitters
- Ionotropic receptors
- Metabotropic receptors
- G-protein–coupled receptors (GPCRs)
- Phosphoinositide cascade
- Inositol trisphosphate (IP$_3$)
- Diacylglycerol (DAG)
- Cyclic mononucleotides
- Myasthenia gravis
- Muscarinic acetylcholine receptors (mAChRs)
- N-methyl-D-aspartate (NMDA) receptors (NMDARs)
- α-Amino-3-hydroxy-5-methyl-4-isoxazole-propionate (AMPA) receptors (AMPARs)
- Metabotropic glutamate receptors (mGluRs)
- Excitatory synapse
- Inhibitory synapse
- End-plate current (EPC)
- Excitatory postsynaptic current (EPSC)
- Mg^{2+} block of NMDAR channels
- Receptor desensitization
- Central nervous system (CNS)
- Inhibitory postsynaptic current (IPSC)
- Inhibitory postsynaptic potential (IPSP)
- Temporal and spatial summation of postsynaptic potentials
- Afterhyperpolarization
- Parkinson's disease
- Selective serotonin reuptake inhibitors (SSRIs)
- Long-term potentiation (LTP)
- Long-term depression (LTD)
- Purinergic neurotransmission
- Soluble guanylyl cyclase
- Nitric oxide synthase (NOS)
- Hemoxygenase-2 (HO2)
- Retrograde synaptic transmission
- 2-Arachidonoylglycerol (2-AG)
- N-arachidonoyl ethanolamide (AEA)
- Neurotransmitter reuptake
- Antidepressants
- Dendritic spine

STUDY PROBLEMS

1. The typical linear dimension of a synaptic cleft is less than 0.5 μm. The diffusion coefficient for a small molecule neurotransmitter such as glutamate is $D \approx 5 \times 10^{-6}$ cm^2/sec. After the neurotransmitter molecules are released from the presynaptic terminal, they spread diffusionally and escape from the synaptic cleft. Using the quantitative description of diffusion from Chapter 2, estimate the time scale for neurotransmitter clearance from the synapse.

2. Based on what you know about synaptic transmission in the CNS, explain why it is so difficult to treat most behavioral disorders with single pharmacological agents.

3. You are investigating the ionic selectivity of an ion channel opened by activation of a specific group of ionotropic receptors. You use the ionic concentrations shown in the following table and determine that the reversal potential for activation of the receptors is −10 mV.

Ion	Intracellular (mM)	Extracellular (mM)
K$^+$	140	5
Na$^+$	10	145
Ca^{2+}	0.0001	2
Cl$^-$	6	106

   a. Give two possible ionic selectivities for the channel that could give rise to the observed reversal potential.

   b. How would you distinguish between these possibilities?

4. In NMDAR-dependent induction of LTP:

   a. Consider the role of AMPARs in the induction of NMDAR-dependent LTP by high-frequency stimulation. If AMPARs are blocked by a specific antagonist, and the presynaptic terminal is stimulated to release glutamate while the postsynaptic neuron is depolarized to 0 mV by voltage-clamp, would LTP be induced?

b. If the postsynaptic neuron is loaded with a high concentration of Ca^{2+} buffer to block increases in $[Ca^{2+}]_i$, would the normal high-frequency stimulation protocol induce LTP?

BIBLIOGRAPHY

Agnati LF, Zoll M, Strömberg I, Fuxe K: Intercellular communication in the brain: wiring versus volume transmission, *Neuroscience* 69:711, 1995.

Albuquerque EX, Pereira EFR, Alkondon M, Rogers SW: Mammalian nicotinic acetylcholine receptors: from structure to function, *Physiol Rev* 89:73, 2009.

Bjorness TE, Greene RW: Adenosine and sleep, *Curr Neuropharmacol* 7:238, 2009.

Garthwaite J: Concepts of neural nitric oxide-mediated transmission, *Eur J Neurosci* 27:2783, 2008.

Haas HL, Sergeeva OA, Selbach O: Histamine in the nervous system, *Physiol Rev* 88:1183, 2008.

Jorgensen EM: GABA. *WormBook* doi/10.1895/wormbook.1.14.1, http:/www.wormbook.org. Accessed August 31, 2005.

Kano M, Ohno-Shosaku T, Hashimotandi Y, et al: Endocannabinoid-mediated control of synaptic transmission, *Physiol Rev* 89:309, 2009.

Katona I, Freund TF: Endocannabinoid signaling as a synaptic circuit breaker in neurological disease, *Nat Med* 14:923, 2008.

Purves D, Augustine GJ, Fitzpatrick DJ, et al, editors. *Neuroscience*, ed 3, Sunderland, MA, 2004, Sinauer.

Salio C, Lossi L, Ferrini F: Neuropeptides as synaptic transmitters, *Cell Tissue Res* 326:583, 2006.

Seybold VS: The role of peptides in central sensitization. In Canning BJ, Spina D, editors: *Handbook of experimental pharmacology*. Vol 194, *Sensory nerves*, Berlin, Germany, 2009, Springer-Verlag, p 451.

Südhof TC: Neurotransmitter release. In Südhof TC, Starke K, editors: *Handbook of experimental pharmacology*. Vol 184, *Pharmacology of neurotransmitter release*, Berlin, Germany, 2008, Springer-Verlag, p 1.

Trudeau L-E: Glutamate co-transmission as an emerging concept in monoamine neuron function, *J Psychiatry Neurosci* 29:296, 2004.

Violin JD, Lefkowitz RJ: β-Arrestin-biased ligands at seven-transmembrane receptors, *Trends Pharmacol Sci* 28:416, 2007.

Section V

Molecular Motors and Muscle Contraction

14

MOLECULAR MOTORS AND THE MECHANISM OF MUSCLE CONTRACTION

OBJECTIVES

1. Understand the common principles that apply to all molecular motors: myosin, kinesin, and dynein.

2. Describe the structure of a skeletal muscle cell and the organization of its contractile elements, and compare and contrast this with the structure of cardiac and smooth muscle.

3. Understand the sliding filament mechanism of muscle contraction.

4. Understand the coupling between the mechanical motions of the myosin motor and the steps involved in ATP hydrolysis during cross-bridge cycling.

5. Describe how Ca^{2+} interacts with the regulatory proteins troponin and tropomyosin to activate contraction in skeletal and cardiac muscle.

6. Describe how Ca^{2+} activates contraction in smooth muscle by promoting the phosphorylation of myosin regulatory light chain.

MOLECULAR MOTORS PRODUCE MOVEMENT BY CONVERTING CHEMICAL ENERGY INTO KINETIC ENERGY

Movement is one of the defining characteristics of all living creatures. Motility is an essential feature of many biological activities, such as the beating of cilia and flagella, cell movement, cell division, development and maintenance of cell architecture, and muscle contraction, the main topic of this and the next two chapters. Moreover, the normal functioning of all cells requires the directional transport, *within* the cell, of numerous substances and organelles, such as vesicles, mitochondria, chromosomes, and macromolecules (e.g., mRNA and protein).

The Three Types of Molecular Motors Are Myosin, Kinesin, and Dynein

All types of cellular motility are driven by **molecular motors** that produce unidirectional movement along structural elements in the cell. The structural elements are either filaments composed of **actin** monomers or **microtubules**, which are polymers of the protein tubulin. Three distinct types of molecular motors that move along these structures have been described: **myosin, kinesin,** and **dynein.** Myosin is a motor that moves along actin filaments. There are many classes of myosins. Myosin II, which is found in all muscles, produces muscle contraction. Myosin V transports vesicles and organelles along actin filaments. Kinesin and dynein transport organelles along microtubules.

211

Kinesins are also involved in spindle formation and chromosome separation during mitosis and meiosis, as well as in mRNA and protein transport. Dyneins mediate the beating of cilia, the movement of flagella, and vesicular trafficking.

Several principles apply to the operation of all molecular motors. Molecular motors convert chemical energy into kinetic energy (movement). The chemical energy is stored in the high-energy phosphate bond of ATP. The motors (myosin, kinesin, and dynein) all have ATPase activity. The binding of ATP, its hydrolysis, and the subsequent release of products are important steps in the generation of movement. In all cases, movement is produced through repetitive cycles of interaction between the motor and either an actin filament or a microtubule. The mechanism whereby ATP hydrolysis is coupled to the conformational and structural changes that produce movement has been elucidated through extensive biochemical, biophysical, and structural studies of muscle contraction and kinesin-based vesicle transport. The mechanism is discussed in a later section.

SINGLE SKELETAL MUSCLE FIBERS ARE COMPOSED OF MANY MYOFIBRILS

Muscle cell types are classified primarily according to their structural and functional properties. An understanding of the detailed ultrastructure of single muscle cells provides insight into their functional properties. *Skeletal muscle cells* (skeletal myocytes) are attached to the skeleton by tendons and are under voluntary control. Their primary function is to shorten and generate force to produce movement of skeletal levers. The other two types of muscle, cardiac and smooth, are described later in this chapter.

Skeletal muscle is composed of many individual *muscle fibers*, each of which is an elongated cell. Each cell is 10 to 100 μm in diameter and may reach several centimeters in length. Electron micrographs reveal that a single skeletal muscle fiber is composed of bundles of filaments, called **myofibrils**. The myofibrils lie parallel to one another and run along the long axis of the cell (Figure 14-1). Surrounding each myofibril is an extensive membrane-enclosed intracellular compartment called

the SR, which plays a key role in activating muscle contraction. Enlarged portions of the SR, the terminal cisterns, are closely apposed to finger-like invaginations of the sarcolemma (muscle PM) called **transverse tubules (T-tubules)** (Figure 14-1). The T-tubule membrane is continuous with the surface membrane. In contrast, the SR membrane is physically distinct, and electrically isolated, from the sarcolemma. The relevance of this point will become clear when we consider the roles of the T-tubule and the SR in excitation-contraction coupling in Chapter 15. Viewed perpendicular to its long axis, a skeletal muscle cell has a striped appearance, with alternating light and dark bands (Figure 14-2); this has led to its classification as *striated muscle.*

THE SARCOMERE IS THE BASIC UNIT OF CONTRACTION IN SKELETAL MUSCLE

Sarcomeres Consist of Interdigitating Thin and Thick Filaments

The banding pattern in striated muscle is produced by the regular arrangement of **thick** and **thin filaments** in the myofibrils. The light bands are I bands, which contain thin (actin) filaments that extend in both directions from a thin dense line, called the Z line (Figure 14-3). The region of myofibril between two adjacent Z lines is called a **sarcomere.** The dark bands, called A bands, contain thick (myosin) filaments arranged in parallel (Figure 14-3). At the center of the A band is a dense line called the M line. The thin filaments extend into the A bands, but are not present in the central H zone,* which therefore appears lighter. The regular arrangement of thick and thin filaments is clearly shown in a cross section of a myofibril taken in the region of the A band where the filaments overlap (Figure 14-3). The thick filaments interdigitate with thin filaments so that each thick filament is

---

* The darker bands are called A bands because they are anisotropic; the I bands are isotropic. Anisotropic material has different refractive indices for different planes of polarized light; isotropic material has a single refractive index. The Z line takes its name from the first letter of *Zwischenscheibe* (intervening disk, in German). The H in H zone stands for *heller* (lighter, in German).

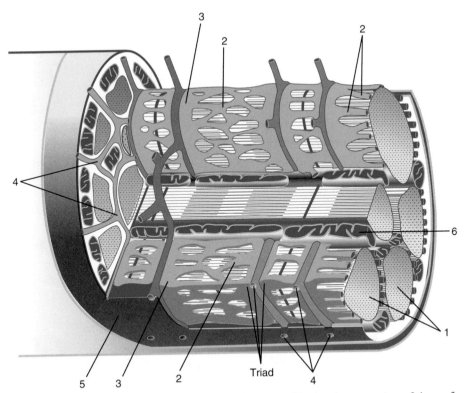

FIGURE 14-1 ■ Ultrastructure of a mammalian skeletal muscle cell. In this drawing a portion of the surface membrane (*5*) has been removed to reveal the parallel arrangement of myofibrils (*1*). The cut ends of the myofibrils reveal that they are composed of arrays of thick and thin filaments. Each myofibril is surrounded by elements of the sarcoplasmic reticulum (*2*) with their terminal cisterns (*3*). The T-tubules (*4*) are invaginations of the surface membrane that form a network of tubules extending into the center of the cell. Note that the lumen of the T-tubule is continuous with the extracellular space (see Figure 15-3); the *triad* is the conjunction of a T-tubule with a pair of SR terminal cisterns (see Figure 15-5). Numerous mitochondria (*6*) lie between myofibrils. (*Modified from Krstic RV: Ultrastructure of the mammalian cell, New York, 1979, Springer-Verlag.*)

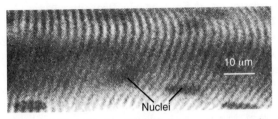

FIGURE 14-2 ■ Skeletal muscle cells have striations. A short segment of a single muscle fiber from human gastrocnemius (calf) muscle clearly shows the alternating light and dark bands that characterize striated muscle. Several nuclei are also visible. (*From Berne RM, Levy MN, Koeppen BM, et al, editors:* Physiology, *ed 4, New York, 1998, Mosby.*)

surrounded by a hexagonal array of thin filaments. This precise filament geometry is maintained by various cytoskeletal proteins that link filaments within a sarcomere and also link the sarcomeres of adjacent myofibrils. One of these important cytoskeletal proteins, **α-actinin,** is a major component of the Z line structure to which the thin filaments attach. **Titin** is a giant muscle protein (~3.8 million daltons) that has an important role in muscle elasticity (see Chapter 16). One end of the titin molecule is inserted into the Z line; the other end forms a portion of the thick filament and inserts into the M line.

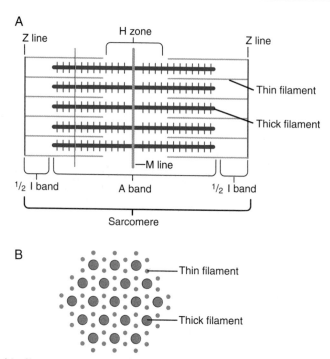

FIGURE 14-3 ■ Thick and thin filaments are arranged in regular arrays in the myofibril. **A,** Schematic drawing of a longitudinal section of a single sarcomere, which is the region of myofibril between two adjacent Z lines (see text for details). **B,** Diagram of a cross section of the myofibril through the A band at the position indicated by the *thin gray line* in **A.** In this region, thick and thin filaments overlap, and each thick filament is surrounded by a hexagonal array of thin filaments. *(Modified from Huxley HE, Hanson J: The molecular basis of contraction in cross-striated muscles. In Bourne GH, editor: The structure and function of muscle, vol 1, New York, 1960, Academic Press.)*

Thick Filaments Are Composed Mostly of Myosin

With a molecular weight of approximately 470,000 daltons, myosin II is a large protein consisting of two heavy chains and two pairs of different light chains—a **myosin essential light chain** and a **myosin regulatory light chain (RLC)**. The myosin molecule has a long, rod-like tail with two globular heads (Figure 14-4). The rod-like portion of the molecule contains an "arm" adjacent to each globular head. At each end of the arm is a flexible region that acts as a hinge, allowing rotation at that point. Many myosin molecules align to form a thick filament (Figure 14-5). The tail regions of the molecules are bundled to form the body of the thick filament. The globular heads and arm regions project out from the bundle. The heads of the myosin molecules can bind to the thin filaments to form **cross-bridges** between the two filaments.

The myosin heads in each half of the thick filament are oriented in opposite directions; the heads are not present in the central region (Figure 14-5).

Thin Filaments in Skeletal Muscle Are Composed of Four Major Proteins: Actin, Tropomyosin, Troponin, and Nebulin

Actin is a globular protein (G-actin) with a molecular weight of 41,700 daltons. G-actin monomers aggregate to form strands resembling a string of pearls. The thin filament consists primarily of two helical strands of G-actin wound around each other (Figure 14-6). The 600-kDa protein molecule **nebulin** runs along the thin filament and forms a template that limits the length of the actin filaments. The thin filament also contains the regulatory proteins **tropomyosin** and **troponin**. Tropomyosin is a long, rod-shaped protein dimer with a molecular weight of approximately 66 kDa. This molecule lies

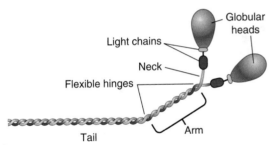

FIGURE 14-4 ■ Structure of myosin. Myosin is composed of two identical heavy chains and two different pairs of light chains ("essential" and regulatory light chains). Each heavy chain has a globular head attached to an elongated, rod-like tail. The two tails are twisted together; most of each tail is buried in the thick filament. The arm projects out from the thick filament and has flexible hinges at both ends. The arm and the globular head can form a cross-bridge to a thin (actin) filament. One essential light chain and one regulatory light chain are associated with each heavy chain near the globular head. The light chains play a role in the regulation of contraction. *(Modified from Berne RM, Levy MN, Koeppen BM, Stanton BA, editors:* Physiology, *ed 4, New York, 1998, Mosby.)*

along both sides of the thin filament in grooves formed by the two strands of actin molecules (Figure 14-6). Each tropomyosin molecule binds to seven actin monomers in one of the strands. Troponin, which is bound to tropomyosin, is a complex of three proteins: **troponin T (TnT), troponin C (TnC),** and **troponin I (TnI).** The roles of tropomyosin and troponin in the calcium (Ca^{2+})-dependent regulation of skeletal muscle contraction are discussed later in this chapter.

MUSCLE CONTRACTION RESULTS FROM THICK AND THIN FILAMENTS SLIDING PAST EACH OTHER (THE "SLIDING FILAMENT" MECHANISM)

The **sliding filament** mechanism of muscle contraction was deduced from the changes in striation pattern of skeletal muscle observed during contraction. Before contraction, a relatively wide I band and H zone are visible (Figure 14-7A). When stimulated to contract, muscle shortening is accompanied by sarcomere shortening (Figure 14-7B). After the muscle shortens, the width of the A band is unchanged but the widths of the I band and H zone decrease. The change in sarcomere length results from a change in the degree of overlap between thick and thin filaments as they slide past one another. In a resting muscle, only partial overlap occurs between thick and thin filaments (Figure 14-8A). The region of the thin filaments that does *not* overlap the thick filaments corresponds to the I band, whereas the region of the thick filaments that does *not* overlap the thin filaments constitutes the H zone. When the muscle shortens during a contraction, the region of overlap between thick and thin filaments increases (Figure 14-8B). In this contracted state, the H zones and I bands are narrower because the nonoverlapped portions of thick and thin filaments are both shorter. The A band corresponds to the entire length of the thick filament. Because the filament length is constant, the length of the A band remains constant during changes in muscle length.

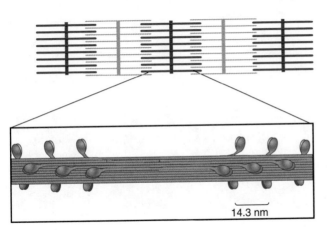

FIGURE 14-5 ■ Structure of the thick filament. *Top,* Schematic drawing of the structure of the sarcomere. *Bottom,* The proposed structure of the thick filament. The body of the thick filament is formed from the tail regions of a large number of myosin molecules. The arms and heads of the myosin molecules project out from the thick filament at regular intervals. Successive projections are rotated 120 degrees around the thick filament. Three pairs of myosin heads project out at intervals of 14.3 nm along the thick filament.

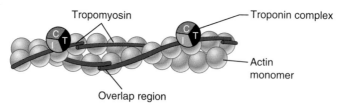

Tropomyosin

Troponin complex

Actin monomer

Overlap region

FIGURE 14-6 ■ The thin filament consists of two helical strands of actin monomers. Double-stranded tropomyosin molecules (drawn as a single strand) lie in each of the two grooves formed by the actin strands. Each tropomyosin molecule binds to seven actin monomers in one strand. The ends of successive tropomyosin molecules overlap slightly, and near this overlap region a troponin complex is bound to tropomyosin. The troponin (Tn) complex consists of three proteins, TnC (C), TnI (I), and TnT (T).

In the region of filament overlap, short connections, or cross-bridges, project from the thick filaments toward the thin filaments (Figure 14-8). The molecular basis for filament sliding involves cross-bridge movement. The cross-bridges attach to and pull on the thin filaments to cause sliding of the thick and thin filaments past each other. This increases overlap between the filaments and shortens the sarcomere.

THE CROSS-BRIDGE CYCLE POWERS MUSCLE CONTRACTION

Cross-bridge movement produces filament sliding in the following way (Figure 14-9). The myosin head attaches to an actin filament to form a cross-bridge. The head then rotates toward the myosin tail, thus pulling on the thin filament and causing it to move relative to the thick filament. The head detaches and rotates back to its original orientation, and the cycle can repeat. This mechanism is analogous to rowing a boat. The oar is dipped into and pulled through the water (myosin binding and rotation; the "power stroke"), the oar is pulled out of the water and pushed back to its original position (myosin detachment and "recocking" of the head), and a new stroke can begin. These mechanical steps are coupled to the hydrolysis of ATP, which is catalyzed by the myosin head during its interaction with actin (i.e., myosin is an ATPase). The cyclical sequence of steps (Figure 14-9), called the **cross-bridge cycle,** illustrates the mechanism by which the

FIGURE 14-7 ■ Electron micrographs of frog sartorius muscle at different degrees of shortening. Muscle shortening in **B** is greater than in **A**. Sarcomere shortening accompanied muscle shortening: the sarcomere length in **B** (the shorter muscle) is shorter than in **A**. The shorter sarcomere length is the result of a shorter I band and H zone. The length of the A band (i.e., the length of the thick filament) is constant. *(From Huxley HE: Structural evidence concerning the mechanism of contraction in striated muscle. In Paul WM, Daniel EE, Kay CM, et al, editors:* Muscle: proceedings of a symposium held at the Faculty of Medicine, University of Alberta, *Oxford, 1964, Pergamon.)*

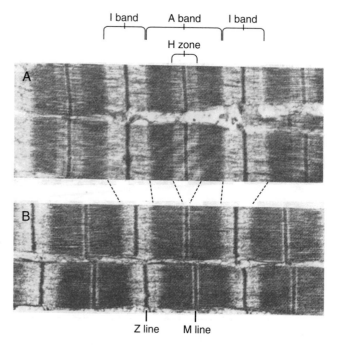

A. Relaxed muscle

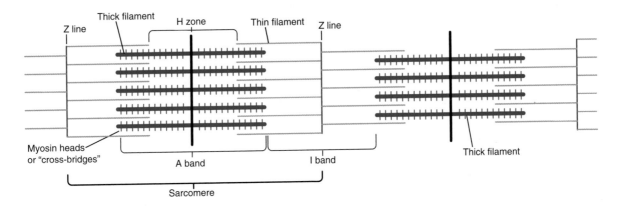

B. Contracted muscle

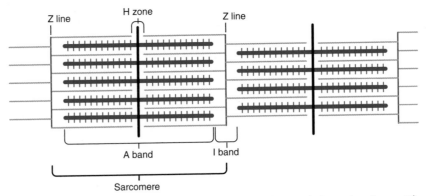

FIGURE 14-8 ■ The sliding filament mechanism. **A,** Schematic drawing of a muscle in a relaxed state. There is little overlap between thin and thick filaments. The projections from the thick filaments are the arm and globular head regions of the myosin molecules that form cross-bridges to the thin filaments. **B,** Schematic drawing of the same muscle during a contraction. The thin filament has slid along the thick filament so that they overlap each other to a greater extent. Note that in the contracted state, both the H zone and the I band are narrower but the A band is unchanged.

muscle cell uses chemical energy stored in ATP to generate force.

In resting muscle, cross-bridges are not attached and the myosin heads are oriented at an angle of 90 degrees to the thin filaments. In this state, the myosin head is phosphorylated and ADP is bound to the head. Actin-myosin interactions are prevented because the myosin binding sites on the thin filament are covered by tropomyosin. When $[Ca^{2+}]_i$ increases upon muscle activation (see Chapter 15), the tropomyosin molecules move to expose the myosin binding sites on actin filaments. The myosin head then binds to the thin filament with low affinity, to form a weakly attached cross-bridge. Pi is then released to allow a high-affinity (strong) attachment between actin and myosin. This is accompanied by a 45-degree rotation of the myosin head. This bending of the cross-bridge generates a force on the thin filament that causes it to slide relative to the thick filament (the power stroke). Subsequent dissociation of ADP and binding of ATP causes the cross-bridge to detach. Hydrolysis of ATP results in a return to the

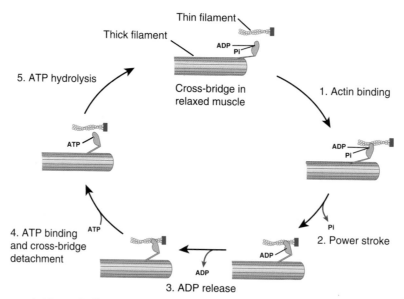

FIGURE 14-9 ■ The cross-bridge cycle illustrates the coupling between ATP hydrolysis and movement. *Top,* The state of the contractile proteins in a relaxed muscle cell. With the products of ATP hydrolysis (ADP and Pi) bound to the myosin, it has high affinity for actin. When $[Ca^{2+}]_i$ is elevated, the regulatory proteins enable myosin to bind to actin *(step 1).* In *step 2,* the myosin head rotates by 45 degrees when Pi is released, thus causing filament sliding. This is the "power stroke," or the force-generating step in the cross-bridge cycle. *Step 3* is the dissociation of ADP with the cross-bridges still attached. In the absence of ATP, cross-bridges are locked in this state of "rigor" (rigor mortis). In *step 4,* ATP binds to the myosin head. In this state, myosin has low affinity for actin and the cross-bridge detaches. In *step 5,* the final step, ATP is hydrolyzed, thereby rephosphorylating the myosin head and restoring it to an angle of 90 degrees.

myosin-ADP-Pi complex, with the myosin head again poised at 90 degrees with respect to the thin filament. This cycle can repeat itself as long as $[Ca^{2+}]_i$ remains elevated. Some of the steps in the cross-bridge cycle have been visualized using atomic force microscopy to study myosin V molecules moving along actin filaments (Box 14-1).

IN SKELETAL AND CARDIAC MUSCLES, Ca²⁺ ACTIVATES CONTRACTION BY BINDING TO THE REGULATORY PROTEIN TROPONIN C

A necessary step in the activation of contraction in all types of muscle cells is an increase in $[Ca^{2+}]_i$. The process by which muscle cell activation leads to

an increase in $[Ca^{2+}]_i$ is elucidated in Chapter 15. Here we describe the mechanism by which an increase in $[Ca^{2+}]_i$ initiates contraction. In both skeletal and cardiac muscle, Ca^{2+} activation of contraction involves the regulatory proteins troponin and tropomyosin.

The ends of consecutive tropomyosin molecules overlap each other, and one troponin complex binds to each tropomyosin molecule near the overlap region (see Figure 14-6). TnT links the troponin complex to tropomyosin, and TnI plays an inhibitory role in actin-myosin interaction. In skeletal and cardiac muscles, Ca^{2+} initiates contraction by binding to TnC (the "actin switch").

In a resting (relaxed) skeletal muscle cell $[Ca^{2+}]_i$ is approximately 100 nM. At this low $[Ca^{2+}]_i$, TnI is tightly bound to actin, and tropomyosin covers the

BOX 14-1
VISUALIZING MYOSIN MOVEMENT THROUGH ATOMIC FORCE MICROSCOPY

Myosin V is a two-headed motor that transports vesicles and organelles along actin filaments. A myosin V molecule walks in 36-nm steps in one direction along the actin filament. Myosin V is composed of two chains, each containing a globular head and an elongated tail, with the two tails twisted together as in myosin II (see Figure 14-4). In myosin V the untwisted "neck" region of the myosin molecule, between the arm and the head, is approximately three times longer than in myosin II. This accounts for the long step length of 36 nm. High-speed atomic force microscopy (AFM) has been used to record the dynamic behavior of myosin V walking along an actin filament. The cycle of events involved in myosin V movement is illustrated in Figure B-1. In the initial state of the cycle, both ADP–bound heads are attached to actin (Figure B-1a):

the leading myosin head (H2) is in an angled conformation and the trailing head (H1) is in a straight conformation. Following ADP release and ATP binding, the trailing head detaches (Figure B-1b). Then the leading head spontaneously switches to the straight conformation, thus allowing the trailing head to swing forward and search for a new binding site on actin. The ATP is hydrolyzed and the ADP-Pi–bound (new) leading head binds to actin; Pi is released immediately, and the new leading head switches to the angled conformation. At this point, the myosin V is again in the original two-head-bound form, and it has advanced 36 nm along the actin filament (Figure B-1c). A movie showing these movements as observed by AFM can be viewed at www. nature.com/nature/journal/v468/n7320/extref/nature09450-s3.mov.

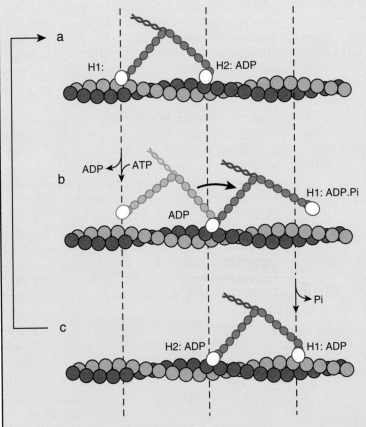

FIGURE B-1 ■ Mechanism of myosin V walking along an actin filament. The distance between two vertical dashed lines represents a spacing of 36 nm along the actin filament. In the initial state (a), both ADP-bound heads of myosin V (H1 and H2) are bound to actin. The leading head (H2) is in the angled conformation and the trailing head (H1) is in the straight conformation. Following release of ADP and binding of ATP (b), the trailing head detaches. The leading head then switches to the straight conformation and causes the trailing head to swing forward. ATP is hydrolyzed, and the ADP-Pi-bound new leading head searches for a fresh actin binding site. The new leading head attaches to actin, Pi is immediately released, and this head then adopts the angled conformation (c). (Modified from Kodera N, Yamamota D, Ishikawa R, et al: Nature 468:72, 2010.)

myosin binding sites on actin and, thus, prevents cross-bridge formation. When an AP causes $[Ca^{2+}]_i$ to rise transiently to micromolar levels (see Chapter 15), Ca^{2+} binds to TnC. This weakens the bond between TnI and actin so that tropomyosin can move laterally on the thin filament to expose the myosin binding sites on actin. Myosin heads can then attach to the actin filament to form cross-bridges and generate force.

THE STRUCTURE AND FUNCTION OF CARDIAC MUSCLE AND SMOOTH MUSCLE ARE DISTINCTLY DIFFERENT FROM THOSE OF SKELETAL MUSCLE

Cardiac Muscle Is Striated

The heart functions as a pump and is designed for continuous, rhythmic activity over the life of an individual. Several unique structural and functional properties of cardiac muscle cells (cardiac myocytes) are important in this continuous pumping activity. The contractile mechanism in an individual cardiac myocyte is very similar to that in skeletal muscle, however. Like skeletal muscle, cardiac muscle is striated (Figure 14-10) because the contractile elements in adjacent myofilaments are aligned in register to one another. As in skeletal muscle, Ca^{2+} initiates contraction by binding to TnC, thus causing tropomyosin to expose myosin binding sites on actin. Shortening is then produced by cross-bridge cycling, which causes the filaments to slide.

Cardiac Muscle Cells Require a Continuous Supply of Energy

Cardiac muscle requires an uninterrupted supply of ATP to support the continuous, repetitive contraction-relaxation cycles. The main source of this ATP is oxidative phosphorylation in mitochondria, which are very abundant in cardiac myocytes (Figure 14-10). The mitochondria keep ATP production in pace with its continuous utilization by the contractile machinery and other metabolic activities. This requires an uninterrupted supply of oxygen, which is provided by the extensive capillary network throughout the heart. Every cardiac muscle cell is in contact with a capillary to ensure continuous delivery of oxygen and nutrients

to, and removal of metabolic waste from, each cardiac myocyte.

To Enable the Heart to Act as a Pump, Myocytes Comprising Each Chamber Must Contract Synchronously

The heart consists of four chambers: right and left atria and right and left ventricles. The cardiac myocytes in the wall of each chamber must contract synchronously so that the chamber can eject its contents efficiently. As in skeletal myocyte, an AP in the cardiac myocyte sarcolemma initiates a contraction (see Chapter 15). Heart muscle contracts synchronously because the AP spreads rapidly from cell to cell through gap junctions (see Chapter 12). The gap junctions are located in **intercalated disks**, which are dense regions of the sarcolemma at the ends of cardiac myocytes (Figure 14-10). Thus, the heart behaves as an electrical syncytium* in which the cardiac myocytes are electrically coupled from end to end, in contrast to skeletal myocytes, which do not have gap junctions and are electrically isolated. The intercalated disks in the heart also transmit force from cell to cell. This connection of multiple cardiac myocytes functionally mimics a single, long skeletal muscle fiber.

Smooth Muscles Are Not Striated

Smooth (nonstriated) muscles are contained within the walls of hollow organs, such as the gastrointestinal tract, arteries and veins, urinary bladder, bronchi, and ureters. These muscles are under involuntary control and are regulated by autonomic nerves and by hormones.

The name smooth muscle derives from the finding that these muscles do not exhibit the striation or banding pattern observed in skeletal and cardiac muscles. Smooth muscles do, however, contain actin and myosin filaments, and they contract by a sliding filament mechanism.

The special feature of smooth muscle is the organization of actin and myosin filaments into small bundles

---

*In general, a syncytium is a large multinucleated cell formed by fusion of smaller cells. An electrical syncytium refers to a group of cells interconnected by gap junctions that allow ionic currents to flow freely from cell to cell.

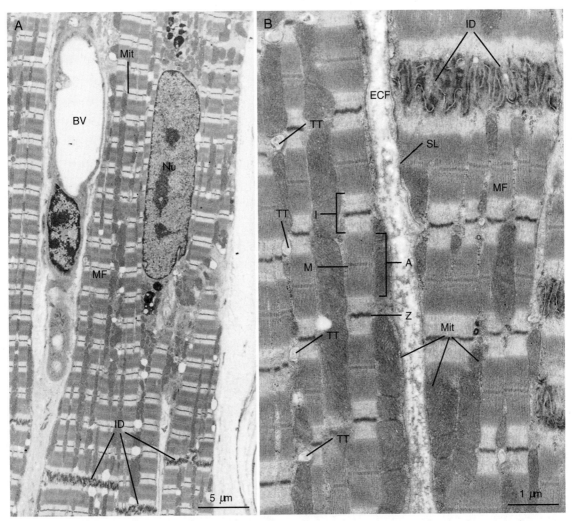

FIGURE 14-10 ■ Structure of a cardiac muscle cell. **A,** Low-magnification electron micrograph of a ventricular myocyte. Typical features of cardiac myocytes include myofibrils (MF) with clearly visible striations, numerous mitochondria (Mit) arranged in columns between the myofibrils, intercalated disks (ID), and an elongated nucleus (Nu). A blood vessel (BV) lies between two cells. **B,** Details of the structure shown at higher magnification. The figure shows portions of two cells separated by extracellular fluid (ECF) between the cells. The surface membrane, or sarcolemma (SL), is folded many times at the intercalated disk (ID). Mitochondria (Mit) are located in columns between myofibrils (MF) or are clustered next to the sarcolemma. The structure of the sarcomere is similar to that of skeletal muscle and includes the A band (A), I band (I), Z line, and M line. T-tubules (TT) are larger in diameter than in skeletal muscle. *(From Berne RM, Levy MN, Koeppen BM, et al, editors:* Physiology, *ed 4, New York, 1998, Mosby.)*

that run obliquely across the long axis of the cells (Figure 14-11). The filaments are organized into sarcomeres and exhibit partial overlap, just as in skeletal and cardiac muscle (see the magnified diagram at the upper right of Figure 14-11). Because individual bundles of contractile filaments have different orientations, the

sarcomeres in adjacent bundles are not in register, hence the lack of striations.

The actin filaments are attached to **dense bodies** (Figure 14-11), which are equivalent to Z line structures of skeletal and cardiac muscle. The dense bodies are composed primarily of α-actinin, which

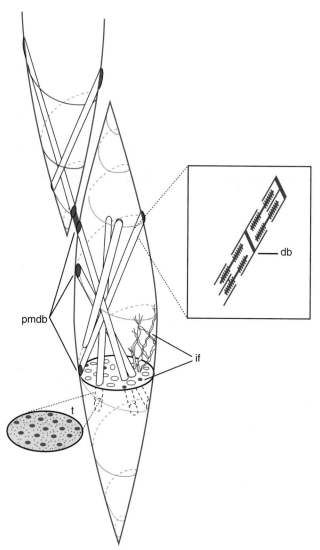

FIGURE 14-11 ■ Organization of the contractile apparatus in smooth muscle. The contractile units are arranged in columns containing thick (myosin) and thin (actin) filaments (t shows a cross section of a column); as shown in the *box,* the interdigitation of the thick and thin filaments is similar to that in skeletal muscle. The columns are shown in a helical arrangement. The actin filaments are anchored to dense bodies (db), and the dense bodies are held in place by a cytoskeletal scaffold composed of intermediate filaments (if). Some dense bodies are inserted into the plasma membrane (pmdb). The insertion points in adjacent cells may abut one another, as depicted in the *upper left portion* of the figure. (*Modified from Small JV, Sobieszek A:* Int Rev Cytol *64:241, 1980.*)

is also a major component of Z line structures. Dense bodies are held in position by cytoskeletal elements, the **intermediate filaments,** which are thicker than actin filaments, but thinner than myosin filaments. Two major components of the intermediate filaments are the cytoskeletal proteins desmin and vimentin.

The two ends of each contractile bundle are attached to *PM-associated dense bodies* located on opposite sides of the smooth muscle cells (Figure 14-11). The PM attachment sites in neighboring cells often abut one another to maximize the transmission of force from cell to cell (Figure 14-11). Because the contractile bundles are oriented obliquely to the long axis of the smooth muscle cells (Figure 14-11), some contractile force is transmitted laterally; most of the force is transmitted longitudinally, however. Thus, in a cylindrical organ such as an artery or the intestine, smooth muscle cells that are oriented circumferentially around the lumen constrict the lumen when they contract. In contrast, contraction of longitudinally oriented smooth muscle cells, as in the longitudinal muscle layer of the intestine, causes the organ to shorten. In saccular organs such as the urinary bladder, both lateral and longitudinal forces cause relatively uniform contraction of the organ wall that leads to a reduction in organ volume.

In Smooth Muscle, Elevation of Intracellular Ca^{2+} Activates Contraction by Promoting the Phosphorylation of the Myosin Regulatory Light Chain

In smooth muscle, as in skeletal and cardiac muscle, elevation of $[Ca^{2+}]_i$ activates contraction, but the mechanism whereby Ca^{2+} activates the contractile apparatus is different. Smooth muscles lack troponin. Instead, the Ca^{2+} binds to the protein, **calmodulin,** and the Ca^{2+}-calmodulin complex then binds and activates **myosin light chain kinase (MLCK)**. Activated MLCK phosphorylates the smooth muscle RLC (Figure 14-12). Phosphorylation induces a conformational change in the myosin—this is the "switch" that enables the cross-bridges to form between actin and myosin (Figure 14-13),

thereby generating force and shortening the muscle cell. Thus, in smooth muscle the switch that turns on cross-bridge cycling is located on the thick filament (myosin), whereas in skeletal and cardiac muscle the switch (TnC) is located on the thin filament (actin).

Another enzyme, **myosin light chain phosphatase (MLCP)**, catalyzes the dephosphorylation of myosin RLCs (Figure 14-12). This halts actin-myosin cross-bridge cycling (Figure 14-13). This dephosphorylation may occur either when the actin is attached to the myosin head or when it is detached (Figure 14-13).

As long as the RLC is phosphorylated, cross-bridge cycling will continue. As in skeletal muscle, the cross-bridge cycle is associated with ATP hydrolysis (Figure 14-13). This ATP is needed *in addition to* the ATP involved in the phosphorylation of the myosin RLC. If the cross-bridges are still attached when the myosin RLC is dephosphorylated, detachment of the cross-bridges will be slowed (see Chapter 16). This slow detachment contributes to the maintenance of **tonic contraction** and force in smooth muscle. This tonic force (**tone**) is maintained with little expenditure of energy because the cross-bridges cycle very slowly, so that little ATP is consumed. Ca^{2+} is required not only to initiate the contraction but also to maintain tone. If, after an initial transient increase, $[Ca^{2+}]_i$ falls to less than the contraction threshold, no new cross-bridges will form, whereas previously formed cross-bridges continue to detach slowly. The several chemical reactions directly involved in the contractile activation and relaxation of smooth muscle contraction are diagrammed in Figures 14-12 and 14-13.

The sensitivity of smooth muscles to Ca^{2+} can be modulated so that a given $[Ca^{2+}]_i$ level can produce more or less contractile force. Some important mechanisms that regulate contraction without a change in $[Ca^{2+}]_i$ (so-called **Ca^{2+}-independent regulatory mechanisms**) are described in Box 14-2. These mechanisms, which reflect the complex regulation of contraction and relaxation in smooth muscles, are important targets for pharmacological intervention.

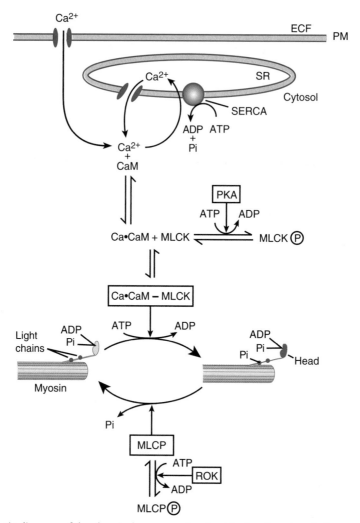

FIGURE 14-12 ■ Schematic diagram of the chemical processes involved in the phosphorylation and dephosphorylation of smooth muscle myosin RLC. The Ca^{2+} that triggers contraction comes from either the ECF or the SR. The Ca^{2+} combines with calmodulin (CaM) to form a complex with myosin light chain kinase (Ca/CaM+MLCK). Ca/CaM+MLCK catalyzes the phosphorylation of the myosin RLC. This initiates the myosin-actin interaction (see Figure 14-13) that generates force and causes shortening. Relaxation results from dephosphorylation of RLC catalyzed by myosin light chain phosphatase (MLCP). The sensitivity of the contractile apparatus to Ca^{2+} (Box 14-2) is regulated, in part, by cyclic AMP–dependent protein kinase (PKA) and by Rho-associated protein kinase (ROK). Activation of PKA reduces Ca^{2+} sensitivity, while activation of ROK increases Ca^{2+} sensitivity.

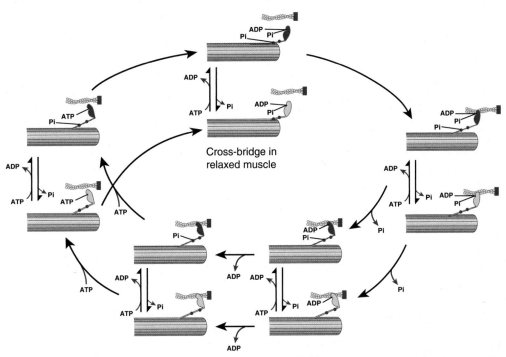

Cross-bridge in
relaxed muscle

FIGURE 14-13 ■ Cross-bridge cycling in smooth muscle. A rise in $[Ca^{2+}]_i$ activates MLCK, which begins the cycle by phosphorylating the myosin RLC (*at the top*). This enables myosin to attach to actin to form the actin-myosin cross-bridge (*at the right*). Pi then dissociates from the myosin head, which then rotates to produce force (*bottom right*). ADP is released, and ATP then binds (*circling clockwise to the left*) and promotes cross-bridge detachment. The ATP bound to the myosin head is hydrolyzed to ADP and Pi (*at the top*) so that a new cycle can begin. The RLC can be dephosphorylated by MLCP at any time during the cycle (represented by lighter-colored myosin heads). When RLC is dephosphorylated, a detached myosin head cannot reattach to actin (as indicated by the break in the cycle) until the RLC is rephosphorylated. *(Modified from Murphy RA: Smooth muscle. In Berne RM, Levy MN, Koeppen BM, et al, editors:* Physiology, *ed 4, New York, 1998, Mosby.)*

BOX 14-2

SMOOTH MUSCLE CONTRACTION CAN BE REGULATED BY MODULATING Ca²⁺ SENSITIVITY

The contraction of smooth muscle depends on Ca^{2+} (see text and Figures 14-12 and 14-13). Contraction is also regulated by modulating the sensitivity of the contractile apparatus to Ca^{2+}. For example, stimulation by a neurostransmitter such as norepinephrine, acting on β-adrenergic receptors, stimulates adenylyl cyclase to generate cAMP. The cAMP activates PKA, which catalyzes the phosphorylation of MLCK (see Figure 14-12). Phosphorylated MLCK (MLCK-P) cannot bind the Ca^{2+}-calmodulin complex. Thus, cAMP effectively reduces the sensitivity of the contractile apparatus to Ca^{2+} and thereby decreases force production and relaxes tonic smooth muscle without altering $[Ca^{2+}]_i$.

Smooth muscles are also regulated by the Rho/Rho-kinase pathway. Various smooth muscle agonists, such as norepinephrine, serotonin (5-HT), and histamine, increase the GTP–bound (active) form of the small GT-Pase, RhoA. RhoA-GTP then activates Rho-associated protein kinase (ROK), which, in turn, phosphorylates

and thus inhibits MLCP (see Figure 14-12). The net effect is sensitization of the smooth muscle to Ca^{2+}; that is, the activating effect of Ca^{2+}-dependent phosphorylation of myosin RLC is prolonged because dephosphorylation is inhibited. Thus, the Rho/Rho-kinase pathway contributes to the tonic contraction of smooth muscles. In contrast, stimulation of cGMP-dependent protein kinase, for example, by endothelium-derived nitric oxide, promotes the phosphorylation of RhoA. Phosphorylated RhoA cannot bind GTP and, therefore, cannot activate ROK. Thus, MLCP can remain active and continue to dephosphorylate the RLCs. In arteries, this desensitizes the smooth muscle to Ca^{2+} and causes arterial dilation. These Rho/ROK-mediated effects on smooth muscle contraction and relaxation can occur without changes in $[Ca^{2+}]_i$. Although they are commonly referred to as Ca^{2+}-independent regulatory mechanisms, these mechanisms actually exert their effect by tuning the sensitivity of the contractile apparatus to Ca^{2+}.

SUMMARY

1. The three types of muscle are skeletal, cardiac, and smooth. In all three, contraction is produced by a myosin motor that converts chemical energy into kinetic energy through the hydrolysis of ATP.

2. A single skeletal muscle fiber contains many myofibrils, which are bundles of myofilaments that lie parallel to one another and run along the long axis of the cell. The SR surrounds each myofibril.

3. The region of myofibril between two Z lines, the contractile unit, is called a sarcomere.

4. There is an ordered arrangement of thick and thin filaments within the myofibril. Thick filaments, composed mostly of myosin, interdigitate with thin filaments. Thin filaments are composed of actin, tropomyosin, troponin and nebulin.

5. Each myosin molecule has a long tail and two globular heads. The heads of myosin molecules

bind to actin in the thin filaments and form cross-bridges between the two filaments.

6. Sarcomere length shortens during muscle contraction. The shortening is caused by thick and thin filaments sliding past one another. Cross-bridge movement is the molecular basis for filament sliding.

7. The cross-bridge cycle is the mechanism by which the chemical energy stored in the high-energy phosphate bond in ATP is converted into force generation.

8. When skeletal muscle is activated, myosin heads bind to actin in the thin filaments, thus forming cross-bridges. In the force-generating step of the cross-bridge cycle, the myosin head rotates 45 degrees as Pi is released. Subsequent dissociation of ADP and binding of ATP causes the cross-bridge to detach.

9. In skeletal and cardiac muscle, Ca^{2+} initiates contraction by binding to troponin C. This causes a movement of tropomyosin that exposes myosin binding sites on actin and permits cross-bridges to form.

10. Smooth muscles do not exhibit the regular striations observed in skeletal and cardiac muscles. The bundles of contractile elements run obliquely with respect to the longitudinal axis of the muscle cell and are not parallel to one another.

11. The diameters and volumes of many hollow organs are controlled by smooth muscles. Thus, many smooth muscles must contract tonically to maintain these diameters and volumes.

12. The contraction of smooth muscles is regulated by the autonomic nerves and by hormones.

13. Smooth muscles are activated by elevation of the $[Ca^{2+}]_i$. This promotes the phosphorylation of myosin RLC, which enables the actin-myosin cross-bridges to form.

14. Smooth muscle contraction is also regulated by mechanisms that alter the sensitivity of the contractile apparatus to Ca^{2+}.

KEY WORDS AND CONCEPTS

- Molecular motor
- Actin
- Microtubule
- Myosin
- Kinesin
- Dynein
- Myofibril
- Transverse tubule (T-tubule)
- Thick and thin filaments
- Sarcomere
- α-Actinin
- Titin
- Cross-bridge
- Nebulin
- Tropomyosin
- Troponin T (TnT), troponin C (TnC), and troponin I (TnI)
- Sliding filament
- Cross-bridge cycle
- Intercalated disks
- Smooth (nonstriated) muscle
- Dense bodies
- Intermediate filaments
- Calmodulin
- Myosin light chain kinase (MLCK)
- Myosin regulatory light chain (RLC)
- Myosin essential light chain
- Myosin light chain phosphatase (MLCP)
- Tonic contraction (tone)
- Ca^{2+}-independent regulatory mechanisms

STUDY PROBLEMS

1. Compare and contrast the mechanisms by which elevation of $[Ca^{2+}]_i$ activates the contractile proteins in skeletal, cardiac, and smooth muscles.

2. Contrast the functions of skeletal, cardiac, and smooth muscle, and discuss how muscle cell structure and organization are related to these functions.

3. Why do you think that mechanisms that regulate the Ca^{2+}-sensitivity of the contractile machinery are especially important in smooth muscle?

BIBLIOGRAPHY

Alberts B, Johnson A, Lewis J, et al: *The molecular biology of the cell*, ed 5, New York, NY, 2008, Garland Science.

Bagby RM: Organization of contractile/cytoskeletal elements. In Stephens NL, editor: *Biochemistry of smooth muscle*, vol 1, Boca Raton, FL, 1983, CRC Press.

Gordon AM, Huxley AF, Julian FJ: The variation in isometric tension with sarcomere length in vertebrate muscle fibres, *J Physiol* 184:170, 1966.

Huxley HE: Structural evidence concerning the mechanism of contraction in striated muscle. In Paul WM, Daniel EE, Kay CM, et al, editors: *Muscle: proceedings of a symposium held at the Faculty of Medicine, University of Alberta*, Oxford, England, 1964, Pergamon.

Huxley HE, Hanson J: The molecular basis of contraction in cross-striated muscles. In Bourne GH, editor: *The structure and function of muscle*, vol 1, New York, NY, 1960, Academic Press.

Kodera N, Yamamoto D, Ishikawa R, et al: Video imaging of walking myosin V by high-speed atomic force microscopy, *Nature* 468:72, 2010.

Krstic RV: *Ultrastructure of the mammalian cell*, New York, NY, 1979, Springer-Verlag.

Leeson CR, Leeson TS: *Histology*, ed 3, Philadelphia, 1976, WB Saunders.

Murphy RA: Smooth muscle. In Berne RM, Levy MN, Koeppen BM, et al, editors: *Physiology*, ed 4, St Louis, MO, 1998, Mosby.

Ruegg JC: Vertebrate smooth muscle. In *Calcium in muscle contraction*, Berlin, Germany, 1992, Springer-Verlag.

Small JV, Sobieszek A: The contractile apparatus of smooth muscle, *Int Rev Cytol* 64:241, 1980.

15

EXCITATION-CONTRACTION COUPLING IN MUSCLE

OBJECTIVES

1. Compare and contrast the mechanisms of excitation-contraction coupling in skeletal, cardiac, and smooth muscle cells.

2. Understand that in skeletal muscle:

   a. All of the Ca^{2+} required for contraction is released from the sarcoplasmic reticulum (SR) through Ca^{2+} release channels.

   b. Movement of a voltage sensor couples sarcolemmal depolarization to the opening of SR Ca^{2+} release channels.

   c. Essentially all of the released Ca^{2+} is pumped back into the SR by the SR Ca^{2+} pump (SERCA).

3. Understand that in cardiac muscle, intracellular Ca^{2+} release channels are opened by Ca^{2+} ions in a process known as Ca^{2+}-induced Ca^{2+} release (CICR).

4. Describe the role of SERCA, the Na^+/Ca^{2+} exchanger (NCX), and the plasma membrane Ca^{2+} pump (PMCA) in the relaxation of cardiac muscle.

5. Understand that smooth muscles are highly diversified and that they can be activated by depolarization, or by agonists through a process known as pharmacomechanical coupling.

6. Understand the different roles of the inositol trisphosphate receptors (IP_3Rs) and ryanodine receptors (RyRs) in smooth muscle.

SKELETAL MUSCLE CONTRACTION IS INITIATED BY A DEPOLARIZATION OF THE SURFACE MEMBRANE

Skeletal muscle fibers are innervated by α motor neurons, which are large neurons (cell body diameter up to 70 μm) that originate in the ventral horn of the spinal cord. As discussed in Chapter 12, a single AP in an α motor neuron causes sufficient ACh release at a single NMJ to trigger an AP that is propagated along an entire muscle fiber.

Skeletal muscle contraction is triggered by *depolarization* of the muscle fiber membrane beyond a critical level, the *mechanical threshold*. Under normal physiological conditions an AP causes this depolarization. The relationship between V_m and the amount of force, or "tension," generated by skeletal muscle is presented in Figure 15-1. The mechanical responses shown in Figure 15-1, which are produced by depolarizations lasting several seconds, are called **contractures.** As V_m becomes more positive than approximately −55 mV (the mechanical threshold), force increases very steeply with further depolarization. The mechanism by which depolarization of the sarcolemma causes contraction of the muscle cell is termed **excitation-contraction coupling** (E-C coupling). The fact that force increases with V_m indicates that the process of E-C coupling involves a **voltage**

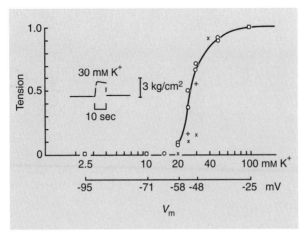

FIGURE 15-1 ■ Relationship between force and membrane potential. The V_m of single muscle fibers was depolarized to different levels by increasing $[K^+]_o$. V_m was measured with an intracellular electrode and the force (tension) developed by the myocyte was measured with a force transducer. The *inset* shows the force resulting from a 10-second depolarization in 30 mM $[K^+]_o$. The graph is a plot of the normalized tension (i.e., the tension at that V_m, divided by the maximum tension) as a function of $[K^+]_o$ or V_m. Depolarizations to voltages more positive than approximately -55 mV result in force production (in the form of contractures). The force increases steeply over a very narrow range of V_m. (*Data from Hodgkin AL, Horowicz P: J Physiol 153:386, 1960.*)

sensor in the sarcolemma that couples depolarization to contraction.

Skeletal Muscle Has a High Resting Cl⁻ Permeability

In skeletal muscle, the resting V_m (~-90 mV) is much more negative than in neurons (~-70 mV) owing to the relatively high permeability of the sarcolemma to both K^+ and Cl^-. Because of the high resting Cl^- permeability, a relatively large stimulus is normally necessary to bring the V_m to mechanical threshold. In some skeletal muscle diseases, loss-of-function mutations in skeletal muscle Cl^- channels drastically reduce the Cl^- permeability of the sarcolemma. As a result, the skeletal muscle AP threshold can be reached more easily and the muscle becomes hyperexcitable (Box 15-1).

A Single Action Potential Causes a Brief Contraction Called a Twitch

The temporal relationship between the skeletal muscle AP and the force generated by the muscle is illustrated in Figure 15-2. The skeletal muscle AP is similar to the nerve axon AP (see Chapter 7) in that it is generated by the activity of voltage-gated Na^+ and K^+ channels. The development of force by the myocyte begins several milliseconds after the AP. The transient contraction produced by a myocyte in response to a single AP is called a **twitch**. During the twitch, the contractile force rises to a peak in 30 to 50 milliseconds and then declines over the next 50 to 100 milliseconds. The duration of the twitch is approximately 100-fold longer than the duration of the AP.

How Does Depolarization Increase Intracellular Ca²⁺ in Skeletal Muscle?

In Chapter 14 we learned that an increase in $[Ca^{2+}]_i$ activates contraction in all types of muscle cells. Therefore, the central question in E-C coupling is, "How does depolarization of the sarcolemma bring about an increase in $[Ca^{2+}]_i$?" A related, but more subtle, question is, "How does the $[Ca^{2+}]_i$ increase fast enough throughout the muscle fiber so that myofibrils deep inside the fiber can contract synchronously with those close to the surface?"

BOX 15-1

BECKER'S MYOTONIA AND THOMSEN'S DISEASE ARE CAUSED BY MUTATIONS IN A Cl⁻ CHANNEL

Two inherited forms of nondystrophic myotonia congenita* involve the Cl⁻ conductance (g_{Cl}) of the skeletal muscle sarcolemma: Thomsen's disease, with autosomal dominant inheritance; and the more severe Becker's myotonia, with autosomal recessive inheritance. Both forms are characterized by attacks of muscle stiffness that are caused by a delay in muscle relaxation after stimulation. Under conditions in which normal muscle responds with a single AP, myotonic muscle cells respond to a single stimulation with repetitive APs. Thus, a single α motor neuron AP evokes a twitch in normal skeletal muscle but a *tetanus* in myotonic muscle. A tetanus is a sustained contraction that is evoked in normal skeletal muscle only when it is stimulated

by a train, or burst, of closely spaced APs (see Chapter 16). The rigidity and delayed relaxation of myotonic muscle are caused by the tetanic response to a *single* stimulus.

The hyperexcitability of myotonic muscle is caused by a mutation in the gene encoding the predominant skeletal muscle Cl⁻ channel, CLCN1. This mutation reduces the Cl⁻ conductance of the skeletal muscle sarcolemma. Because of the reduction in g_{Cl} and, thus, an increase in membrane resistance, a stimulus will produce a larger change in V_m in myotonic muscle than in normal muscle (see Ohm's Law in Chapter 6). As a result, the current required to reach AP threshold is less, thereby making the muscle hyperexcitable.

*Myotonia is muscle rigidity caused by delayed muscle relaxation. It is typical of several distinct diseases, including myotonic dystrophy, which is a relatively common neuromuscular disorder characterized by abnormal expression of CTG-trinucleotide repeats. The diseases discussed in this box are nondystrophic myotonias (i.e., they are not associated with muscle wasting).

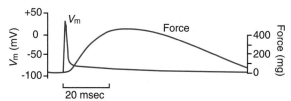

FIGURE 15-2 ■ The temporal relationship between the AP and a twitch contraction. The trace labeled V_m shows the time course of the skeletal muscle AP. The twitch contraction (trace labeled force) begins a few milliseconds after the AP. Peak twitch force is reached in approximately 30 milliseconds, and the muscle then relaxes over the next 50 milliseconds. After the peak of the AP, the sarcolemma initially repolarizes rapidly. This is followed by a prolonged period of much slower repolarization called an afterdepolarization, which is caused by accumulation of K⁺ in the lumen of the T-tubule during fast repolarization. This transiently increases E_K.

DIRECT MECHANICAL INTERACTION BETWEEN SARCOLEMMAL AND SARCOPLASMIC RETICULUM MEMBRANE PROTEINS MEDIATES EXCITATION-CONTRACTION COUPLING IN SKELETAL MUSCLE

In Skeletal Muscle, Depolarization of the T-Tubule Membrane Is Required for Excitation-Contraction Coupling

Is diffusion of a soluble factor from the surface membrane (e.g., Ca^{2+} ions entering the cell through voltage-gated Ca^{2+} channels) to the interior of the cell sufficiently fast to explain the rapid activation of skeletal muscle? A molecule would take about a second to diffuse to the center of a 100-μm diameter skeletal muscle cell (see Chapter 2). This is much too slow to account for the development of force during a twitch contraction, which begins just a few milliseconds after the AP (Figure 15-2).

How, then, does the AP in the sarcolemma rapidly activate the myofibrils in the center of the cell? The critical clue was the discovery that "hot spots" distributed over the sarcolemma (later shown to be T-tubule openings)

FIGURE 15-3 ■ The extensive network of T-tubules in skeletal muscle. A Golgi stain was used to infiltrate the T-tubule system from the extracellular space. The resulting black areas in this electron micrograph clearly delineate the extensive network of tubules. Longitudinal elements of the T-tubules interconnect transverse elements. The enlarged areas of the tubules are regions of functional coupling between the T-tubule and sarcoplasmic reticulum membranes. *(Courtesy of D. Appelt and C. Franzini-Armstrong.)*

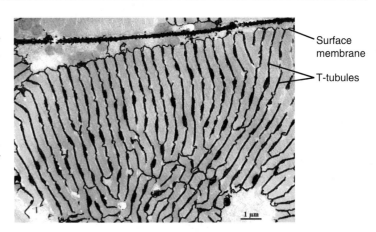

Surface membrane

T-tubules

provide a pathway for depolarization to spread from the surface into the interior of the fiber. T-tubules form an intricate network that extends throughout the skeletal muscle cell (Figure 15-3). Thus, the T-tubule lumen is a narrow extension of the extracellular space into the interior of the muscle cell. The T-tubule membrane contains voltage-gated Na^+ and K^+ channels, and APs from the surface are propagated along the T-tubule membrane into the interior of the cell.

In Skeletal Muscle, Extracellular Ca^{2+} Is *Not* Required for Contraction

The T-tubule membrane of skeletal muscle contains receptors for the dihydropyridine derivatives that block L-type voltage-gated Ca^{2+} channels (see Chapter 8). Voltage clamp studies in skeletal muscle demonstrate that inward Ca^{2+} currents are generated by depolarization (Figure 15-4) and that these currents are blocked by dihydropyridines. Thus, the **dihydropyridine receptors (DHPRs)** in the T-tubule membrane are L-type Ca^{2+} channels. Therefore, we could reasonably expect that the Ca^{2+} entry through these channels raises $[Ca^{2+}]_i$ and activates contraction, but this is *not* the case. Skeletal muscle continues to contract normally when bathed in a solution containing *no* Ca^{2+} ions. This finding indicates that Ca^{2+} entry is not essential for skeletal muscle contraction and that *all* the Ca^{2+} required for activating the contractile machinery is derived from intracellular sources. This is in marked contrast to the mechanism of E-C coupling in cardiac muscle, where Ca^{2+} entry through voltage-gated Ca^{2+} channels is required (see later). We shall see, however, that even though Ca^{2+}

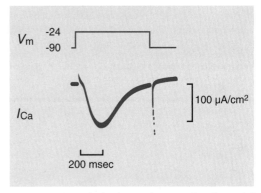

V_m −24 −90

I_{Ca}

100 μA/cm²

200 msec

FIGURE 15-4 ■ Depolarization activates a slow inward Ca^{2+} current (I_{Ca}) in skeletal myocytes. A voltage clamp was used to record I_{Ca} after current flow through other channels was eliminated. In response to a voltage clamp step from −90 to −24 mV, an inward I_{Ca} developed slowly, peaking in approximately 200 milliseconds. Surprisingly, although L-type Ca^{2+} channels generate I_{Ca} in skeletal muscle, their activation is nearly two orders of magnitude slower than in other cell types (see Figure 8-2). *(Data from Sanchez JA, Stefani E: J Physiol 37:1, 1983.)*

entry through DHPRs is not required for E-C coupling in skeletal muscle, the DHPRs *do* play a crucial role in this mechanism.

In Skeletal Muscle, the Sarcoplasmic Reticulum Stores All the Ca^{2+} Needed for Contraction

The SR is an extensive intracellular membrane enclosed system of sacs and tubules that envelopes the myofibrils in skeletal muscle cells (see Figure 14-1).

That the SR surrounds all myofibrils has long suggested that it plays an important role in skeletal muscle contraction. The SR is a specialized ER with copious expression of three proteins: SERCA (see Chapter 11); the **ryanodine receptor (RyR),** which is the *Ca^{2+} release channel* in the SR membrane; and **calsequestrin.** Virtually all the Ca^{2+} released during a contraction is rapidly and efficiently transported back into the SR by the extremely abundant SERCA in the SR membrane. The rapid removal of Ca^{2+} from the cytoplasm causes relaxation. Because essentially all the Ca^{2+} is recycled in this way, extracellular Ca^{2+} is not required in the short term, and skeletal muscle contraction can proceed even in the absence of extracellular Ca^{2+}. The capacity of the SR to store Ca^{2+} is greatly enhanced by calsequestrin, the Ca^{2+}-binding protein that is highly expressed in the SR lumen and acts as a Ca^{2+} buffer.

The SR Ca^{2+} release channels are called RyRs because they avidly bind ryanodine, a paralyzing plant alkaloid. There are three RyR isoforms: RyR1 in adult skeletal muscle; RyR2 in cardiac muscle; and RyR3 in embryonic skeletal muscle, brain (in ER), and other tissues. Each Ca^{2+} release channel is a tetramer of four identical RyR protein subunits. Malignant hyperthermia is a genetic disease of skeletal muscle caused by mutations in RyR1 (Box 15-2).

The Triad Is the Structure That Mediates Excitation-Contraction Coupling in Skeletal Muscle

The SR and T-tubule membranes in skeletal muscle are closely apposed at specialized junctions called **triads** (Figure 15-5). The triad consists of a T-tubule that is flanked on either side by enlarged terminal sacs of the SR, called **terminal cisterns**. The RyRs are located in the terminal cisterns. A large cytoplasmic domain of the RyR spans the narrow gap between the T-tubule and SR membranes. These domains are visible by electron microscopy and have been termed *junctional feet* (Figure 15-5). At the triads, the T-tubule membrane contains clusters of DHPRs that are located in precise register with the RyR subunits in the underlying SR membrane (Figure 15-6). The DHPRs are grouped into tetrads, and each of the four DHPR molecules in the T-tubule membrane is aligned with, and apposed to,

---

BOX 15-2

MALIGNANT HYPERTHERMIA IS CAUSED BY A SUSTAINED INCREASE OF INTRACELLULAR Ca^{2+} IN SKELETAL MUSCLE

Malignant hyperthermia (MH) is a skeletal muscle disorder characterized by hypermetabolism. MH is triggered by exposure to volatile, halogenated anesthetics, such as halothane. In children undergoing anesthesia, the incidence is 1 in 3000 to 15,000—about one case in a busy metropolitan hospital every other year. Although relatively rare, MH is life-threatening.

MH is characterized by an uncontrolled, sustained increase in $[Ca^{2+}]_i$ in skeletal muscle. This results in hypermetabolism: as cross-bridges continuously cycle and SERCA attempts to restore normal $[Ca^{2+}]_i$, ATP is rapidly consumed in both processes. The maintained high $[Ca^{2+}]_i$ produces muscle rigidity. The rapid ATP consumption leads to a dramatic increase in aerobic and anaerobic metabolism, an increase in CO_2 production, and a sustained rise in body temperature. If untreated, this clinical syndrome is usually fatal. The skeletal muscle relaxant dantrolene is a lifesaving therapy. This drug, which inhibits SR Ca^{2+} release through the RyRs, rapidly reduces the uncontrolled $[Ca^{2+}]_i$ rise in skeletal muscle and suppresses the hypermetabolism that threatens the patient with MH.

MH is genetically heterogeneous; most cases involve a point mutation in the gene encoding the skeletal muscle SR Ca^{2+} release channel (RyR1). The altered channels exhibit several abnormalities that enhance Ca^{2+} release from the SR. For example, MH muscle is more sensitive than normal muscle to depolarization and to caffeine, both of which open RyRs. The standard test for MH susceptibility involves measuring contracture in a muscle sample on exposure to caffeine. Compared with normal muscle, muscle from patients with MH produces larger caffeine contractures and the contractures are triggered at a lower caffeine concentration.

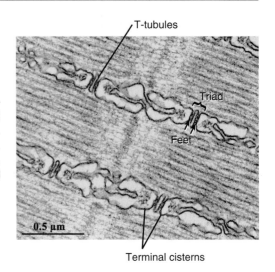

FIGURE 15-5 ■ A triad in skeletal muscle consists of a T-tubule sandwiched between two terminal cisterns of the SR. This longitudinal electron microscopic section shows the three elements of the triad. The flattened, oval elements are cross-sectional views of T-tubules. On either side of each T-tubule is a terminal cistern of the SR. Junctional "feet" *(arrows)* span the space between the T-tubule and SR membranes. *(Courtesy of C. Franzini-Armstrong.)*

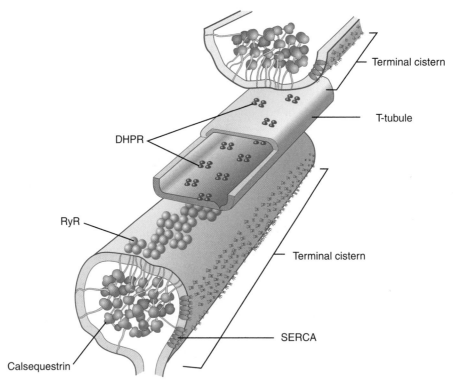

FIGURE 15-6 ■ The three-dimensional structure of a triad. Four RyR subunits in the SR membrane form the Ca^{2+} release channel. These subunits are the "foot" proteins that connect the SR terminal cistern membrane to the cytoplasmic face of the T-tubule. In the T-tubule membrane, DHPRs are grouped into tetrads, and each member of the tetrad is apposed to an RyR subunit. Every second RyR tetramer, however, lacks the opposing tetrad of DHPRs. Calsequestrin and SERCA are also shown (see text). *(Modified from Block BA, Imagawa T, Campbell KP, et al: J Cell Biol 107:2587, 1988.)*

one subunit of the RyR tetramer in the SR membrane. Alternate RyR tetramers are, however, unapposed by DHPRs (Figure 15-6). This precise spatial organization of RyRs and DHPRs, which does not occur in cardiac muscle, is the structural basis of the E-C coupling mechanism in skeletal muscle.

In Skeletal Muscle, Excitation-Contraction Coupling Is Mechanical

In response to depolarization of the T-tubule membrane, the RyRs open to allow Ca^{2+} to flow out of the SR, down its concentration gradient, into the cytoplasm, where it binds to TnC and activates contraction. The central question in E-C coupling is this: How does depolarization of the T-tubule membrane open the SR Ca^{2+} release channel? The finding that the SR membrane is electrically isolated from the T-tubule membrane rules out the possibility that depolarization spreads directly from the T-tubule to the SR. Two classes of indirect coupling models have been proposed. First, T-tubule membrane depolarization could generate a *chemical* messenger that diffuses to and activates the RyRs. Alternatively, the coupling mechanism could be a *mechanical* link between a voltage sensor in the T-tubule membrane and the gating mechanism of the RyR in the SR membrane. The available evidence supports the mechanical coupling model in skeletal muscle.

The DHPR is homologous to the voltage-gated Na^+ channel. Like the Na^+ channel (see Chapter 5), the DHPR has a positively charged, membrane-spanning segment (S4 region) that acts as the voltage sensor. In the T-tubules of skeletal muscle the DHPR also acts as the voltage sensor for E-C coupling. Importantly, its function as a voltage sensor is *independent* of its ability to conduct Ca^{2+} ions. The movement of the positive charges in the S4 segment of the DHPR in response to depolarization generates a transient capacitive current that can be measured with a voltage clamp (Box 15-3). This charge movement is an essential first step in E-C coupling in skeletal muscle. The relationships among voltage-dependent activation of DHPRs, charge movement, and E-C coupling have been demonstrated in animals expressing mutated DHPRs (Box 15-4). These studies suggest that the movement of the voltage sensor in the DHPR is mechanically linked to the opening of the RyR Ca^{2+} release channel (Figure 15-7). The main elements of E-C coupling in skeletal muscle are illustrated in Figure 15-8.

Skeletal Muscle Relaxes When Ca^{2+} Is Returned to the Sarcoplasmic Reticulum by SERCA

As $[Ca^{2+}]_i$ increases during contraction, Ca^{2+} ions bind not only to TnC, but also to SERCA, which immediately begins to pump Ca^{2+} back into the SR. SERCA has higher affinity for Ca^{2+} than does TnC. Thus, after the RyRs have closed, SERCA restores $[Ca^{2+}]_i$ to its resting level and brings about relaxation. Essentially all the Ca^{2+} that is released from the SR during a contraction is pumped back into the SR by SERCA.

BOX 15-3

NONLINEAR CHARGE MOVEMENT REFLECTS MOVEMENT OF THE VOLTAGE SENSOR IN SKELETAL MUSCLE

In skeletal muscle, the voltage clamp can be used to measure a nonlinear component of the capacitive current that is qualitatively similar to the Na^+ channel gating current (see Box 7-3). A depolarizing voltage step generates an outward current. This current reflects the movement of charged amino acid residues in the S4 segment of the DHPR that act as the voltage sensor for E-C coupling in skeletal muscle. Several lines of evidence indicate that this charge movement is related to E-C coupling. For example, the magnitude of the charge movement increases with the size of the depolarization and this occurs over the same range of potentials where muscle force increases (see Figure 15-1). Furthermore, Ca^{2+} release from the SR is tightly correlated with charge movement. Finally, no experimental or physiological conditions have been found that eliminate charge movement while maintaining depolarization-induced SR Ca^{2+} release in skeletal muscle.

BOX 15-4

A DYSGENIC MUSCLE MODEL REVEALS THAT DIHYDROPYRIDINE RECEPTORS ARE THE VOLTAGE SENSORS FOR EXCITATION-CONTRACTION COUPLING IN SKELETAL MUSCLE

An animal model has provided some of the evidence that DHPRs act as the voltage sensors for E-C coupling in skeletal muscle. A lethal autosomal recessive mutation in mice, termed muscular dysgenesis *(mdg),* is a loss-of-function mutation in the skeletal muscle DHPR gene and is functionally expressed as a failure of E-C coupling. The α-subunit of the DHPR is missing and the channel does not conduct inward Ca^{2+} current. Expression of other proteins remains unchanged, and the skeletal muscle cells are normal in all other respects.

The nonlinear charge movement (see Box 15-3) is also absent in *mdg* skeletal muscle, a finding suggesting that the DHPRs are the source of the charge movement. When *mdg* skeletal muscle cells are injected with cDNA encoding the normal skeletal muscle DHPR, normal E-C coupling, inward Ca^{2+} current, *and* the nonlinear charge movement are all restored. These findings support the idea that the DHPR is the voltage sensor that generates the charge movement, which is then coupled to the opening of the RyRs of skeletal muscle.

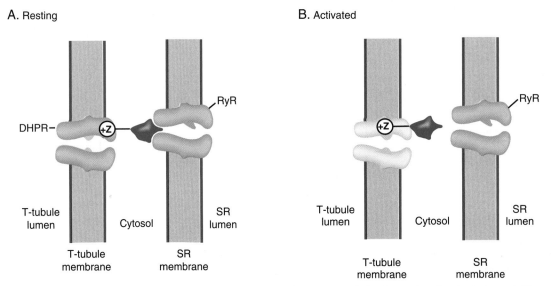

A. Resting

B. Activated

FIGURE 15-7 ■ Coupling of charge movement to the opening of an RyR in skeletal muscle. The T-tubule and SR membranes separate the T-tubule lumen, the cytosol, and the SR lumen. **A,** The voltage sensor in the DHPR, with charge of +Z, is attached to a plug that moves to open or close the RyR. At rest, the negative V_m drives the voltage sensor to the inner face of the T-tubule membrane, and the plug blocks the RyR. **B,** On depolarization, as V_m becomes more positive, the voltage sensor moves outward, pulling the plug to open the RyR channel. *(Modified from Chandler WK, Rakowski RF, Schneider MF:* J Physiol *254:285, 1976.)*

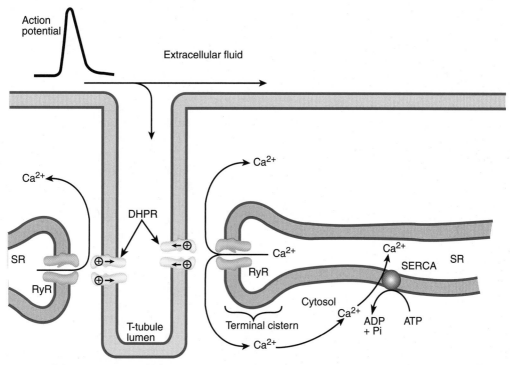

FIGURE 15-8 ■ Cellular components of E-C coupling and relaxation in skeletal muscle. An AP initiated at the neuromuscular junction is normally the first event in E-C coupling. The AP rapidly propagates over the sarcolemma and into the T-tubules. When the T-tubule is depolarized, the voltage sensors ($\oplus$) in the DHPRs move and open the RyRs, thus permitting Ca^{2+} to flow out of the SR into the cytosol to trigger contraction. The Ca^{2+} released from the SR is subsequently returned to the SR by SERCA to relax the muscle.

Ca^{2+}-INDUCED Ca^{2+} RELEASE IS CENTRAL TO EXCITATION-CONTRACTION COUPLING IN CARDIAC MUSCLE

Despite general similarities, cardiac and skeletal muscle E-C coupling mechanisms differ in important respects. Contraction of a cardiac cell is initiated by an AP that rapidly propagates over the surface of the cell and into the T-tubules. In cardiac, as in skeletal muscle, the sarcolemmal and T-tubule AP leads to the opening of the RyRs in the SR, thus, allowing Ca^{2+} to flow out of the SR, bind to TnC, and activate contraction. The mechanism by which the AP is coupled to the opening of RyRs in cardiac myocytes is, however, entirely different from the charge-coupled (electromechanical) mechanism just described for skeletal muscle. Moreover, initiation

of the cardiac sarcolemmal AP does not occur through neuromuscular transmission, as it does in skeletal muscle. Instead, the sinoatrial node of the heart contains special pacemaker cells (**sinoatrial pacemaker cells**) that spontaneously generate APs. Each AP is conducted from myocyte to myocyte throughout the heart because all the cells are electrically coupled through gap junctions (see Chapter 12 and Figure 14-10).

In Cardiac Muscle, Communication Between the Sarcoplasmic Reticulum and Sarcolemma Occurs at Dyads and Peripheral Couplings

In cardiac cells the T-tubule system is less extensive than in skeletal muscle, and the T-tubule diameter is much larger (see Figure 14-10). The SR in cardiac muscle is less regular and less extensive than in skeletal muscle. The sarcolemma and the SR membrane are functionally coupled and, as in skeletal muscle, the cytoplasmic domains of RyRs span the narrow gap (<20 nm) between the two membranes. In cardiac myocytes the functional interaction between the sarcolemma and the SR membrane occurs either at a **dyad** formed between a T-tubule and a single flattened terminal cistern of the SR, or at a **peripheral coupling** formed between the sarcolemma and a subsarcolemmal cistern of the SR. At these sites, the sarcolemmal DHPRs are close to the RyRs in the SR, but these proteins do not exhibit the precise spatial organization observed in skeletal muscle.

Cardiac Excitation-Contraction Coupling Requires Extracellular Ca^{2+} and Ca^{2+} Entry Through L-Type Ca^{2+} Channels (Dihydropyridine Receptors)

The activation of cardiac contraction requires *both* Ca^{2+} entry from the extracellular space and Ca^{2+} release from the SR (Figure 15-9); removal of extracellular Ca^{2+} abolishes cardiac contraction. The cardiac sarcolemma contains DHPRs that open rapidly (2 to 3 milliseconds) during the AP. This is fast enough to allow Ca^{2+} influx through the DHPR channels to activate contraction directly. This Ca^{2+} influx is also crucial in activating RyRs, which mediate Ca^{2+} release from the SR. This process, known as **Ca^{2+}-induced Ca^{2+} release (CICR)**, is the predominant component of cardiac E-C

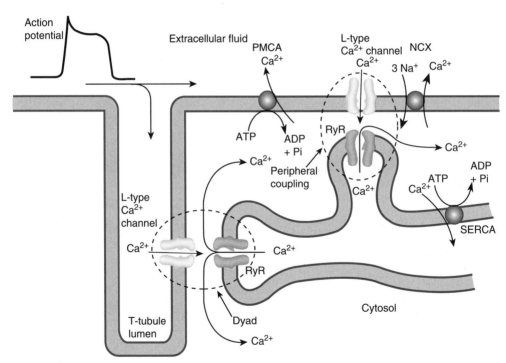

FIGURE 15-9 ■ Cellular components of E-C coupling and relaxation in cardiac muscle. The cardiac AP opens DHPRs in both surface and T-tubule membranes, thus allowing Ca^{2+} to enter the cell. At dyads and peripheral couplings (enclosed by *dashed ellipses*), Ca^{2+} entering through DHPRs binds to and thereby opens RyR channels on the SR. Of the Ca^{2+} required to activate contraction in cardiac muscle, ~80% comes from the SR and ~20% comes from the extracellular space. During relaxation (diastole), ~80% of the Ca^{2+} is transported into the SR, whereas the NCX and PMCA extrude ~15% and ~5%, respectively.

coupling—it accounts for approximately 80% of the Ca^{2+} required for contraction (Box 15-5). The close association between DHPRs in the sarcolemma and RyR Ca^{2+} release channels in the SR membrane suggests a functional coupling between these proteins, as illustrated in Figure 15-10. Depolarization of the sarcolemma opens DHPRs and permits Ca^{2+} to enter.

This results in a local $[Ca^{2+}]_i$ increase that activates those RyRs in the immediate vicinity of the DHPRs. The Ca^{2+} released through this small group of RyRs can be visualized as a **Ca^{2+} spark** (Box 15-6). When they are triggered synchronously, many of these Ca^{2+} release events combine to generate a global $[Ca^{2+}]_i$ rise that activates contraction.

BOX 15-5

IN CARDIAC MUSCLE, BUT NOT SKELETAL MUSCLE, Ca^{2+}-INDUCED Ca^{2+} RELEASE IS THE PRIMARY MECHANISM OF EXCITATION-CONTRACTION COUPLING

Comparison of the voltage dependence of I_{Ca}, $[Ca^{2+}]_i$, and the activation of contraction in cardiac (Figure B-1A) and skeletal (Figure B-1B) muscle provides key evidence that Ca^{2+}-induced Ca^{2+} release (CICR) is crucial in cardiac E-C coupling. In cardiac muscle, as V_m moves from the resting level to more positive values, I_{Ca} increases, reaches a maximum, and then decreases in amplitude (see Figure 8-1). I_{Ca} increases because the number of open voltage-gated Ca^{2+} channels increases as the membrane is depolarized. At still more positive potentials, I_{Ca} decreases because of a reduction in the driving force ($V_m - E_{Ca}$). In cardiac muscle, $[Ca^{2+}]_i$ and cell shortening have a voltage dependence that is similar

to I_{Ca} (Figure B-1A): they reach a maximum value at about the same voltage and then decrease in size as the V_m becomes more positive. The finding that $[Ca^{2+}]_i$, and cell shortening are correlated with the amplitude of I_{Ca} is part of the evidence that CICR is the primary mechanism of E-C coupling in cardiac muscle. In skeletal muscle, which does not depend on CICR for E-C coupling, the voltage dependence of $[Ca^{2+}]_i$ and tension do not mirror the voltage dependence of I_{Ca} (Figure B-1B). As V_m becomes depolarized from the resting level, $[Ca^{2+}]_i$ and tension increase. Further increasing V_m does not reduce $[Ca^{2+}]_i$ or tension, but it does reduce I_{Ca} (Figure B-1B).

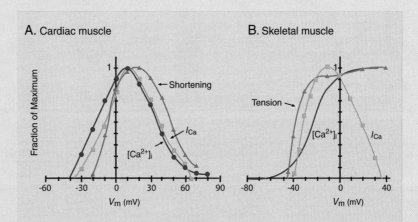

FIGURE B-1 ■ Comparison of the voltage dependence of I_{Ca}, $[Ca^{2+}]_i$, and contraction (shortening or tension) in cardiac and skeletal muscle. **A,** Each parameter was measured in voltage clamp experiments on isolated ventricular muscle cells. Each set of values was normalized and plotted as a function of V_m. **B,** I_{Ca}, $[Ca^{2+}]_i$, and tension were measured in voltage clamp experiments on skeletal muscle cells. *(Modified from Bers DM: Excitation-contraction coupling and cardiac contractile force, Dordrecht, The Netherlands, 1991, Kluwer.)*

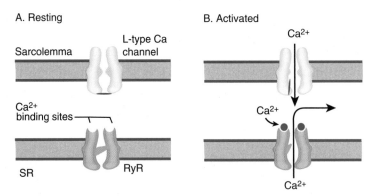

FIGURE 15-10 ■ Functional coupling between DHPRs and RyRs in cardiac muscle. DHPRs in the sarcolemma, at dyads and peripheral couplings, are located directly opposite RyRs in the SR. **A,** In the resting state, the gates on both channels are closed. **B,** When a cardiac AP opens DHPRs, Ca^{2+} entering through these channels binds to, and opens, RyR channels, through which Ca^{2+} is released from the SR into the cytosol. Although two Ca^{2+} binding sites are shown on the RyR, the precise stoichiometry is unknown.

BOX 15-6

Ca^{2+} SPARKS ARE LOCALIZED INCREASES IN INTRACELLULAR $[Ca^{2+}]$ TRIGGERED BY THE OPENING OF A SINGLE DHPR/L-TYPE Ca^{2+} CHANNEL IN CARDIAC MUSCLE

In living cells, $[Ca^{2+}]_i$ can be measured with fluorescent Ca^{2+} indicator dyes. One such indicator, Fluo-3, becomes brightly fluorescent only when it binds Ca^{2+}. When Fluo-3 is monitored in single cardiac myocytes with a laser-scanning confocal microscope, *localized* increases in $[Ca^{2+}]_i$ can be observed as small bursts of fluorescence called *Ca^{2+} sparks*. In resting cardiac cells, Ca^{2+} sparks occur spontaneously at a relatively low frequency: they increase in frequency with membrane depolarization. The Ca^{2+} sparks result from the transient opening of a small cluster of RyRs. A comparison of the properties (e.g., the voltage dependence) of Ca^{2+} sparks with the properties of DHPRs reveals that a Ca^{2+} spark can be triggered by the Ca^{2+} entering through a single, open nearby DHPR. The global increase in $[Ca^{2+}]_i$ that occurs during the cardiac action potential is the result of the synchronized triggering of myriad Ca^{2+} sparks throughout the cell.

Ca^{2+} That Enters the Cell during the Cardiac Action Potential Must Be Removed to Maintain a Steady State

When $[Ca^{2+}]_i$ increases during contraction, TnC and three Ca^{2+} transport proteins all compete for Ca^{2+} ions. The transporters are PMCA, NCX, and SERCA. The excess Ca^{2+} ions in the cytoplasm are transported out of the cell primarily by the NCX, but the PMCA also contributes. Most, however, is resequestered in the SR by SERCA. Thus, both NCX and SERCA lower $[Ca^{2+}]_i$; cause the relaxation of cardiac muscle, and maintain long-term Ca^{2+} balance.

Cardiac Contraction Can Be Regulated by Altering Intracellular Ca^{2+}

The contractile force of cardiac muscle can be regulated by mechanisms that change $[Ca^{2+}]_i$ to meet physiological demands. For example, partial inhibition of Na^+ pumps with cardiotonic steroids indirectly increases the SR Ca^{2+} content and, thus, the amount of Ca^{2+} released during the AP (see Chapter 11). In another example of physiological regulation, activation of β-adrenergic receptors in the cardiac sarcolemma increases DHPR open probability (see Chapter 8). This increases Ca^{2+} influx during the AP, leads to a larger $[Ca^{2+}]_i$ rise, and thereby augments contraction.

SMOOTH MUSCLE EXCITATION-CONTRACTION COUPLING IS FUNDAMENTALLY DIFFERENT FROM THAT IN SKELETAL AND CARDIAC MUSCLES

Smooth Muscles Are Highly Diverse

Contraction in smooth muscles, as in skeletal and cardiac muscles, is triggered by a rise in $[Ca^{2+}]_i$ (see Chapter 14). The mechanisms of smooth muscle activation differ, however, from those in skeletal and cardiac muscles and even vary considerably among the many diverse types of smooth muscle. Some smooth muscles (e.g., intestinal and vascular) are extensively innervated by sympathetic or parasympathetic neurons. These neurons do not form specialized synapses comparable to the NMJs of skeletal muscle. Instead, the neurons have numerous varicosities along their axons. Synaptic vesicles containing neurotransmitter molecules are clustered in the varicosities, which are the sites

of transmitter release (see Chapter 12). Different types of smooth muscle express receptors for one or more neurotransmitters (e.g., ACh, NE, purines, peptides, and amino acids; see Chapter 13). The receptors tend to be clustered in microdomains of the smooth muscle cell sarcolemma close to the neuronal varicosities.

The Density of Innervation Varies Greatly among Different Types of Smooth Muscles

Some smooth muscles, such as those in many blood vessels, are densely innervated by sympathetic neurons. The GI tract has its own **enteric nervous system** that includes sympathetic, parasympathetic, and sensory neurons. This system is needed to coordinate the contractions of the circular and longitudinal smooth muscle layers to achieve the complex motility of the GI tract known as **peristalsis.** Peristalsis is responsible for mixing the contents of the intestinal lumen and for propelling the contents along the GI tract. When neuronal ganglion cells of the colon fail to develop (*aganglionic colon*), the results are intestinal obstruction, the absence of defecation, and massive distention of the colon (*megacolon*) early in infancy (Hirschsprung's disease; Box 15-7).

At the other extreme are some smooth muscles, such as uterine smooth muscle, that have little or no neural innervation. In this case, the smooth muscles

---

BOX 15-7

HIRSCHSPRUNG'S DISEASE (AGANGLIONIC MEGACOLON) RESULTS FROM THE FAILURE OF COLON GANGLION CELLS TO DEVELOP

Congenital aganglionic megacolon (Hirschsprung's disease) is manifested, usually in newborns or young infants, as a massive distention of the colon. This condition is the result of the absence of ganglion cells in a small, distal segment of the colon. This segment of the colon is therefore unable to undergo reflex dilation and remains constricted when the more proximal, normally innervated segments are distended. As a consequence, fecal material accumulates and greatly distends the proximal segments.

In approximately 1 in 5000 live births, colon ganglion cells fail to develop. At least half of the cases can

be attributed to defects in the genes that encode a receptor tyrosine kinase (RET) or endothelin 3 (EDN3) or its receptor, the endothelin B receptor (EDNRB). The endothelins are a family of oligopeptide hormones. EDN3 is synthesized by glial cells and is a neurotrophic factor (i.e., a substance that promotes the survival and development of nerve cells). During embryonic development, RET, EDN3, and EDNRB are all required for the migration of neural crest-derived cells to the intestinal wall and for their subsequent development into the colon ganglion cells of the enteric nervous system.

are activated by circulating and locally released factors. For example, the pregnant uterus at term (i.e., the time of delivery at the end of fetal gestation) is activated by the posterior pituitary gland hormone *oxytocin*, by locally released *prostaglandins*, and by several other substances. These agents, too, act on specific PM receptors to initiate contraction. Circulating and locally released substances, other than neurotransmitters, can also activate smooth muscles that receive abundant neuronal innervation.

Some smooth muscles in the walls of hollow organs can be activated by stretch or increased wall pressure. Activation occurs, in part, by opening stretch-activated cation channels (members of the transient receptor potential [TRP] channel family; see Chapter 8). For example, increased intraluminal pressure slightly depolarizes vascular smooth muscle cells, increases Ca^{2+} entry, and activates contraction in small arteries, although the detailed mechanisms are still incompletely understood. This pressure-activated, sustained vascular constriction, known as **myogenic tone** (Figure 15-11), is needed to maintain a relatively constant blood flow even when blood pressure varies.

Some Smooth Muscles Are Normally Activated by Depolarization

Some smooth muscles (e.g., intestinal smooth muscles) are excitable: activation of contraction is triggered by APs or by rhythmic changes in V_m. This type of smooth muscle is often referred to as **phasic smooth muscle.** The depolarization opens L-type voltage-gated Ca^{2+} channels, and the entering Ca^{2+} then triggers Ca^{2+} release from the SR through RyRs (i.e., CICR). In this case Ca^{2+} entry is essential for triggering contraction, as in cardiac muscle.

In many instances the smooth muscle cells are electrically connected to one another by gap junctions. Such electrical coupling allows depolarization to propagate from cell to cell (as in the heart). This enables many cells to act in a coordinated fashion, as a syncytium, to alter organ lumen diameter or volume. Smooth muscles that function in this way are known as *single-unit smooth muscles* (e.g., intestinal smooth muscles). In contrast, smooth muscles in which the individual myocytes act independently are called *multiunit smooth muscles* (e.g., the iris diaphragm of the eye).

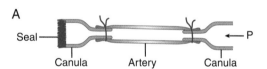

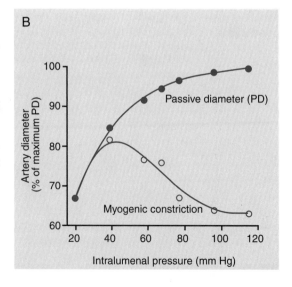

FIGURE 15-11 ■ Myogenic tone in an isolated, cannulated, and pressurized mouse mesenteric small artery. **A,** A small artery is cannulated at both ends. One canula is sealed, and pressure (P) is applied at the other canula. **B,** Pressure-diameter relationships. *Solid circles,* Passive external diameter (PD) measured as a function of intraluminal pressure in Ca^{2+}-free medium; PD at 120 mm Hg is taken as the maximum PD. *Open circles,* Relative diameter (i.e., relative to maximum PD) as a function of pressure in normal medium containing 2.5 mM Ca^{2+}. Myogenic tone is the vasoconstriction observed in the Ca^{2+}-containing solution, relative to that in Ca^{2+}-free solution. *(Courtesy of J. Zhang.)*

Some Smooth Muscles Can Be Activated without Depolarization by Pharmacomechanical Coupling

Smooth muscles that do not normally generate APs are known as **tonic smooth muscles.** These smooth muscles also express L-type Ca^{2+} channels, but they have few, if any, voltage-gated Na^+ channels. This type of smooth muscle (e.g., vascular smooth muscles) often has relatively small resting V_m (e.g., –55 to –40 mV), at which some Ca^{2+} channels are already open. Because the number of open channels depends on V_m, small changes in V_m can significantly alter Ca^{2+} entry by changing the number of open Ca^{2+} channels. Such variations in Ca^{2+} entry may in turn modulate the strength of the tonic contraction that occurs under "resting" (or quasi–steady-state) conditions (Figure 15-11). Thus, incremental changes in V_m may lead to graded changes in tonic contraction (Figure 15-12).

A major mechanism of tonic smooth muscle activation is called **pharmacomechanical coupling.** In contrast to electrical E-C coupling, which involves depolarization, pharmacomechanical coupling involves the activation of smooth muscles by various agonists that often have little or no effect on V_m. These agonists are usually neurotransmitters or hormones; some examples are NE, ACh, serotonin, histamine, NO, vasopressin, angiotensin, and oxytocin. Specificity of response depends on the specific neurotransmitter and hormone receptors expressed by a smooth muscle; most smooth muscles express multiple receptor types. Some transmitters or hormones antagonize contraction and relax smooth muscles. In the intestine, for example, ACh activates contraction, whereas NE promotes relaxation by stimulating adenylyl cyclase and thereby activating Na^+ pumps to lower $[Na^+]_i$ and promote NCX-mediated Ca^{2+} extrusion. Conversely, NE directly

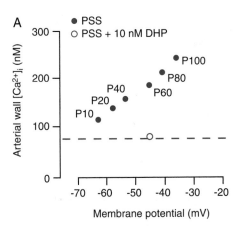

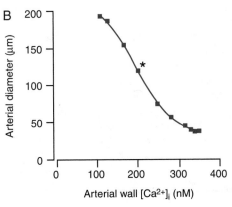

FIGURE 15-12 ■ Graded changes in smooth muscle membrane potential cause graded changes in $[Ca^{2+}]_i$ and, consequently, contraction. **A,** Relationship between arterial wall $[Ca^{2+}]_i$ and myocyte V_m. The intra-arterial pressure was varied between 10 and 100 mm Hg (P10 to P100), and $[Ca^{2+}]_i$ and V_m were both measured at each pressure *(filled circles)*. The normal, in vivo, pressure is 60 mm Hg. Blockade of L-type Ca^{2+} channels by the dihydropyridine (DHP), nisoldipine, prevented Ca^{2+} influx irrespective of V_m *(dashed horizontal line); the open circle* shows the measured $[Ca^{2+}]_i$ at 60 mm Hg in the presence of nisoldipine. PSS, normal physiological salt solution. **B,** Relationship between arterial wall $[Ca^{2+}]_i$ and artery diameter. The arterial pressure was maintained at 60 mm Hg. Changing V_m by varying $[K^+]_o$ affected both $[Ca^{2+}]_i$ and diameter *(squares)*. The *asterisk* marks the data point acquired at the normal resting condition, with $[K^+]_o = 6$ mM. An experimental setup similar to that in Figure 15-11A was used to obtain the data in isolated small cerebral arteries with myogenic tone. *(Data from Knot HJ, Nelson MT: J Physiol 508:199, 1998.)*

activates arterial smooth muscle contraction, whereas ACh promotes relaxation by acting on the vascular endothelium that lines the lumen of blood vessels. ACh triggers the secretion of NO by the endothelial cells. NO, by stimulating the intracellular production of cGMP, relaxes the adjacent arterial smooth muscle. cGMP promotes smooth muscle relaxation by reducing the sensitivity of the contractile proteins to Ca^{2+} (see Box 14-2). Indeed, this is the basis of NO's role in promoting penile erection, as well as the mechanism of action of sildenafil (Viagra) in treating erectile dysfunction (Box 15-8). It is also the basis of action of nitroglycerin and other organic nitrates, which have long been used to treat *angina pectoris,* the chest pain associated with obstructed coronary arteries (Box 15-8). This may be contrasted with the relief of angina pectoris by L-type Ca^{2+} channel blockers that dilate arteries by reducing Ca^{2+} influx into coronary artery myocytes.

In pharmacomechanical coupling, an agonist binds to its specific GPCR to trigger a signaling cascade and the generation of the second messengers, IP_3 and DAG (see Figure 13-2). DAG usually induces transient opening of ROCs (see Chapter 8) that mediate Ca^{2+} entry, which can activate contraction. The IP_3 generated at the PM diffuses to nearby SR, where it binds to and opens the IP_3R, which mediates Ca^{2+} release from the SR into the cytosol. The consequent $[Ca^{2+}]_i$ rise promotes smooth muscle contraction through the activation of MLCK (see Figure 14-12).

Smooth muscles also express RyRs, but their physiological role is less clear than in skeletal and cardiac muscles. When $[Ca^{2+}]_i$ rises as a result of Ca^{2+} entry through the PM and Ca^{2+} release from the SR, CICR mediated by RyRs may help sustain the elevated $[Ca^{2+}]_i$ and, thus, augment contraction.

The SR Ca^{2+} stores that are depleted by release of Ca^{2+} through IP_3R and RyR channels must be replenished. In addition to resequestration of the released Ca^{2+} into the SR by SERCA, SOCs (see Chapter 8) are opened as a result of SR store depletion; this permits extracellular Ca^{2+} to enter and then to be transported rapidly into the SR by SERCA. The mechanism by which depletion of SR Ca^{2+} stores opens SOCs is controversial.

BOX 15-8

NITRIC OXIDE, NITROGLYCERIN, ORGANIC NITRATES, AND SILDENAFIL (VIAGRA) PROMOTE VASODILATION BY ACTIVATING A cGMP–DEPENDENT PROTEIN KINASE (G-KINASE)

Nitric oxide is a product of the oxidation of arginine by the enzyme NO synthase, which is found in endothelial cells, neurons, and certain other cell types. NO is a highly reactive, short-lived (several seconds) hormone neurotransmitter (see Chapter 13). In blood vessels, for example, NO rapidly diffuses from the endothelial cells to the smooth muscle cells. There, NO binds to the heme group of a soluble cytosolic guanylyl cyclase and activates this enzyme, which catalyzes the synthesis of cGMP from GTP. The elevated cGMP level then stimulates a G-kinase, which catalyzes the phosphorylation of RhoA. The net result (see Box 14-2) is activation of MLCP, which dephosphorylates the myosin regulatory light chain and thereby decreases the Ca^{2+} sensitivity of the smooth muscle contractile apparatus. This process induces vasodilation and augments blood flow.

Drugs such as nitroglycerin and isosorbide dinitrate release NO to activate soluble guanylyl cyclase. The consequent vasodilation relieves the chest pain (angina pectoris) that occurs when coronary artery vasospasm (transient vasoconstriction) reduces blood flow and oxygen delivery to the cardiac muscle.

The cGMP signal is terminated when cGMP is hydrolyzed by the enzyme phosphodiesterase. Therefore, phosphodiesterase inhibitors can greatly prolong and amplify the action of cGMP. Sildenafil (Viagra) is such an inhibitor; it relaxes the vascular smooth muscle of the corpus cavernosum and thereby enables it to fill with blood and induce erection (enlargement and stiffening) of the penis.

Ca²⁺ Signaling, Ca²⁺ Sensitivity, and Ca²⁺ Balance in Smooth Muscle May Be Altered under Physiological and Pathophysiological Conditions

Various Ca²⁺ entry mechanisms in smooth muscle have been discussed. Much of the entering Ca²⁺ is rapidly pumped into the SR by SERCA. Indeed, activated smooth muscle cells frequently exhibit net gain of Ca²⁺ in the short term. Sustained net gain of Ca²⁺ would, however, lead to Ca²⁺ overload in the smooth muscle cells. Therefore, in the long term, Ca²⁺ entry must be exactly balanced by Ca²⁺ exit. The two Ca²⁺ extrusion mechanisms in smooth muscle cells, as in other cell types, are the PMCA and the NCX (see Chapters 10 and 11). Although the precise roles of these two transport systems are uncertain, the NCX,

because of its localization at PM-SR junctions (see Chapter 11), apparently plays a key role in regulating the smooth muscle Ca²⁺ concentration in the SR compartments adjacent to the PM. The various smooth muscle Ca²⁺ transport mechanisms are diagrammed in Figure 15-13.

Depending on physiological conditions, smooth muscle Ca²⁺ homeostasis and contractility can be greatly altered. For example, before and during pregnancy, the uterus is poorly responsive to activators such as the hormone oxytocin. Shortly before the onset of labor, however, expression of oxytocin receptors, L-type Ca²⁺ channels, and components of the Rho signaling cascade are increased. Consequently, Ca²⁺ signaling and the Ca²⁺ sensitivity of the contractile apparatus are augmented, so that the uterus can

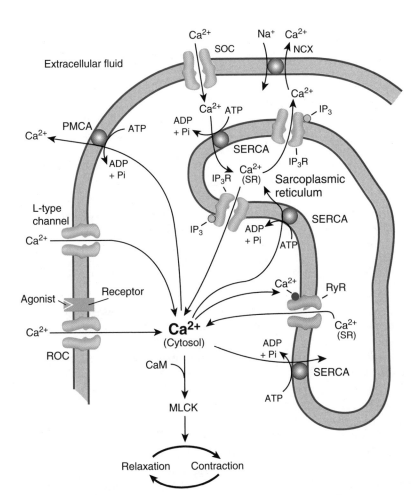

FIGURE 15-13 ■ The various transport systems that participate in the regulation of [Ca²⁺]ᵢ in smooth muscle cells. Ca²⁺ can enter the cells through DHPRs, ROCs, SOCs, and the NCX. Cytosolic Ca²⁺ is extruded from the cells by PMCA and the NCX and pumped into the SR by SERCA. Ca²⁺ is released from the SR through channels gated by IP₃ (i.e., IP₃Rs), and through RyRs that are opened by elevated [Ca²⁺]ᵢ and that mediate CICR.

contract and expel the fetus when the uterus is activated by oxytocin and other agonists.

Another example of functional alteration of smooth muscle occurs in hypertension. In this case, the Ca^{2+} regulatory mechanisms may be reset to maintain an elevated $[Ca^{2+}]_i$ and increased tonic vascular constriction. Evidence indicates that expression of NCX and some TRP channel proteins (components of ROCs) is increased in arterial smooth muscle in several types of hypertension in animal models and humans. Upregulation of these proteins can be induced by ouabain (see Chapter 11). This provides a direct link between Na^+

metabolism and the increased vascular tone and elevated blood pressure.

In sum, the responsiveness of smooth muscles to activators (agonists), as well as the maintenance of tonic contractions, is governed by the relative contributions of (1) the mechanisms that deliver Ca^{2+} to the cytosol, (2) the mechanisms that remove Ca^{2+} from the cytosol (Figure 15-13), and (3) the mechanisms that regulate the Ca^{2+} sensitivity of the contractile machinery (see Box 14-2). These mechanisms can all be modulated under physiological and pathophysiological conditions and can be subject to pharmacological manipulation.

SUMMARY

1. Skeletal muscles are activated by depolarization of the muscle sarcolemma. The process by which depolarization of the surface membrane causes contraction is E-C coupling.

2. E-C coupling in skeletal muscle involves a voltage sensor in the sarcolemma that couples depolarization to contraction; thus, force generation is voltage-dependent.

3. A brief skeletal myocyte contraction in response to a single AP is called a twitch. A contracture is a long-lasting contraction produced by a depolarization that lasts several seconds.

4. A triad is a specialized junction in skeletal muscle consisting of a T-tubule flanked by terminal cisterns of the SR.

5. At the triads, Ca^{2+} release channels, or RyRs, in the SR membrane are in precise spatial register with clusters of DHPRs in the T-tubule membrane.

6. In skeletal muscle, T-tubule depolarization, which is required for E-C coupling, causes movement of the voltage sensor in the DHPRs. The voltage sensor appears to be mechanically linked to the opening of RyR Ca^{2+} release channels in the SR.

7. All the Ca^{2+} required for contraction in skeletal muscle is stored in the SR. Relaxation of skeletal muscle occurs when essentially all the released Ca^{2+} is returned to the SR by SERCA.

8. In cardiac muscle, the functional interactions between SR and the sarcolemma occur at dyads

and peripheral couplings. At these sites, DHPRs are located close to RyRs.

9. In cardiac muscle, the AP is coupled to the opening of RyR Ca^{2+} release channels by CICR. The Ca^{2+} that enters the cell through DHPRs activates the nearby RyRs.

10. The activity of both the NCX and SERCA lowers $[Ca^{2+}]_i$ and causes cardiac muscle to relax.

11. Some smooth muscles are activated by neurotransmitters released from sympathetic and parasympathetic nerves.

12. The enteric nervous system coordinates the contractions of the circular and longitudinal smooth muscle layers in the intestine during peristalsis.

13. Some smooth muscles (e.g., uterine) receive little neuronal innervation. Contraction of these smooth muscles is regulated by circulating hormones and locally released factors.

14. Stretch or increase in wall pressure may also directly activate some smooth muscles, such as those in the arterial wall.

15. Phasic smooth muscles are activated by brief or rhythmic depolarizations that trigger the opening of DHPRs. The resulting Ca^{2+} entry is required to activate contraction.

16. Tonic smooth muscles have a relatively small V_m (~-55 to -40 mV), and contraction can be regulated by small changes in V_m.

17. Smooth muscle contraction can also be activated by Ca^{2+} release from the SR in a process called

pharmacomechanical coupling, which involves little or no change in V_m. Agonists that activate smooth muscle cells usually trigger the production of IP_3 in the cytosol, which binds to IP_3R Ca^{2+} release channels on the SR. The consequent efflux of Ca^{2+} into the cytosol initiates contraction. The role of RyRs in smooth muscle is less well understood.

18. Maintained contraction in tonic smooth muscles is controlled by the balance between Ca^{2+} entry into and removal from the cytosol, as well as by the mechanisms that regulate the Ca^{2+} sensitivity of the contractile apparatus. These mechanisms may be modulated under physiological and pathophysiological conditions.

KEY WORDS AND CONCEPTS

- Contracture
- Excitation-contraction coupling (E-C coupling)
- Voltage sensor
- Twitch
- Dihydropyridine receptor (DHPR)
- Ryanodine receptor (RyR)
- Calsequestrin
- Triad
- Terminal cistern
- Dyads and peripheral couplings in cardiac muscle
- Sinoatrial pacemaker cells
- Calcium-induced calcium release (CICR)
- Calcium sparks
- Enteric nervous system
- Peristalsis
- Myogenic tone
- Tonic and phasic smooth muscles
- Pharmacomechanical coupling

STUDY PROBLEMS

1. E-C coupling is similar in skeletal and cardiac muscle in some respects but different in others. For example, a depolarization of the sarcolemma (e.g., by an AP) initiates E-C coupling in both types of muscle. Describe other similarities and differences.

2. In skeletal muscle, only about half of the RyRs in the SR are apposed to (and presumably linked to) tetrads of DHPRs in the T-tubules. These RyRs are opened by a mechanical link between the voltage sensor in the DHPR and the gating mechanism of the RyR. During E-C coupling, what do you think happens to the RyRs that are not opposed by DHPRs? Do you think they are nonfunctional? Or could they be opened through a different mechanism?

3. What general types of mechanisms influence smooth muscle contraction by altering (a) the availability of Ca^{2+} or (b) the sensitivity of the contractile apparatus to Ca^{2+}? (c) Why do you suppose so many types of mechanisms are available for regulating contraction in smooth muscles? (d) What are the possible clinical (therapeutic) implications?

BIBLIOGRAPHY

Bennett MR: Autonomic neuromuscular transmission at a varicosity, *Prog Neurobiol* 50:505, 1996.

Bers DM: *Excitation-contraction coupling and cardiac contractile force*, Dordrecht, The Netherlands, 1991, Kluwer.

Block BA, Imagawa T, Campbell KP, et al: Structural evidence for direct interaction between the molecular components of the transverse tubule/sarcoplasmic reticulum junction in skeletal muscle, *J Cell Biol* 107:2587, 1988.

Bryant HJ, Harder DR, Pamnani MB, et al: In vivo membrane potentials of smooth muscle cells in the caudal artery of the rat, *Am J Physiol* 249:C78, 1985.

Chandler WK, Rakowski RF, Schneider MF: Effects of glycerol treatment and maintained depolarization on charge movement in muscle, *J Physiol* 254:285, 1976.

Giachini FR, Tostes RC: Does Na^+ really play a role in Ca^{2+} homeostasis in hypertension? *Am J Physiol* 299:H602, 2010.

Hodgkin AL, Horowicz P: Potassium contractures in single muscle fibres, *J Physiol* 153:386, 1960.

Knot HJ, Nelson MT: Regulation of arterial diameter and wall $[Ca^{2+}]$ in cerebral arteries of rat by membrane potential and intravascular pressure, *J Physiol* 508:199, 1998.

Murphy RA: Smooth muscle. In Berne RM, Levy MN, Koeppen BM, et al, editors: *Physiology*, ed 4, St Louis, 1998, Mosby.

Rüegg JC: *Calcium in muscle contraction*, ed 2, Berlin, Germany, 1992, Springer-Verlag.

Sanchez JA, Stefani E: Kinetic properties of calcium channels of twitch muscle fibres of the frog, *J Physiol* 37:1, 1983.

Somlyo AP, Somlyo AV: Signal transduction and regulation in smooth muscle, *Nature* 372:231, 1994.

16

MECHANICS OF MUSCLE CONTRACTION

OBJECTIVES

1. Understand how a skeletal muscle twitch and a tetanus are generated.

2. Understand how muscle contractile force can be varied.

3. Understand the mechanical properties of muscle as described by the length-tension relationship and the force-velocity relationship.

4. Understand the roles of the three main types of phasic skeletal muscle.

5. Understand how the mechanical properties of skeletal muscle, cardiac muscle, and smooth muscle differ.

6. Understand the diversity of smooth muscle types.

7. Understand how the kinetic properties of some smooth muscles contribute to their ability to generate tonic contractions or "tone."

The sliding filament mechanism of contraction and E-C coupling are described in Chapters 14 and 15. We now consider the mechanical properties of muscle.

THE TOTAL FORCE GENERATED BY A SKELETAL MUSCLE CAN BE VARIED

Whole Muscle Force Can Be Increased by Recruiting Motor Units

Skeletal muscle is normally activated through its innervation by α motor neurons. An individual α motor neuron may innervate just a few muscle fibers (myocytes) in extraocular muscles or as many as several thousand in large limb and trunk muscles. A single α motor neuron, together with all the myocytes it innervates, is referred to as a **motor unit**. The motor neurons release ACh at NMJs (see Chapter 12). A single motor neuron AP normally releases sufficient ACh at each NMJ to trigger an AP in all the muscle fibers it innervates. Consequently, all the muscle cells in a motor unit contract synchronously when the motor neuron generates an AP. Thus, the motor unit is the functional force-generating unit in whole muscle. The primary mechanism by which the central nervous system increases whole muscle force is by recruiting (i.e., simultaneously activating) motor units. The amount of force generated by whole muscle, therefore, is directly related to the number of motor units that are activated.

A Single Action Potential Produces a Twitch Contraction

A single AP in a skeletal myocyte triggers Ca^{2+} release from the SR (see Chapter 15) to increase $[Ca^{2+}]_i$ and, thus, causes a **twitch** contraction. A comparison of the time course of $[Ca^{2+}]_i$ elevation and force development during a twitch is shown in Figure 16-1A. Shortly after the AP, $[Ca^{2+}]_i$ rapidly rises; the Ca^{2+} ions bind to TnC and activate the formation of actin-myosin cross-bridges. The resultant increase in force develops over a slower time course than the rise in $[Ca^{2+}]_i$ (Figure 16-1A).

Force generated by skeletal muscle is transmitted to the bones through elastic elements (e.g., tendons). These elastic elements are connected in series with the contractile elements and bones and must be stretched before force can be transmitted. Stretching these **series elastic elements** causes a delay between the activation of actin-myosin cross-bridges and the development of force. The following analogy helps to explain this effect: you hold in your hand a taut, but not stretched, rubber band that is attached to a weight resting on a table. As you raise your hand, the weight is not lifted immediately because the rubber band must first be stretched. As the rubber band is stretched (i.e., the length is increased), a force is developed in the

rubber band, which pulls against the attached weight. The development of this force takes time, and the more elastic the rubber band, the longer the delay in lifting the weight (Box 16-1).

Sufficient Ca^{2+} is released from the SR by a single AP to saturate all the TnC binding sites. These sites are saturated for only a brief time because $[Ca^{2+}]_i$ reaches a peak and then begins to fall rapidly as the Ca^{2+} is pumped back into the SR. Therefore, the muscle does not develop its maximum *possible* force during a twitch because Ca^{2+} dissociates from TnC (as $[Ca^{2+}]_i$ declines) before the series elastic elements are completely stretched. If $[Ca^{2+}]_i$ remains elevated for a longer period, however, muscle force will continue to increase. This occurs physiologically if a second AP is evoked before $[Ca^{2+}]_i$ has returned to the resting level. Note that contractile force can also be enhanced by increasing the amount of Ca^{2+} stored in the SR. This is an important mechanism for altering contractile force in the heart (see later), but not in skeletal muscle.

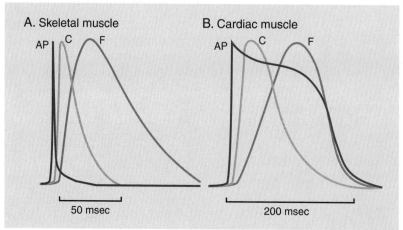

FIGURE 16-1 ■ Time courses of the sarcolemmal AP, average rise in $[Ca^{2+}]_i$, and force development in skeletal and cardiac muscle. **A,** A single stimulus applied to skeletal muscle evokes an AP, which causes Ca^{2+} to be released from the SR (see Chapter 15). $[Ca^{2+}]_i$ (C) rises rapidly after the AP and reaches a level that saturates TnC binding sites, thus permitting actin-myosin cross-bridges to form. Because elastic elements in the muscle need to be stretched, force development (F) is much slower than the $[Ca^{2+}]_i$ rise and the activation of the contractile apparatus. **B,** Time courses of the AP, $[Ca^{2+}]_i$, and force are shown for a cardiac muscle activated by a single stimulus. The force transient lasts much longer than in skeletal muscle because Ca^{2+} channels remain open during the long-duration AP, thus allowing prolonged Ca^{2+} influx and, consequently, prolonged CICR. (*A, Data from Palade P, Vergara J: J Gen Physiol 79:679, 1982. B, Data from Langer G, editor: Calcium and the heart, New York, 1990, Raven Press.*)

BOX 16-1

A MECHANICAL MODEL OF MUSCLE INCLUDES A SPRING IN SERIES WITH THE CONTRACTILE ELEMENTS

Skeletal muscle can be modeled as a system consisting of three mechanical components (Figure B-1). A contractile element is in series with an elastic element, represented by a spring. A parallel elastic element is in parallel with the contractile element and the series elastic element. The series elastic element resides primarily in the tendons. The sarcolemma, intracellular components of the cytoarchitecture, and extracellular connective tissue are part of the parallel elastic element. The series elastic element explains the delay that occurs between the activation of the contractile elements and the development of force: the series elastic element must be stretched before force can be transmitted to a load. Furthermore, if a muscle is rapidly stretched immediately after stimulation and before it begins to develop tension, it will develop more force than if it had not been stretched. The rapid stretch extends the series elastic element, so that the shortening of the contractile elements will pull on the external load sooner. When resting muscle is stretched to relatively long sarcomere lengths, the resting tension increases. This maintained resting tension is evidence of a parallel elastic element because without it the non-elastic contractile elements would passively stretch and eliminate steady-state tension.

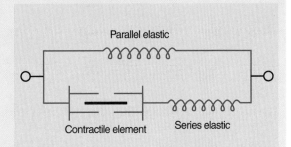

FIGURE B-1 ■ Three-component mechanical model of skeletal muscle. A parallel elastic element is in parallel with a series combination of the contractile element and the series elastic element. The elastic elements are modeled as springs that, when stretched, develop tension according to the relationship $F = k \times \Delta x$, where F is the force (tension), Δx is the change in length of the spring, and k is the spring constant. A stiffer, or stronger, spring has a larger spring constant. When the contractile element shortens, the series elastic element must be stretched before force is transmitted to the external load. Resting tension is supported by the parallel elastic element alone because the contractile elements are not elastic.

Repetitive Stimulation Produces Fused Contractions

The maximum force generated by skeletal muscle in response to two consecutive stimuli depends on the interval between the stimuli (Figure 16-2). If the two stimuli are separated by more than 100 milliseconds, both stimuli will produce twitch contractions (Figure 16-2A). If the second stimulus is given before the first twitch relaxes completely, the force from the second twitch will be added to the force from the first (Figure 16-2B) and the resultant peak force will be larger than the force generated by a single stimulus. This occurs because the second stimulus rereleases Ca^{2+} from the SR before all the Ca^{2+} released by the first stimulus dissociates from TnC. Thus, TnC remains Ca^{2+} bound for a longer period. This gives the contractile elements more

time to stretch the series elastic elements and, thus, generate more force. If the interval between the two stimuli is short enough, the second twitch will become completely fused with the first, thus generating a single force transient with a smooth rising phase (Figure 16-2C). If a train of stimuli is applied at a frequency of 20 to 40 per second, a partially fused contraction will result (Figure 16-2D). This is called a semifused tetanus; a **tetanus*** is a large contraction caused by a high-frequency train of APs in skeletal muscle. The tetanus becomes completely

---

*Tetanus toxin, from the bacterium *Clostridium tetani*, causes tetanic convulsions by blocking inhibitory synaptic receptors in the spinal cord. The resultant disinhibition greatly increases the frequency at which α motor neurons fire APs .

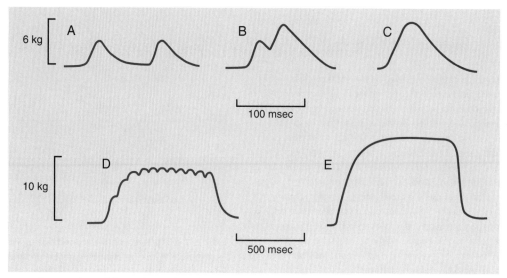

FIGURE 16-2 ■ Summation of force during high-frequency stimulation in mammalian gastrocnemius muscle. The force generated by a whole skeletal muscle in response to repetitive stimulation is shown. **A,** When a pair of stimuli is separated by approximately 100 milliseconds, the muscle responds with two separate twitches of equal size. **B,** When the inter-stimulus interval is reduced to approximately 50 milliseconds, the force from the second twitch is added to the force from the first. **C,** With a sufficiently short interstimulus interval, a single contraction with a smooth rising phase is generated. **D,** A train of stimuli at a rate of 24 sec^{-1} generates a partial tetanus. **E,** A complete, fused tetanus is generated at a stimulus frequency of 115 sec^{-1}. The 100-millisecond time scale and 6-kg force scale apply to **A, B,** and **C,** and the 500-millisecond time scale and 10-kg force scale apply to **D** and **E.** *(Data from Coopers S, Eccles JC: J Physiol 69:377, 1930.)*

fused at higher rates of stimulation (80 to 100 stimuli per second; Figure 16-2E). Thus, force generation in skeletal muscle is regulated through two mechanisms: (1) control of the frequency at which the α motor neuron generates APs and (2) control of the number of α motor neurons (and consequently the number of motor units) activated.

SKELETAL MUSCLE MECHANICS IS CHARACTERIZED BY TWO FUNDAMENTAL RELATIONSHIPS

The **length-tension curve** and the **force-velocity relationship** characterize the mechanical properties of muscle. Note that "tension" and "force" have the same meaning in muscle physiology and are used inter-changeably. The experimental apparatus diagrammed in Figure 16-3 is used to determine these relationships, either for a single muscle fiber or for whole muscle. In this experimental setup, two mechanical variables are

measured as a function of time: contractile force and muscle length. The initial length of the muscle and the attached load can both be varied.

When muscle is activated and filaments begin to slide, two types of mechanical events can occur. The muscle can generate force as it pulls on the tendons, or the muscle can shorten. If the force generated by the muscle is too small to lift the attached load, the muscle will not shorten (Figure 16-4A). In this **iso-metric contraction,** the muscle generates force at a *constant* length. The magnitude of the force varies with the length of the muscle in a manner that is char-acterized by the length-tension curve (Figure 16-5). If the muscle is activated with no weight attached, the contraction will result in shortening with no force generation (see Figure 16-4B). This is an **isotonic contraction**—the muscle shortens while generating a constant amount of force. Macroscopic movement and force generation are the results of summing nu-merous molecular power strokes that are generated

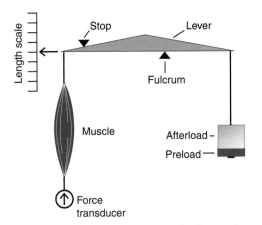

FIGURE 16-3 ■ Apparatus used to study the mechanical properties of muscle. One end of an isolated muscle is attached to a force transducer, which measures total force. The other end is attached to a lever, which is used to measure shortening. Weights attached to the opposite end of the lever are used to set the initial muscle length and to provide loads against which the muscle will contract. A preload weight is first applied to stretch the muscle to a desired length. With the preload in place, a mechanical stop is adjusted to prevent further stretch of the muscle when the afterload is added. Before stimulation, the transducer measures the passive force and the vertical position of the lever indicates the initial muscle length. On stimulation, contraction is activated and the muscle begins to develop force. If the force exceeds the gravitational force of the attached load (preload plus afterload), the muscle will shorten.

by interactions between individual actin and myosin molecules in the form of cross-bridges. The force and movement generated by a single actin-myosin cross-bridge have been measured (Box 16-2).

The Sliding Filament Mechanism Underlies the Length-Tension Curve

The force generated by skeletal muscle depends on the length to which a muscle is stretched. This fundamental property of muscle can be studied by setting the initial length of a single myocyte and measuring the maximum isometric tetanic force generated at that length. Some force is required to stretch relaxed muscle; this is termed **passive tension**. The relationship between passive tension and muscle length is shown in Figure 16-5. When the muscle is stimulated to contract, total tension increases. The difference between total tension generated by contracting muscle and passive tension in relaxed muscle is **active tension** (Figure 16-5). The active length-tension relationship shows an **optimal length** (L_0) at which tension is maximal. In skeletal muscle L_0 is 2.0 to 2.2 μm per sarcomere (Figure 16-5, points B and C).

The active length-tension relationship can be understood in terms of the sliding filament mechanism. At L_0, the maximum number of cross-bridges can form between thick and thin filaments. As the muscle is stretched from 2.2 to 3.65 μm per sarcomere, reduced overlap between thick and thin filaments decreases

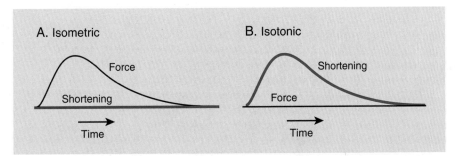

FIGURE 16-4 ■ Isometric and isotonic contractions. **A,** The maximum force generated by the muscle is not enough to lift the load. Thus, no shortening occurs *(blue line)* and the muscle generates force *(black line)* at a constant length. This is an isometric contraction. **B,** With no load attached the muscle does not need to generate force to overcome a load and thus begins to shorten immediately. This is an isotonic contraction, which, in general, refers to a period of muscle shortening at constant force.

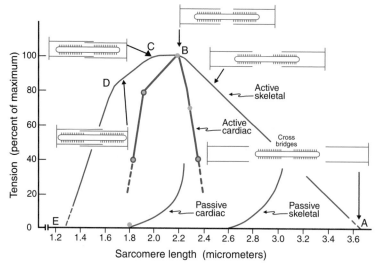

FIGURE 16-5 ■ The length-tension curves in skeletal and cardiac muscle are similar. Actively developed isometric tension and passive, resting tension are plotted against sarcomere length for skeletal and cardiac muscle. The diagrams of the myofilaments show the extent of overlap between thick and thin filaments at five specific sarcomere lengths in skeletal muscle. In skeletal muscle, maximum tension occurs at sarcomere lengths between 2.0 and 2.2 μm (between points B and C). Between 2.2 and 3.65 μm of sarcomere length (between points B and A), tension decreases because of decreasing overlap between thick and thin filaments. Tension also decreases at less than 2.0 μm sarcomere length, in part because of double overlap of thin filaments (between points C and D). In cardiac muscle, active tension falls steeply as sarcomere length changes from 2.2 μm. The increase in passive tension occurs at much shorter sarcomere lengths in cardiac muscle than in skeletal muscle. *(Redrawn from Gordon AM, Huxley AF, Julian FJ: J Physiol 184: 143, 1966; and Braunwald E, Ross J Jr, Sonnenblick EH: Mechanisms of contraction of the normal and failing heart, Boston, 1976, Little, Brown.)*

BOX 16-2

THE FORCE AND DISPLACEMENT GENERATED BY A SINGLE ACTIN-MYOSIN INTERACTION CAN BE MEASURED WITH AN OPTICAL "TRAP"

In an "optical trap" experiment, myosin molecules are bound to a small silica particle, which is then fixed to a glass coverslip. A single actin filament is attached at each end to a plastic bead. The bead is held in place, or trapped, at the focal point of a laser. The operator can move the bead (and thus the attached actin) by moving the laser focal point and can change the force with which the bead (and actin) is held in place by varying the laser intensity. The actin filament is moved close to the myosin-coated particle to permit single actin-myosin interactions. When the actin filament is permitted to move freely during a single cross-bridge cycle, the displacement generated by that single interaction is approximately 10 nm (nm = 10^{-9} m). By adjusting the intensity of the trapping laser so that it is just sufficient to prevent actin filament movement, the isometric force generated by a single cross-bridge can be estimated to be approximately 1 to 3×10^{-12} Newtons. (The Newton is a unit of force and is equal to 1 kg × m/sec². Applying a 1-Newton force to a 1-kg mass imparts an acceleration of 1 m/sec².)

the number of cross-bridges that can form; therefore, active force decreases (Figure 16-5, points *A* and *B*). At sarcomere lengths less than 2.0 μm, force declines for at least two reasons (Figure 16-5, points *C* to *E*). Thin filaments from opposite ends of the sarcomere can overlap, which allows cross-bridges to form in the wrong orientation, and filament geometry may be disrupted as thick filaments hit the Z line and actin filaments are forced to move away from the thick filaments.

Because the shape of the active length-tension relationship is precisely predicted by the sliding filament mechanism, we conclude that *the active force generated by a muscle is proportional to the number of attached cross-bridges.*

In Isotonic Contractions, Shortening Velocity Decreases as Force Increases

The force-velocity relationship for a given muscle can also be determined by use of the apparatus shown in Figure 16-3. Force and length are measured during contractions with different weights attached. When the muscle generates sufficient force to lift the weight, a partly isometric, partly isotonic contraction occurs (Figure 16-6). After activation, the muscle begins to develop force isometrically until the force is sufficient to lift the attached weight ("load"). At this point, force development stops and the muscle begins to shorten isotonically. With a small load, the velocity of shortening (i.e., the rate of change of muscle length with time) is relatively fast (Figure 16-6). With a larger load, more time is spent developing force, thus leaving less time for isotonic shortening. As a result, the velocity of shortening is slower. The overall force-velocity relationship, which is constructed by measuring a series of contractions against different loads, is shown in Figure 16-7. The maximum velocity of shortening (V_0) occurs with no load.

At the molecular level, the rate of cross-bridge cycling determines the rate of muscle shortening. Thus, V_0 is a reflection of the maximum rate of cross-bridge cycling. Cross-bridge cycling requires myosin ATPase activity (see Chapter 14); therefore, V_0 is also directly proportional to the myosin ATPase activity. The inverse relationship between load and shortening velocity can be understood with the help of Figure 16-8. Shortening is caused by rotation of the myosin head; the force generated by this rotation is opposed by the load. If the load is relatively small, myosin head rotation will meet

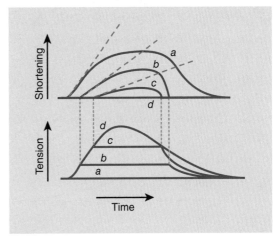

FIGURE 16-6 ■ Partly isometric, partly isotonic contractions. When a contracting muscle can generate enough force to lift an attached load, the contraction has an isometric component and an isotonic component. This figure shows superimposed traces of four contractions of skeletal muscle obtained with the apparatus shown in Figure 16-3. On stimulation in the presence of an afterload, the muscle begins to develop tension isometrically *(lower traces)*. When the muscle tension exceeds the attached load, tension remains constant and the muscle starts to shorten isotonically *(upper traces)*. The maximum velocity of shortening is measured from the slope of a line *(dashed line)* that is tangent to the shortening curve just after shortening has begun. The maximum velocity of shortening is fastest with no load and becomes smaller as the load is increased.

little resistance and will occur at a fast rate. In addition, many myosin heads are unattached and available for cycling. Together, these factors produce a relatively fast shortening velocity. In contrast, if the load is large, head rotation will meet stiff opposition and the rotation speed will be slow. Furthermore, many more cross-bridges are attached at any given moment, and fewer are available for recycling. Thus, shortening velocity is slow.

THERE ARE THREE TYPES OF SKELETAL MUSCLE MOTOR UNITS

The main criteria used to classify motor units are speed of contraction and susceptibility to fatigue. The three types of motor units are **fast-twitch, fatigable**

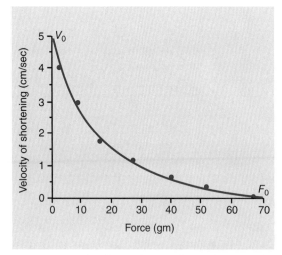

FIGURE 16-7 ■ The force-velocity relationship in skeletal muscle. The velocity of shortening during an isotonic contraction is plotted as a function of the force generated by the muscle during the isotonic shortening. The maximum velocity of shortening (V_0) occurs with no load (isotonic contraction). When the attached load is equal to the maximum force produced by the muscle (F_0), the muscle does not shorten (isometric contraction). *(Data from Hill AV: Proc R Soc Lond B 126:136, 1938.)*

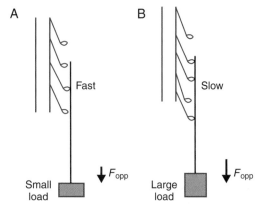

FIGURE 16-8 ■ Myosin head rotation occurs at a slower rate with a larger load attached to the muscle. The load attached to an isotonically contracting muscle represents a force (F_{opp}) that opposes myosin head rotation. **A,** When the muscle is contracting isotonically against a small load, many cross-bridges are unattached and available for cycling. Myosin head rotation occurs rapidly because little resistance is present. **B,** At larger loads, head rotation is slower because the opposing force is larger. *(Redrawn from Berne RM, Levy MN, Koeppen BM, et al, editors: Physiology, ed 4, St Louis, 1998, Mosby.)*

(type FF), fast-twitch, fatigue-resistant (FR), and **slow-twitch, fatigue-resistant (S).** Both FF and FR motor units have twitch contraction times (time to peak twitch force) of 10 to 40 milliseconds, whereas S units take from 60 to more than 100 milliseconds to develop peak twitch force. The contraction time is directly related to the myosin ATPase activity. The myosin isoforms in FF and FR muscle fibers have a higher ATPase activity than the isoform in S fibers. FF motor units generate the largest tetanic forces, S motor units generate the smallest forces, and FR motor units are intermediate. Several factors contribute to the variation in force among motor unit types. These include the *innervation ratio* (i.e., the number of muscle fibers innervated by the motor neuron) and the force output of individual muscle fibers, which depends, in part, on the cross-sectional area of the individual muscle fibers.

The responses of the different types of motor units to repetitive tetanic stimulation is shown in Figure 16-9.

The decrease in force that occurs with repetitive trains of tetanic stimuli is called **fatigue.** This property is quantified by the *fatigue index,* which is defined as the ratio of the tetanic force after 2 minutes of repetitive stimulation to the force generated by the first tetanus. The fatigue index of FF motor units is typically less than 0.25, whereas it is greater than 0.75 in FR and S motor units. Although fatigue is a well-characterized phenomenon, the underlying cellular mechanisms remain obscure.

Whole muscle usually contains mixtures of the three types of motor units. For example, the medial gastrocnemius (calf) muscle of the cat contains FF, FR, and S motor units. The recruitment order of these motor units follows the size principle: motor units that generate the smallest force are recruited first, followed by larger motor units as the requirement for additional force increases. This size-ordered recruitment confers certain mechanical advantages on the operation of the whole muscle. Type S units are used

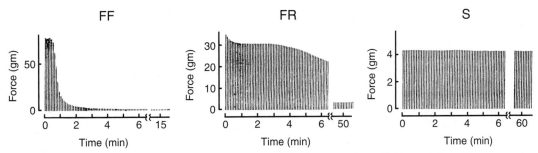

FIGURE 16-9 ■ Response of fast-twitch, fatigable (FF), fast-twitch, fatigue-resistant (FR), and slow (S) motor units to repetitive tetanic stimulations. Individual motor neurons were stimulated to produce a tetanic contraction in a single motor unit, and the tetanic stimulation was repeated every second for up to 60 minutes. This figure shows the change in tetanic force during these repetitive tetanic stimulations. In FF motor units the tetanic force decreases rapidly: after 1 minute the force has dropped to 25% of its initial value. Tetanic force declines much more slowly in FR motor units: after 5 minutes of stimulation, the force has dropped only 30%. These motor units do eventually fatigue, and after 50 minutes of stimulation the force is a small fraction of its initial value. In contrast, the tetanic force in S motor units does not decrease even after 60 minutes of stimulation. Note that FF units generate the greatest force, but are the most easily fatigable. In contrast, S units generate the smallest force, but are fatigue-resistant. *(Data from Burke RE, Levine DN, Tsairis P, et al: J Physiol 234:723, 1973.)*

to maintain posture; FR and FF units are used, respectively, for sustained and burst movements (Box 16-3).

THE FORCE GENERATED BY CARDIAC MUSCLE IS REGULATED BY MECHANISMS THAT CONTROL INTRACELLULAR Ca^{2+}

Cardiac Muscle Generates Long-Duration Contractions

Cardiac muscle contraction is initiated by an AP that triggers Ca^{2+} release from the SR (see Chapter 15). The time courses of the AP, changes in $[Ca^{2+}]_i$, and force development in cardiac muscle are compared in Figure 16-1B. One obvious difference between the cardiac muscle force transient and the skeletal muscle twitch contraction (see Figure 16-1A) is the longer duration of the cardiac contraction. During the long-duration cardiac AP, prolonged Ca^{2+} entry through L-type Ca^{2+} channels, and, thus, prolonged CICR, causes $[Ca^{2+}]_i$ to be elevated for a longer time than in skeletal muscle, thus resulting in a longer contraction.

Contractions in healthy cardiac muscle are always of the twitch type. The summation of twitch forces with repetitive stimulation that occurs during a tetanus in skeletal muscle cannot occur in the heart under

normal physiological conditions because the cardiac AP has approximately the same duration as the force transient (see Figure 16-1B). Thus, even at the shortest possible inter-AP interval, the force transient from the first AP has decayed completely when the succeeding AP begins. Therefore, summation cannot occur.

Total Force Developed by Cardiac Muscle Is Determined by Intracellular Ca^{2+}

Cardiac muscle contractile force can be altered by factors that are intrinsic and extrinsic to the myocytes. The major intrinsic factor is the length-tension relationship, which is described in the next section. Many extrinsic factors affect cardiac contractility by altering $[Ca^{2+}]_i$. For example, β-adrenergic agonists increase the inward Ca^{2+} current activated by an AP and, thus, increase cardiac contractility (see Chapter 8). This is referred to as a **positive inotropic effect.*** L-type Ca^{2+} channel blockers (see Chapter 8) reduce Ca^{2+} entry and, thus, decrease contractility (a negative inotropic effect). Cardiotonic steroids reduce the Na^+ gradient, which reduces Ca^{2+} extrusion through the NCX and thereby increases $[Ca^{2+}]_{SR}$ (see Chapter 11). Activation

---

*In the heart, an inotropic effect is a change in contractility, whereas a chronotropic effect is a change in heart rate.

BOX 16-3

SIZE-ORDERED RECRUITMENT OF MOTOR UNITS PROVIDES MECHANICAL ADVANTAGES IN THE OPERATION OF WHOLE MUSCLE

The medial gastrocnemius (calf) muscle in the cat is a mixture of fast-twitch, fatigable (FF); fast-twitch, fatigue-resistant (FR); and slow (S) motor units. The order in which motor units are recruited depends on the magnitude of the force generated by each motor unit. Consequently, the total muscle force is a nonlinear function of the fraction of motor units activated (Figure B-1). First to be recruited are S units, which generate small, slow contractions and are fatigue-resistant advantageous during sustained activity at low force levels, such as maintaining posture. For actions requiring more force,

FR units are recruited next. FR motor units contract quickly, generate moderate force, and are relatively resistant to fatigue. These properties are advantageous for sustained activity requiring moderate force, such as running. Together, FR and S motor units constitute 60% of the pool, but generate only approximately 25% of the maximum force. Demand for larger forces is met by activating FF motor units, which generate large forces rapidly, but they fatigue quickly and are used infrequently. They are recruited when a short burst of strenuous activity is required, as in sprinting or jumping.

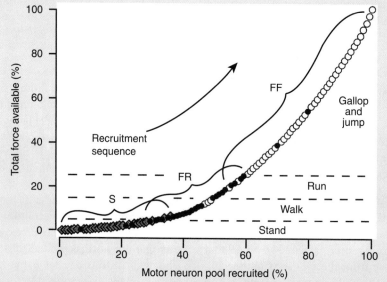

FIGURE B-1 ■ Size-ordered recruitment of motor units in the gastrocnemius muscle. The total force available from the medial gastrocnemius muscle (the chief muscle of the calf of the leg) is plotted as a function of the percentage of the motor neuron pool recruited under the assumption that the order of recruitment is strictly by motor unit size (smallest first). The initial 30% of the pool is dominated by type S units, the next 30% is dominated by FR units, and the final 40% is dominated by FF units. The *diamonds, filled circles,* and *open circles* represent S, FR, and FF motor units, respectively. *(From Karpati G, Hilton-Jones D, Griggs RC, editors:* Disorders of voluntary muscle, *Cambridge, 2001, Cambridge University Press.)*

then results in larger Ca^{2+} release and, thus, higher $[Ca^{2+}]_i$, leading to enhanced contractility (a positive inotropic effect).

MECHANICAL PROPERTIES OF CARDIAC AND SKELETAL MUSCLE ARE SIMILAR BUT QUANTITATIVELY DIFFERENT

Cardiac and Skeletal Muscles Have Similar Length-Tension Relationships

The length-tension relationship in isolated cardiac muscle is similar to that observed in skeletal muscle (see Figure 16-5). The active isometric force is maximal when the initial sarcomere length is approximately 2.2 μm. At this sarcomere length in cardiac muscle, the overlap of thick and thin filaments is optimal, which enables the maximum number of cross-bridges to form. The active force decreases at both longer and shorter sarcomere lengths.

Although the length-tension relationships in skeletal and cardiac muscle are similar, at least two important differences exist. In cardiac muscle, changes in filament overlap account only partly for the changes in active tension that occur with changes in sarcomere length. At muscle lengths less than L_0 (i.e., the ascending limb of the length-tension curve), an increase in muscle length increases the Ca^{2+} sensitivity of the myofilaments, which partially accounts for the increase in active tension. The second difference is in passive tension: cardiac muscle is stiffer than skeletal muscle. Thus, when cardiac muscle is passively stretched beyond L_0, the passive tension increases to high levels (see Figure 16-5). This prevents sarcomere length from exceeding about 2.3 μm and prevents the decreased overlap of thick and thin filaments that would otherwise occur at longer sarcomere lengths. Consequently, healthy cardiac muscle functions almost exclusively on the ascending limb of the length-tension curve.

The Contractile Force of the Intact Heart Is a Function of Initial (End-Diastolic) Volume

The cardiac length-tension relationship (see Figure 16-5) helps us to understand how the heart works. Figure 16-10 shows the relationship between ventricular pressure (which is a direct function of the force generated by

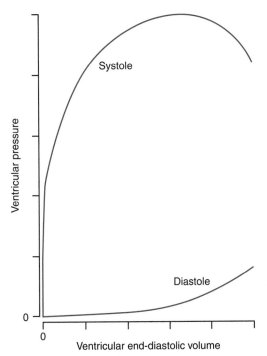

FIGURE 16-10 ■ Pressure-volume curve in the heart. The curve labeled Diastole shows the relationship between the ventricular pressure and the ventricular volume in a relaxed heart (i.e., during diastole). The curve labeled Systole shows the relationship between peak systolic ventricular pressure during contraction as a function of end-diastolic volume. *(Data from Patterson SW, Piper H, Starling EH: J Physiol 48:465, 1914.)*

ventricular myocytes) and the ventricular end-diastolic volume (which is a measure of the initial length of the myocytes). The curve labeled *Diastole* is the relationship between the ventricular pressure and the intraventricular volume in a relaxed heart (i.e., during diastole). This is analogous to the passive length-tension curve in isolated cardiac muscle (see Figure 16-5). The curve labeled *Systole* shows the maximum pressure developed by the ventricle during contraction (systole) for each level of filling. This is the **Frank-Starling relationship,** which is analogous to the active length-tension relationship of cardiac muscle shown in Figure 16-5. An important consequence of this relationship for cardiac performance is that an increase in ventricular filling pressure (i.e., passive tension) will increase the end-diastolic volume of the ventricle and thereby increase the output of the heart.

The normal heart operates on the ascending limb of the Frank-Starling relationship (Figure 16-10), which corresponds to the ascending limb of the active length-tension relationship (see Figure 16-5).

Shortening Velocity Is Slower in Cardiac Than in Skeletal Muscle

The V_0 during isotonic contractions in cardiac muscle occurs at zero load, as in skeletal muscle. As the load increases, the shortening velocity decreases until a load is reached at which the muscle can no longer shorten and the contraction is isometric (Figure 16-11). Although the shape of the curve is similar to that for skeletal muscle (see Figure 16-7), the velocity of shortening is slower in cardiac muscle. In cardiac muscle, a positive inotropic intervention, such as stimulation of β-adrenergic receptors with norepinephrine, increases the velocity of shortening at all loads and increases the isometric contraction force, F_0 (Figure 16-11).

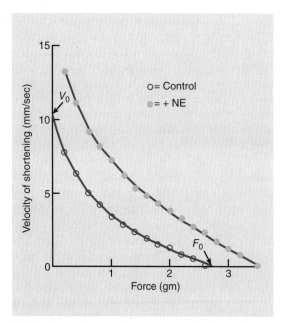

FIGURE 16-11 ■ Force-velocity relationship in cardiac muscle. The velocity of shortening during isotonic contractions in isolated cardiac muscle decreases as the load increases. The addition of norepinephrine (+NE) increases the velocity of shortening at all loads. F_0, maximum force; V_0, maximum velocity of shortening. (*Data from Braunwald E, Ross J Jr, Sonnenblick EH:* Mechanisms of contraction of the normal and failing heart, *Boston, 1976, Little, Brown.*)

DYNAMICS OF SMOOTH MUSCLE CONTRACTION DIFFER MARKEDLY FROM THOSE OF SKELETAL AND CARDIAC MUSCLE

Three Key Relationships Characterize Smooth Muscle Function

In smooth muscle, as in skeletal and cardiac muscles, many critical aspects of contractile function are characterized by the three relationships mentioned previously: the processes that relate cell activation to $[Ca^{2+}]_i$ elevation and force development, the force-velocity relationship, and the length-tension relationship. Nevertheless, some key quantitative differences in these relationships distinguish smooth muscles from skeletal and cardiac muscles.

The Length-Tension Relationship in Smooth Muscles Is Consistent with the Sliding Filament Mechanism of Contraction

As in skeletal muscle, smooth muscle has an optimal length (L_0) at which the largest active force can be developed (Figure 16-12). When the initial length of a smooth muscle differs from L_0, the maximum active force declines.

The shape of the smooth muscle length-tension relationship, as well as its similarity to that of skeletal

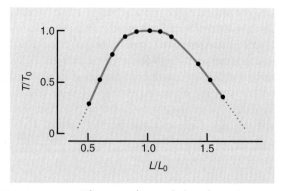

FIGURE 16-12 ■ The smooth muscle length-tension curve for a rat small mesenteric artery. The length (*L*) refers to the internal circumference of the artery. L_0 is the length at which maximum active tension is developed. *T* is the tension (force) at any length, and T_0 is maximum tension developed at the optimal length (L_0). (*Data from Mulvany MJ, Warshaw DM:* J Gen Physiol *74:85, 1979.*)

muscle (see Figure 16-5), implies that the sliding filament mechanism of muscle contraction also applies to smooth muscle. For several reasons, however, the smooth muscle length-tension relationship cannot be correlated directly with sarcomere length, as it is in skeletal muscle (see Figure 16-5). First, as noted in Chapter 14, in smooth muscle, the myofibrils and, thus, the sarcomeres are not all arrayed in register; therefore, muscle length cannot be expressed in terms of sarcomere length. Second, as discussed in Chapter 14, smooth muscle contractile force can be altered by an increase or decrease in the Ca^{2+} sensitivity of the contractile apparatus without a change in $[Ca^{2+}]_i$ or myosin RLC phosphorylation. Third, the ability of smooth muscle cells to accommodate to stretch and to generate *tonic* force complicates the effort to correlate active tension with smooth muscle "sarcomere" length.

Studies at the molecular level, which are not subject to the complications just described, reveal that smooth muscle actin-myosin interactions are similar to those in skeletal muscle. With phosphorylated smooth muscle myosin molecules tethered to a surface, single actin filaments can be moved along the myosin heads as a result of ATP-dependent cross-bridge cycling. Thus, there is no doubt that the basis of force development in smooth muscle, as in skeletal muscle, is the sliding of thin actin filaments along the thick myosin filaments.

The Velocity of Shortening Is Much Lower in Smooth Muscle Than in Skeletal Muscle

Despite the similarity in the mechanisms of force generation in smooth muscle and skeletal muscle, some critical kinetic differences exist. This is exemplified by the force-velocity relationship illustrated in Figure 16-13. The shapes of the skeletal and smooth muscle curves are similar (Figure 16-13A and B). However, over the entire range of forces, from zero force (i.e., $F = 0$, shortening with no load) to maximal force (just sufficient to prevent shortening at $F = F_0$), the velocity of contraction of tonic smooth muscle is less than one-tenth that of skeletal muscle. This is true even when the smooth muscle myosin RLC is maximally phosphorylated.

When smooth muscle cells are maximally activated and $[Ca^{2+}]_i$ is maximally elevated, approximately 60% of the RLC is phosphorylated. This level of phosphorylation is associated with maximal steady force development (Figure 16-13D). Interestingly, smooth muscles and skeletal muscles generate comparable **stress**, which is the active force per unit cross-sectional area. This may seem surprising in view of the finding that smooth muscles contain only about one-fifth as much myosin as do skeletal muscles. How can these observations be explained? One possibility is that smooth muscle myosin may generate much more force than skeletal muscle myosin. Alternatively, the explanation may lie in the structural arrangement of the contractile apparatus. These ideas are explored next.

Single Actin-Myosin Molecular Interactions Reveal How Smooth and Skeletal Muscles Generate the Same Amount of Stress despite Very Different Shortening Velocities

Formation of a single actin-myosin cross-bridge leads to a *unitary displacement step* of the actin by approximately 10 nm (1 nm = 10^{-9} m) along either skeletal muscle myosin or smooth muscle myosin. In other words, the unitary displacement step is the same length in both types of muscle. At the molecular level, shortening (i.e., filament sliding) velocity is defined as the length of the unitary displacement step (d) divided by the length of time (t_s) that the myosin head remains attached to actin (Figure 16-14); that is, velocity = d/t_s. Thus, with identical unitary displacement steps, the lower shortening velocity of smooth muscle implies a longer *step duration* (t_s) in smooth muscle.

The force that a muscle produces under isometric conditions can also be described at the molecular level. This force is related to both the *unitary force* (F_{uni}; Figure 16-14) generated by a single actin-myosin cross-bridge and its *duty cycle ratio* (t_s/t_{cycle}), which is the fraction of the total cross-bridge cycle time (t_{cycle}; Figure 16-14) that myosin is attached to actin and generating force (t_s). Interestingly, both skeletal and smooth muscle myosins are able to generate similar unitary force per attached cross-bridge (~1 to 3×10^{-12} Newtons, or 1 to 3 pN; see Box 16-2). In contrast, the duty cycle ratio (Figure 16-14) for tonic smooth muscle myosin is fourfold greater than in skeletal muscle. This difference in the duty cycle ratio of the actin-myosin interaction appears to account for the

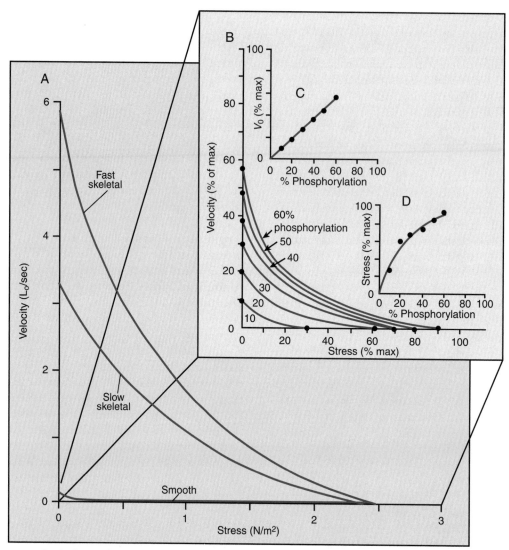

FIGURE 16-13 ■ Smooth muscle stress-velocity relationship; comparison with skeletal muscle. **A,** Relationship between the shortening velocity (in muscle lengths per second [L_0/sec]) and stress (in Newtons/m²) for human fast and slow skeletal muscles and smooth muscle. **B,** Shortening velocity is expressed as a percentage of the maximum unloaded shortening velocity (i.e., stress = 0). Note that the shape of the curve is similar to those for skeletal muscle in **A. C,** Relationship between the percentage of myosin RLC phosphorylation and the unloaded shortening velocity, V_0 (i.e., the y-intercepts in **B**). **D,** Relationship between the percentage of RLC phosphorylation and the percentage of maximal developed stress (i.e., the x-intercepts in **B**). *(Redrawn from Berne RM, Levy MN, editors:* Principles of physiology, *St Louis, 1990, Mosby.)*

ability of smooth muscle to generate comparable stress to skeletal muscle despite having only one-fifth as much myosin. Smooth muscle's lower shortening velocity and enhanced force generation (per mole of myosin) are, therefore, the result of differences in the kinetics of the cross-bridge cycle. The rate-limiting step in the cross-bridge cycle is cross-bridge detachment: the slower the detachment, the slower the filament sliding (and contraction). This detachment, which is limited by the slow rate of ADP dissociation from the myosin head,

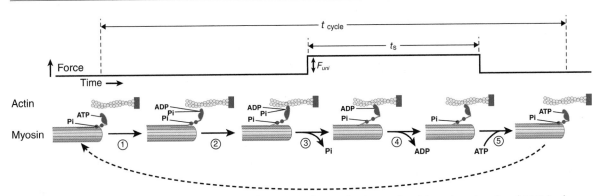

FIGURE 16-14 ■ The actin-myosin cross-bridge cycle in smooth muscle. A myosin with a phosphorylated RLC is shown. Starting with ATP bound to the unattached myosin, in step 1 the ATP is hydrolyzed to ADP + Pi. This recocks the myosin head and enables step 2, the attachment of the myosin head to actin (i.e., the formation of the cross-bridge). In step 3, Pi is released and force is generated (the "power stroke") as the myosin head bends back (to the *left* in this diagram). In step 4, the rate-limiting step, the ADP dissociates. This dissociation is much slower in smooth muscle than in skeletal muscle and is even slower in smooth muscles if the myosin RLC is dephosphorylated while the cross-bridge is still attached (not shown; see Boxes 16-4 and 16-6). Step 5, ATP binding, is rapid and causes cross-bridge detachment. The cycle is then repeated. The time line at the *top* shows the portion of the cycle (t_s) during which force is developed. F_{uni}, unitary force; T_{cycle}, cross-bridge cycle time.

influences both shortening velocity and force generation. The slower detachment rate in smooth muscle also contributes to smooth muscle's high duty cycle ratio. The result is that, at any given moment, more myosin heads are attached to actin and generating force in smooth muscle than in skeletal muscle.

Velocity of Smooth Muscle Shortening and the Amount of Stress Generated Depend on the Extent of Myosin Light Chain Phosphorylation

Myosin RLC phosphorylation triggers activation of smooth muscle contraction (see Figure 14-12). Therefore, it is not surprising that the velocity of shortening in the absence of a load (Force = 0) is directly proportional to the extent of RLC phosphorylation (see Figure 16-13C). When the smooth muscle is maximally activated with no load, shortening velocity is a measure of the maximal rate of cross-bridge cycling.

In contrast, the amount of stress that can be generated, a measure of the number of attached actin-myosin cross-bridges, is a nonlinear function of the level of RLC phosphorylation (see Figure 16-13D). The maximum number of cross-bridges is formed under isometric conditions, when maximum stress is developed. Thus, as in skeletal muscle, smooth muscle shortening velocity is determined by the rate of cross-bridge cycling, and stress is a function of the number of attached cross-bridges. In smooth muscles, but not skeletal muscles, however, both the average cross-bridge cycling rate and the number of attached cross-bridges depend on the extent of RLC phosphorylation.

The Kinetic Properties of the Cross-Bridge Cycle Depend on the Myosin Isoforms Expressed in the Myocytes

Expression of two C-terminal (SM1, SM2) and two N-terminal (SM-A, SM-B) alternative splice-variants of the smooth muscle myosin heavy chain, all products of a single gene, influences contraction kinetics. For example, the SM-B isoform, with a seven amino acid insert near the N-terminus, is predominantly expressed in **phasic smooth muscles** (e.g., esophageal smooth muscles). This may explain the greater myosin ATPase activity, lower ADP affinity, and faster shortening velocity than in **tonic smooth muscles** (e.g., vascular smooth muscles), which express predominantly SM-A (no insert).

BOX 16-4

PROSTATE GLAND HYPERTROPHY OBSTRUCTS URINE OUTFLOW FROM THE URINARY BLADDER AND INDUCES A CHANGE IN THE EXPRESSION OF MYOSIN ISOFORM IN BLADDER SMOOTH MUSCLE

A common form of urinary bladder outflow obstruction occurs in older men as a result of the progressive enlargement with age of the prostate gland that surrounds the urethra. This condition, benign prostatic hypertrophy (BPH), is actually neither benign nor a hypertrophy (cell enlargement). The prostate enlargement is primarily the result of hyperplasia (an increase in cell number), and it may cause serious problems, such as acute urinary retention (i.e., a sudden inability to urinate), which requires immediate medical intervention.

In mice, urethral obstruction increases myosin SM1 isoform expression and reduces the SM2/SM1 ratio—perhaps a compensation for the slow and incomplete bladder emptying. These effects of the obstruction are reversible. Removal of the obstruction (in the case of prostate hypertrophy, surgical extirpation of some prostate tissue) leads to an increase in the SM2/SM1 ratio, improves the urinary stream, and promotes more complete emptying of the urinary bladder.

SM1 and SM2 isoforms are both expressed in all smooth muscles, but the SM2/SM1 ratio differs in different smooth muscles. In genetically engineered mice, increasing SM1 expression increases shortening velocity and isometric force development, whereas increasing SM2 expression has the opposite effects. The interrelationship between the C-terminal and N-terminal variants and smooth muscle kinetics has not been elucidated.

Pathological conditions (or even normal aging) may alter myosin isoform expression and, thus, the kinetic properties of smooth muscles. A good example is the slowed urinary bladder contraction and incomplete bladder emptying that results from partial obstruction of the urethra secondary to prostate hypertrophy; this is associated with increased expression of the SM1 isoform (Box 16-4).

THE RELATIONSHIPS AMONG INTRACELLLULAR Ca²⁺, MYOSIN LIGHT CHAIN PHOSPHORYLATION, AND FORCE IN SMOOTH MUSCLES IS COMPLEX

The time courses of changes in $[Ca^{2+}]_i$, myosin RLC phosphorylation, and mechanical force in a tonic smooth muscle during brief and sustained stimulation are compared in Figure 16-15. The response of the smooth muscle to a brief stimulation is similar to, but

slower than, that of skeletal muscle after a single AP. Brief stimulation of the smooth muscle evokes a rapid, transient rise in $[Ca^{2+}]_i$. This is followed, after a short delay, by phosphorylation of RLC and, after a further delay, by a transient increase in mechanical force. The force reaches a peak at about the time that $[Ca^{2+}]_i$ has nearly recovered to its resting level, and the RLC phosphorylation and force also then slowly return to their resting levels.

When a tonic smooth muscle, such as arterial smooth muscle, is activated by a large, prolonged stimulus, $[Ca^{2+}]_i$ transiently rises to a peak synchronously in all cells (Figure 16-15B, and Figure 16-16A to C). $[Ca^{2+}]_i$ then declines and begins to oscillate, but the oscillations are asynchronous from cell to cell (Figure 16-16A and B, show the behavior of single cells). Thus, although $[Ca^{2+}]_i$ averaged over many cells maintains a steady level substantially above the resting level (see Figure 16-15B; Figure 16-16C), $[Ca^{2+}]_i$ actually fluctuates markedly in individual cells (Figure 16-16A and B).

Tonic Smooth Muscles Can Maintain Tension with Little Consumption of ATP

A characteristic of tonic smooth muscles, such as those in the walls of arteries or the lower esophageal sphincter, is their ability to remain at least partly contracted for very long periods. This tonic smooth

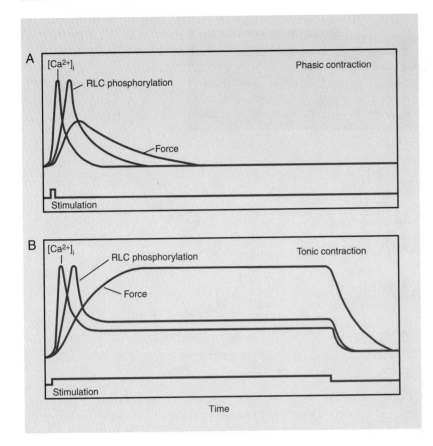

FIGURE 16-15 ■ Time courses of average $[Ca^{2+}]_i$, myosin RLC phosphorylation, and force development in smooth muscle with a brief (phasic) stimulus (**A**) and a sustained (tonic) stimulus (**B**). $[Ca^{2+}]_i$ shown is the spatially averaged $[Ca^{2+}]_i$ from a large number of smooth muscle cells in the arterial wall (see Figure 16-16). In tonic smooth muscle (e.g., arterial smooth muscle), with prolonged stimulation (**B**), $[Ca^{2+}]_i$ and RLC phosphorylation peak quickly and then decline to much lower, sustained levels, whereas force rises to a maximum sustained level. *(Data from Berne RM, Levy MN, editors: Principles of physiology, St Louis, 1990, Mosby.)*

muscle contraction is maintained despite minimal consumption of ATP and, thus, a low cross-bridge cycling rate. How is this possible?

ATP is consumed during active cross-bridge cycling as a result of the myosin ATPase (see Chapter 14). Skeletal muscle is approximately four fold more *efficient* than arterial smooth muscle, in which the **muscle efficiency** is the amount of mechanical work (i.e., shortening with a load) that can be performed per mole of ATP consumed. In contrast, arterial smooth muscle is at least 300-fold more *economical* than skeletal muscle in terms of using ATP to *maintain* stress. The **economy of contraction** is the amount of stress that can be maintained at a given rate of ATP consumption. Underlying the maintenance of this stress or tone in smooth muscle may be the low rate of ADP dissociation from attached cross-bridges with a dephosphorylated myosin RLC (Box 16-5; see Figure 16-14).

Perspective: Smooth Muscles Are Functionally Diverse

The preceding discussion indicates the wide range of properties exhibited by smooth muscles. Urinary bladder smooth muscle is an example of a phasic smooth muscle that is relatively relaxed most of the time; the bladder muscle relaxes more as the bladder fills with urine. This smooth muscle then contracts when neuronally activated to empty the bladder. Most gastric and intestinal smooth muscles, which are also phasic, usually contract rhythmically, so that they are rarely fully relaxed.

Among the tonic smooth muscles, sphincters such as the lower esophageal sphincter occupy the opposite end of the spectrum. This sphincter is closed (i.e., the smooth muscle is contracted) most of the time; the smooth muscle relaxes during brief periods to permit a food bolus to move from the lower esophagus into

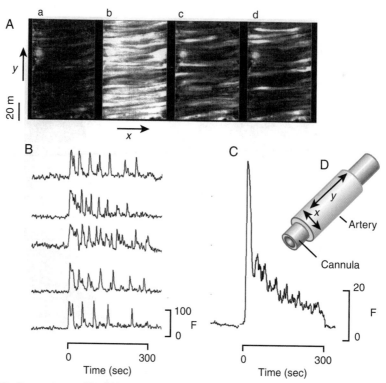

FIGURE 16-16 ■ **A** to **D**, Comparison of $[Ca^{2+}]_i$ in individual arterial smooth muscle cells and the average $[Ca^{2+}]_i$ in a large group of smooth muscle cells. The increase in $[Ca^{2+}]_i$ was evoked by sustained application of a relatively high concentration (5 μM) of the α_1-adrenergic receptor agonist phenylephrine (PE) in a rat small mesenteric artery (~150 μm diameter). The artery was stretched on a glass cannula (**D**) to prevent vasoconstriction. The myocytes form a single circular layer in the artery wall around the lumen. The lumen runs vertically (y-axis) and is perpendicular to the long axis of the cells (x-axis). Cells in the artery were loaded with a Ca^{2+}-sensitive fluorescent dye whose brightness (F) increases with $[Ca^{2+}]_i$. **A,** Fluorescence images of the myocytes in one wall of the artery before (a) and during (b to d) PE application; b, c, and d were acquired at approximately 20, 70, and 120 seconds after PE application, respectively. Fluorescence is very low at resting $[Ca^{2+}]_i$ (~0.1 μM) and increases markedly when $[Ca^{2+}]_i$ peaks (~1 μM). Approximately 25 myocytes are visible in b, which shows that $[Ca^{2+}]_i$ rose synchronously in virtually all cells on PE addition. **B,** Fluorescence changes in five representative cells. **C,** Fluorescence change averaged over all cells imaged in **A**; this curve is comparable to the $[Ca^{2+}]_i$ curve in Figure 16-15B. *Time scale bars at the bottom of* **B** *and* **C** *marked the duration of PE application.* (From Zang W-J, Balke CW, Wier WG: Cell Calcium 29:327, 2001.)

the stomach. Arterial smooth muscle, another tonic muscle, is usually partially contracted and may further contract or relax to alter local blood flow.

To carry out their diverse activities, the many different smooth muscles must express different myosin isoforms and different complements of agonist receptors, ion channels, and ion carriers and pumps. These membrane proteins not only play key roles in conferring the cell-specific properties, but also may be good targets for drug development.

BOX 16-5

MAINTENANCE OF FORCE WITH MINIMAL ATP HYDROLYSIS CAN BE EXPLAINED BY SLOW DETACHMENT OF CROSS-BRIDGES CONTAINING DEPHOSPHORYLATED MYOSIN LIGHT CHAIN

A critical question in smooth muscle physiology is, "How is force maintained in tonic smooth muscles with very little consumption of ATP?" Smooth muscle myosin in which the RLC is phosphorylated undergoes the cross-bridge cycle and consumes ATP continuously (see Chapter 14). When the RLC becomes dephosphorylated while the cross-bridge is still attached, however, detachment is greatly slowed. Thus, force can be maintained without continually detaching and reattaching the actin-myosin cross-bridges and concomitant hydrolysis of ATP. The slowed cycling rate is apparently attributable to a large increase in the affinity of myosin for ADP and a reduction in affinity for ATP that occur when the attached cross-bridges are dephosphorylated. The force is maintained because the actin cannot detach from the myosin until ADP dissociates and ATP binds to the myosin.

SUMMARY

1. Shortly after a single AP in a skeletal muscle cell, $[Ca^{2+}]_i$ rises, causing a transient, twitch contraction. Contractile force develops with slower time course than the rise in $[Ca^{2+}]_i$ because elastic elements must be stretched before external force can be generated.

2. A single α motor neuron and all the skeletal muscles it innervates are collectively referred to as a motor unit. The force generated by whole muscle can be increased by recruiting motor units. Muscle force can also be increased by fast repetitive stimulation.

3. When skeletal muscle is activated and filament sliding begins, the muscle can generate force, or it can shorten, or both can occur. In an isometric contraction the muscle generates force at a constant length. In an isotonic contraction the muscle shortens while generating a constant amount of force. Most contractions are part isometric and part isotonic.

4. Skeletal muscle mechanics are characterized by two relationships: the length-tension curve and the force-velocity relationship.

5. The relationship between length and tension is consistent with the sliding filament mechanism. Active force is maximal at the optimal length, L_0, where overlap between thick and thin filaments is maximal. Active tension declines at lengths both shorter and longer than L_0 because fewer cross-bridges can form.

6. The active force generated by a muscle is proportional to the number of attached cross-bridges.

7. During isotonic shortening, the velocity of shortening is inversely related to the load. The maximum velocity of shortening (V_0) occurs with no load.

8. The rate of muscle shortening reflects the rate of cross-bridge cycling and is directly proportional to myosin ATPase activity.

9. The three main motor unit types are fast-twitch, fatigable (FF); fast-twitch, fatigue-resistant (FR); and slow (S). The time to peak twitch force is shorter in FF and FR motor units than in S units. FF units generate the largest tetanic forces, S units generate the smallest forces, and FR units are intermediate.

10. Cardiac muscle generates long-duration twitch-type contractions caused by prolonged Ca^{2+} entry through voltage-gated Ca^{2+} channels that open during the AP.

11. The total force generated by cardiac muscle is directly related to the level of $[Ca^{2+}]_i$ attained during the contraction. Agents that elevate $[Ca^{2+}]_i$ and augment force, such as cardiotonic steroids, are said to produce a positive inotropic effect.

12. Cardiac and skeletal muscle cells have qualitatively similar length-tension relationships, but the velocity of shortening is much slower in cardiac than in skeletal muscle.

13. The kinetic properties of smooth muscle contraction can be defined in terms of three relationships: the temporal sequence that relates cell activation to the rise in $[Ca^{2+}]_i$ and force development, the force-velocity relationship, and the length-tension relationship.

14. Like skeletal muscle, smooth muscle has an optimal length, L_0, at which active tension is maximal; active tension declines at shorter and longer lengths. This is consistent with the sliding filament mechanism.

15. Smooth muscles and skeletal muscles generate about the same amount of active force per unit cross-sectional area (= stress). Nevertheless, shortening velocity is approximately 10-fold slower in smooth muscles than in fast skeletal muscles.

16. Single actin-myosin cross-bridges formed by smooth muscle and skeletal muscle myosin molecules generate similar unitary forces and have similar unitary displacement steps during a single cross-bridge cycle. The duration of force maintenance by cross-bridges formed by smooth muscle myosin is much longer than the duration of skeletal muscle cross-bridges.

17. In smooth muscle, but not skeletal muscle, the cross-bridge cycling rate and the number of attached cross-bridges depend on myosin RLC phosphorylation. Maximum stress is developed when the maximum number of cross-bridges is formed under isometric conditions.

18. The velocity of smooth muscle shortening is directly proportional to the extent of RLC phosphorylation. In the absence of a load, shortening velocity is a measure of the maximal rate of cross-bridge cycling.

19. Shortening velocity depends on the myosin isoform. Phasic smooth muscles shorten more rapidly than tonic smooth muscles because of differences in myosin isoform expression.

20. Pathological conditions may alter myosin isoform expression and thus the speed of shortening (e.g., partial obstruction of the urethra may slow urinary bladder smooth muscle contraction).

21. The brief contractions observed in phasic smooth muscles, as well as in tonic smooth muscles after brief stimulation, are associated with transient increases in $[Ca^{2+}]_i$, RLC phosphorylation, and tension development.

22. During prolonged stimulation of tonic smooth muscle cells, $[Ca^{2+}]_i$ oscillates after the initial transient rise, whereas RLC phosphorylation, after an initial peak, declines to a much lower steady level. During this time, tension rises slowly to a high, sustained level (= tonic tension or tone).

23. Smooth muscle tone is maintained with little consumption of ATP because of the slow cross-bridge cycling rate. The reason is a slow rate of dissociation of ADP from attached cross-bridges when RLC is dephosphorylated.

KEY WORDS AND CONCEPTS

- Motor unit
- Twitch
- Series elastic elements
- Tetanus
- Length-tension curve
- Force-velocity relationship
- Isometric and isotonic contractions
- Active and passive tension
- Optimal length (L_0)
- Fast-twitch, fatigable (FF); fast-twitch, fatigue-resistant (FR); slow (S) motor units
- Fatigue
- Positive inotropic effect
- Frank-Starling relationship
- Stress = force per unit cross-sectional area
- Phasic smooth muscles
- Tonic smooth muscles
- Muscle efficiency = amount of mechanical work performed per mole of ATP consumed
- Economy of contraction = stress maintained at a given rate of ATP consumption

STUDY PROBLEMS

1. Consider the partly isometric, partly isotonic contraction shown in the figure below (similar to the ones shown in Figure 16-6), in which the initial muscle length is 2.8 μm per sarcomere. On this figure, identify the following four phases of the contraction: isometric force development, isotonic shortening, isotonic relaxation, and

isometric relaxation. Then, on the skeletal muscle length-tension curve (see Figure 16-5), indicate each of these phases of the contraction.

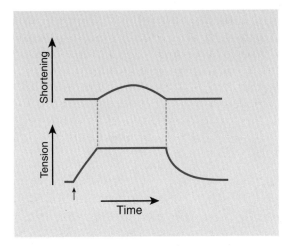

Graphs show the time course of the change in tension (*lower trace*) and the muscle shortening (*top trace*) during a partly isometric, partly isotonic contraction.

2. Stimulation of β-adrenergic receptors on ventricular myocytes increases contractility in these cells, by increasing the maximum force at all muscle lengths along the ascending limb of the length-tension curve. What effect would this increase in contractility have on the change in ventricular volume following contraction?

3. Compare the primary mechanisms used to change the amount of force generated by whole skeletal muscle with those used by cardiac ventricular muscle.

4. How do the mechanisms that control shortening velocity and cross-bridge cycling rate enable some smooth muscles to maintain tone without fatigue? Why is this an important property that is observed in many smooth muscles, but is undesirable in skeletal or cardiac muscles?

BIBLIOGRAPHY

Arafat HA, Kim GS, DiSanto ME, et al: Heterogeneity of bladder myocytes in vitro: modification of myosin isoform expression, *Tissue Cell* 33:219, 2001.

Babu GJ, Warshaw DM, Periasamy M: Smooth muscle myosin heavy chain isoforms and their role in muscle physiology, *Microsc Res Tech* 50:532, 2000.

Blinks JR, Rudel R, Taylor SR: Calcium transients in isolated amphibian skeletal muscle fibres: detection with aequorin, *J Physiol* 277:291, 1978.

Braunwald E, Sonnenblick EH, Ross J: Mechanisms of cardiac contraction and relaxation. In Braunwald E, editor: *Heart disease: a textbook of cardiovascular medicine*, ed 4, Philadelphia, PA, 1992, WB Saunders.

Burke RE: The structure and function of motor units. In Karpati G, Hilton-Jones D, Griggs RC, editors: *Disorders of voluntary muscle*, Cambridge, UK, 2001, Cambridge University Press.

Burke RE, Levine DN, Tsairis P, et al: Physiological types and histochemical profiles in motor units of the cat gastrocnemius, *J Physiol* 234:723, 1973.

Gordon AM, Huxley AF, Julian FJ: The variation in isometric tension with sarcomere length in vertebrate muscle fibers, *J Physiol* 184:143, 1966.

Guilford WH, Warshaw DM: The molecular mechanics of smooth muscle myosin, *Comp Biochem Physiol B Biochem Mol Biol* 119:451, 1998.

Hill AV: The heat of shortening and the dynamic constants of muscle, *Proc R Soc Lond B* 126:136, 1938.

Miriel VA, Mauban JR, Blaustein MP, et al: Local and cellular Ca^{2+} transients in smooth muscle of pressurized rat resistance arteries during myogenic and agonist stimulation, *J Physiol* 518:815, 1999.

Morano I: Tuning smooth muscle contraction by molecular motors. *J Mol Med* 81:41, 2003.

Murphy RA: Muscle cells of hollow organs, *News Physiol Sci* 3:124, 1988.

Murphy RA: Smooth muscle. In Berne RM, Levy MN, Koeppen BM, et al, editors: *Physiology*, ed 4, St Louis, MO, 1998, Mosby.

Rovner AS, Fagnant PM, Lowey S, et al: The carboxy-terminal isoforms of smooth muscle myosin heavy chain determine thick filament assembly properties, *J Cell Biol* 156:113, 2002.

Tyska MJ, Warshaw DM: The myosin power stroke, *Cell Motil Cytoskeleton* 51:1, 2002.

EPILOGUE

This book provides a conceptual view of the currently understood molecular mechanisms responsible for maintaining cellular solute and solvent homeostasis. We also discuss the mechanisms of electrical signaling, synaptic physiology and muscular contraction. To appreciate how far the study of cellular physiology has come, consider that, in 1950, none of the mechanisms described in this book was understood at the molecular level. The cellular physiological processes were then known only at the phenomenological (descriptive) level. For example, in the 1950s electrical signals were just starting to be recognized in terms of "ionic conductances," and transport processes were identified only on the basis of solute and water fluxes. The sliding filament mechanism of muscle contraction was proposed in 1954.

Cellular physiology was revolutionized by the identification of the specific proteins responsible for transport phenomena. The application of molecular biological methods led to the cloning, expression, and molecular manipulation of these proteins. Structural analysis of the proteins by x-ray crystallography, exemplified by K^+ channels, the Na^+ pump, and SERCA, is illuminating the precise mechanisms by which these molecules are gated and transfer solutes across a membrane. Analysis of single-molecule function, made possible by the patch clamp for ion channels and the laser trap for molecular motors, has greatly facilitated the elucidation of physiological mechanisms.

The importance of understanding molecular mechanisms is illustrated by a brief history of the physiology of gastric acid secretion and the therapeutics of gastric hyperacidity (heartburn), peptic ulcers, and gastric reflux disease, as introduced in Chapter 11. The clinical problem was known to the early Greeks and Romans (e.g., Diocles of Carystos and Aurelius Cornelius Celsus), who recognized sour eruptions and epigastric hunger pains. The "modern" era began with the English physician and physiological chemist Prout, who demonstrated in 1824 that the stomach secretes hydrochloric acid. In the midnineteenth century the Austrian physician Rokitansky and the French physician Cruveilhier first prescribed chalk and alkaline substances to treat gastric ulcers. Numerous newer antacid formulations such as Tums and Maalox came into widespread clinical use during the twentieth century. These "first-generation" therapeutic agents were the mainstay of pharmacological therapy for peptic ulcers, heartburn, and gastroesophageal reflux. An important side effect of this therapy is the alkalinization of the urine, which greatly increases the tendency to form kidney stones.

In 1902 the Russian physician Pavlov demonstrated that the vagus nerve plays a key role in the neural control of gastric secretion. Nineteen years later, the French physician Latarjet treated peptic ulcers by surgical ablation of the branch of the vagus nerve that innervates the stomach (vagotomy). This procedure was not immediately accepted but came into more frequent use when it was reintroduced by the Chicago surgeon and physiologist Dragstedt in 1943. Vagotomy is a major surgical procedure that, despite relieving ulcer pain and reducing gastric acidity, is accompanied by some unpleasant gastrointestinal side effects. These include nausea and the feeling of fullness in the stomach as a result of gastric retention. Thus the use of vagotomy in the therapy of peptic ulcer disease has been abandoned.

Knowledge of the details of neural control of gastric acid secretion, specifically the involvement of histamine as a neurotransmitter that activates gastric parietal cells by histamine type 2 (H_2) receptors, is relatively recent. Black and colleagues first described H_2 receptors in 1972. This mechanistic discovery led to the development, in the 1970s, of H_2-receptor blockers such as cimetidine (Tagamet) and ranitidine (Zantac). These "second-generation" compounds became a mainstay in the treatment of gastric hyperacidity. The H_2 antagonists are effective and well tolerated, and they have few side effects. They do not, however, totally suppress gastric acid secretion, and because they are short acting, they are usually administered twice daily.

The discovery of the parietal cell proton pump (H^+,K^+-ATPase) by Forte and Sachs and their colleagues in the 1970s resulted in the development, by Wallman and associates, of specific proton pump inhibitors such as omeprazole (Prilosec) and lansoprazole (Prevacid). These "third-generation" therapeutic agents were introduced into clinical use in the late 1980s. They are particularly effective because they directly block the acid secretion mechanism, are well tolerated, and have few side effects. Moreover, they are long-acting and can be administered in a single daily dose.

The preceding historical synopsis highlights the rationale for studying cellular physiological mechanisms as a basis for understanding therapeutics and current clinical practice. The aim is to develop the most specific therapies with the fewest side effects. Indeed, with basic molecular mechanisms being elucidated at an accelerating rate, we can expect the understanding of cellular physiology and its clinical applications to advance at an ever increasing pace.

BIBLIOGRAPHY

Black JW, Duncan WAM, Dunant CJ, et al: Definition and antagonism of histamine H_2-receptors, *Nature* 236:385, 1972.

Harvey SC: Gastric antacids and digestants. In Goodman LS, Gilman AG, editors: *The pharmacological basis of therapeutics*, ed 4, New York, 1970, Macmillan.

Hoogerwerf WA, Pasricha PJ: Agents used for control of gastric acidity and treatment of peptic ulcers and gastroesophageal reflux disease. In Hardman JG, Limbird LE, editors: *Goodman & Gilman's the pharmacological basis of therapeutics*, ed 10, New York, 2001, McGraw-Hill.

Modlin IM: From Prout to the proton pump: a history of the science of gastric acid secretion and the surgery of peptic ulcer, *Surg Gynecol Obstet* 170:81, 1990.

Sachs G: The gastric proton pump: the H^+,K^+-ATPase. In Johnson LR, editor: *Physiology of the gastrointestinal tract*, ed 2, New York, 1987, Raven Press.

APPENDIX A
Abbreviations, Symbols, and Numerical Constants

Abbreviations	Meaning
$\alpha_{GTP}, \alpha_{GDP}$	GTP- and GDP-bound α-subunit of heterotrimeric G-protein
ABC	ATP-binding cassette
ACh	Acetylcholine
AChE	Acetylcholinesterase
AChR	Acetylcholine receptor
ADH	Antidiuretic hormone (vasopressin)
ADP	Adenosine-5′-diphosphate
AE1	Anion exchanger type 1
AEA	*N*-arachidonoylglycerol
2-AG	2-Arachidonoylglycerol
AMPAR	α-Amino-3-hydroxy-5-methyl-4-isoxazole-propionate receptor
AP	Action potential
AQP-2	Aquaporin-2
AR	Adrenergic receptor
ATP	Adenosine-5′-triphosphate
BoTox	Botulinum toxin
CaM	Calmodulin
CaMKII	Ca^{2+}/CaM-dependent protein kinase II
cAMP	3′,5′-cyclic adenosine monophosphate
CFTR	Cystic fibrosis transmembrane conductance regulator
CICR	Ca^{2+}-induced Ca^{2+} release
cGMP	3′5′-cyclic guanosine monophosphate
CNS	Central nervous system
CO	Carbon monoxide
DAG	Diacylglycerol
DHPR	Dihydropyridine receptor
E-C coupling	Excitation-contraction coupling
ECF	Extracellular fluid
EPC	End-plate current
Epi	Epinephrine
EPP	End-plate potential
EPSC	Excitatory postsynaptic current
EPSP	Excitatory postsynaptic potential

Abbreviations	Meaning
ER	Endoplasmic reticulum
FF motor unit	Fast-twitch, fatigable motor unit
FR motor unit	Fast-twitch, fatigue-resistant motor unit
GABA	γ-Aminobutyric acid
GDP	Guanosine-5′-diphosphate
GHK	Goldman-Hodgkin-Katz
GHRH	Growth hormone–releasing hormone
GI	Gastrointestinal
GLUT	Glucose transporter
GlyR	Glycine receptor
GPCR	G-protein–coupled receptor
GRK	G-protein-coupled receptor kinase
GTP	Guanosine-5′-triphosphate
hERG	Human EAG-related gene
HO2	Hemoxygenase-2
5-HT	5-Hydroxytryptamine (serotonin)
ICF	Intracellular fluid
IP_3	Inositol 1,4,5-trisphosphate
IPSC	Inhibitory postsynaptic current
IPSP	Iinhibitory postsynaptic potential
KAR	Kainic acid (kainate) receptor
K_{ATP} channel	ATP-sensitive K^+ channel
K_{Ca} channel	Ca^{2+}-activated K^+ channel
Kir channel	Inward rectifier K^+ channel
LTD	Long-term depression
LTP	Long-term potentiation
mAChR	Muscarinic acetylcholine receptor
MEPP	Miniature end-plate potential
mEPSC	Miniature excitatory postsynaptic current
mGluR	Metabotropic glutamate receptor
MLCK	Myosin light chain kinase
MLCP	Myosin light chain phosphatase
MRP	Multidrug resistance protein
MW	Molecular weight
nAChR	Nicotinic acetylcholine receptor
NCX	Na^+/Ca^{2+} exchanger

Abbreviations	Meaning
NE	Norepinephrine
NHE	Na^+/H^+ exchanger
NKCC	Na^+-K^+-$2Cl^-$ cotransporter
NMDAR	N-methyl-D-aspartate receptor
NMJ	Neuromuscular junction
NO	Nitric oxide
NOS	Nitric oxide synthase
NPY	Neuropeptide Y
NSF	N-ethylmaleimide-sensitive factor
Pi	Phosphate
PIP_2	Phosphatidylinositol 4,5-bisphosphate
PKA	Protein kinase A
PKC	Protein kinase C
PLC	Phospholipase C
PM	Plasma membrane
PMCA	Plasma membrane Ca^{2+}-ATPase
PSD	Postsynaptic density
PTP	Post-tetanic potentiation
RBC	Red blood cell
Rim	Rab3-interacting molecule
RLC	Regulatory light chain
ROC	Receptor-operated channel
RyR	Ryanodine receptor
S motor unit	Slow-twitch, fatigue-resistant motor unit
SERCA	Sarcoplasmic/endoplasmic reticulum Ca^{2+}-ATPase
SGLT	Na^+-glucose cotransporter
SNAP	Soluble NSF-attachment protein
SNARE	Soluble NSF-attachment protein receptor
SOC	Store-operated channel
SR	Sarcoplasmic reticulum
SSRI	Serotonin-selective reuptake inhibitor
SV	Synaptic vesicle
TnC	Troponin C
TnI	Troponin I
TnT	Troponin T
TRP channel	Transient receptor potential channel
TTX	Tetrodotoxin
VAMP	Vesicle-associated membrane protein
V-ATPase	Vacuolar H^+-ATPase
VGCC	Voltage-gated Ca^{2+} channel
VIP	Vasoactive intestinal peptide

Symbols	Quantity Represented
β	Partition coefficient
γ	Single-channel conductance
$\Delta C/\Delta x$	Concentration gradient
λ	Length constant
μ_s	Electrochemical potential of solute S
π	Osmotic pressure
σ	Reflection coefficient
τ	Time constant
τ_m	Membrane time constant
τ_p	Action potential propagation time constant
C	Concentration
C, c	Capacitance
D	Diffusion coefficient
d_{RMS}	Root-mean-squared displacement
E_{ion}	Equilibrium potential of an ion (e.g., E_K is the K^+ equilibrium potential)
E_{rev}	Reversal potential
g	Conductance
I, i	Current
J	Flux
J_v	Volume flow
K_f	Filtration constant
L_0	Optimal length of muscle
L_p	Hydraulic conductivity
P	Hydrostatic pressure
p_o	Open probability of channel
P_S	Permeability coefficient of solute S
q	Charge
R, r	Resistance
$[S]$	Concentration of solute S
T	Absolute (Kelvin) temperature (= Celsius temperature + 273.15)
t	Time
V	Potential difference
V_0	Maximal velocity of muscle shortening
V_m	Membrane potential
z	Ionic charge

Numerical Constants

Quantity	Symbol	Value
Avogadro's number	N_A	6.022×10^{23} mole^{-1}
Elementary charge	e	1.602×10^{-19} coulombs
Faraday's constant	F	96,485 coulombs/mole
Universal gas constant	R	0.08205 liter·atmosphere·mole^{-1}·Kelvin^{-1} or 1.987 Joule·mole^{-1}·Kelvin^{-1}

APPENDIX B
A Mathematical Refresher

EXPONENTS

Definition of Exponentiation

Multiplying the number 3 by itself four times gives

$$3 \times 3 \times 3 \times 3 = 81$$

which can be written more simply with the shorthand notation

$$3^4 = 81$$

The shorthand notation 3^4 is read as "3 raised to the fourth power" or, more simply, "3 to the fourth power." The number 3 in the foregoing example is usually referred to as the base, and the number 4 is called the exponent. In the natural sciences the most frequently used base is an irrational number given the symbol e. An irrational number is just a number that cannot be written as a fraction. The square root of $2 (\sqrt{2})$ and the number pi (π) are examples of irrational numbers.

The number e is referred to as the base of the natural exponential function; the reason for this will become apparent in a later section. The value of e to three decimal places is 2.718.

The most general representation of exponentiation is

$$a^m = c$$

where a and m can be arbitrary numbers. A few numerical examples are

$$2^{13} = 8192; \quad 7^3 = 343; \quad 3.7^4 = 187.4161;$$

$$10^{3.32} = 2089.2961\ldots; \quad \text{and} \quad 2.13^{4.71} = 35.21015\ldots$$

The first three examples just given are readily verified by hand calculation. The last two examples could also be verified by hand with the information we will provide shortly, but the task is incomparably easier with the use of a calculator.

Multiplication of Exponentials

The rule for multiplying exponentials by combining exponents follows from the definition of exponentials:

$$a^m \times a^n = a^{(m + n)} \qquad [1]$$

This rule is easily verified by checking an example:

$$3^2 \times 3^4 = (3 \times 3) \times (3 \times 3 \times 3 \times 3) = 3^6 = 3^{(2 + 4)}$$

Meaning of the Number 0 as Exponent

All the other properties of exponentials follow directly from the rule for combining exponents. First, the rule allows us to deduce what a to the 0th power (a^0) means. To make the point concrete, take $a = 3$. Equation [1] allows us to write

$$3^2 \times 3^0 = 3^{(2 + 0)} = 3^2$$

This expression is true if, and only if, $3^0 = 1$. In general, any non-0 number raised to the 0th power is equal to 1:

$$a^0 = 1 \text{ (as long as } a \neq 0) \qquad [2]$$

The reason for excluding 0 from the definition will be clear shortly.

Negative Numbers as Exponents

So far we have dealt with positive exponents. What does a negative exponent mean? Equation [1] also allows us to deduce the answer. If we again use $a = 3$ as the base, Equation [1] implies that

$$3^2 \times 3^{-2} = 3^{[2 + (-2)]} = 3^0 = 1$$

This expression is true if, and only if, $3^{-2} = 1/3^2$. In general,

$$a^{-m} = \frac{1}{a^m} \text{ (as long as } a \neq 0) \qquad [3]$$

275

Division of Exponentials

The definition in Equation [3] extends the rule for combining exponents (Equation [1]) to include division of exponentials:

$$\frac{a^m}{a^n} = a^m \times a^{-n} = a^{(m-n)} \qquad [4]$$

We now see why in the definition of the 0th power (Equation [2]), the base a could not be 0. That is because $a = 0$ in Equation [4] would force a division of 0 by 0—an operation that has no meaning.

Exponentials of Exponentials

Knowing how to combine exponents allows us to see what happens when an exponential is raised to another power, for example, $(7^2)^3$:

$$(7^2)^3 = 7^2 \times 7^2 \times 7^2 = 7^{(2+2+2)} = 7^6$$

In general,

$$(a^m)^n = a^{(m \times n)} \qquad [5]$$

Fractions as Exponents

So far, only exponents that are whole numbers have been discussed. We now investigate the case in which the exponent is a fraction (i.e., $a^{m/n}$). Again, taking a concrete example in which $a = 7$, we ask what is meant by $7^{1/2}$. Applying the multiplication rule (Equation [1]) gives

$$7^{1/2} \times 7^{1/2} = 7^{(1/2 + 1/2)} = 7^1 = 7$$

which immediately shows that $7^{1/2} = \sqrt{7}$ (the square root of 7). In general, a to the $1/n$ is the nth root of a:

$$a^{\frac{1}{n}} = \sqrt[n]{a} \qquad [6]$$

A direct consequence of combining Equations [6] and [4] is that

$$a^{\frac{m}{n}} = a^{\left(m \times \frac{1}{n}\right)} = \left(a^m\right)^{\frac{1}{n}} = \sqrt[n]{a^m} = (a^{\frac{1}{n}})^m = \left(\sqrt[n]{a}\right)^m \qquad [7]$$

Because any rational decimal number can always be written as a fraction (e.g., $1.5 = 3/2$, $0.47 = 47/100$), decimal numbers in the exponent can be dealt with precisely as if they were fractions.

Equations [1] to [7] constitute essentially all the properties of exponentials that are important for computation. For convenience, all these properties are summarized in Box B-1.

LOGARITHMS

Definition of the Logarithm

The logarithm can be thought of as an inverse way to look at the exponential process. The general way of representing a logarithm is

$$\log_a c = m$$

This expression is read as "the logarithm 'to the base a' of the number c is equal to m." The meaning of this expression is "m is the power to which a must be raised in order to get the number c." However, this is really just another way of representing the exponential process $a^m = c$, which we had just examined (we may consider the logarithm the *inverse* of the exponential). The logarithm is the answer to the question, "To what power must a be raised to give the value c ($a^? = c$)?" The answer is the definition of the logarithm:

$$a^{\log_a c} = c$$

Examples corresponding to some of those in the previous section would be

$$\log_2 8192 = 13; \quad \log_7 343 = 3; \quad \log_{3.7} 187.4161 = 4;$$

$$\log_{10} 2089.2961 \ldots = 3.32, \quad \log_{2.13} 35.21015 \ldots = 4.71$$

BOX B-1
PROPERTIES OF EXPONENTIALS

$$a^m \times a^n = a^{(m+n)}$$

$$\frac{a^m}{a^n} = a^m \times a^{-n} = a^{(m-n)}$$

$$a^0 = 1 \text{ (as long as } a \neq 0)$$

$$a^{-m} = \frac{1}{a^m} \text{ (as long as } a \neq 0)$$

$$(a^m)^n = a^{(m \times n)}$$

$$a^{\frac{1}{n}} = \sqrt[n]{a}$$

$$a^{\frac{m}{n}} = a^{\left(m \times \frac{1}{n}\right)} = (a^m)^{\frac{1}{n}} = \sqrt[n]{a^m} = (a^{\frac{1}{n}})^m = (\sqrt[n]{a})^m$$

Historically, the number 10 is the most commonly used base for logarithms. For this reason, base-10 logarithms are referred to as common logarithms. In the natural sciences the irrational number e is usually used as the base for logarithms. Just as the exponential function using e as the base is called the natural exponential function, a logarithm using the base e is called the natural logarithm and is given the symbol ln. Thus the natural logarithm of x is symbolized as ln$x$.

The definitions given previously, along with the properties of exponentials, imply three rules governing calculations involving logarithms.

Logarithm of a Product
The logarithm of a product is given by

$$\log_a (b \times c) = \log_a b + \log_a c \qquad [8]$$

Examining the definition of a logarithm immediately shows where this rule comes from:

$$b = a^{\log_a b} \text{ and } c = a^{\log_a c}$$

so

$$b \times c = a^{\log_a b} \times a^{\log_a c} = a^{(\log_a b + \log_a c)}$$

(the last step is a consequence of the rule for combining exponents—Equation [1]). Therefore

$$\log_a (b \times c) = \log_a b + \log_a c$$

Incidentally, the product rule (in conjunction with the rule for taking the logarithm of an exponential—Equation [9]) also gives the logarithm of a quotient:

$$\log_a \frac{b}{c} = \log_a [b \times c^{-1}] = \log_a b + \log_a (c^{-1})$$

$$= \log_a b + (-1)\log_a c = \log_a b - \log_a c$$

Logarithm of an Exponential
The logarithm of an exponential is given by

$$\log_a c^m = m \times \log_a c \qquad [9]$$

This rule is a consequence of the rule governing logarithm of a product (Equation [8]) combined with the definition of an exponential:

$$\log_a c^m = \log_a (c \times c \times \ldots \times c)_{m \text{ times}}$$
$$= (\log_a c + \log_a c + \ldots + \log_a c)_{m \text{ times}} = m \times \log_a c$$

A special case of the logarithm of an exponential occurs frequently and is sometimes given as a separate rule:

$$\log_a \frac{1}{c} = \log_a c^{-1} = -1 \times \log_a c = -\log_a c \qquad [10]$$

Changing the Base of a Logarithm
For any number c, its logarithms using two different bases, a and b, are related by the expression

$$\log_a c = (\log_a b) \times \log_b c \qquad [11]$$

This rule is the result of combining the definition of a logarithm

$$c = b^{\log_b c}$$

with the rule for taking the logarithm of an exponential (Equation [9])

$$\log_a c = \log_a b^{(\log_b c)} = (\log_b c) \times \log_a b$$

which is the rule shown in Equation [11].

The rules of logarithms are summarized in Box B-2.

SOLVING QUADRATIC EQUATIONS

Any equation that can be put into the form

$$ax^2 + bx + c = 0 \qquad [12]$$

can be solved for the value of x through the *quadratic formula:*

$$x = \frac{-b \pm \sqrt{b^2 - 4ac}}{2a} \qquad [13]$$

The $\pm$ sign means that x may take on two different values, one corresponding to using the $+$ sign in the numerator, and one corresponding to using the $-$ sign.

BOX B-2
PROPERTIES OF LOGARITHMS

$$\log_a (b \times c) = \log_a b + \log_a c$$

$$\log_a \frac{b}{c} = \log_a b - \log_a c$$

$$\log_a c^m = m \times \log_a c$$

$$\log_a \frac{1}{c} = -\log_a c$$

$$\log_a c = (\log_a b) \times \log_b c$$

Solving the quadratic equation $4x^2 - 9 = 0$ illustrates the technique. Comparing this equation with the general form of the quadratic equation given above shows the coefficients to be $a = 4$, $b = 0$, and $c = -9$. The two solutions that will satisfy the equation are given by the quadratic formula:

$$x = \frac{-(0) \pm \sqrt{0^2 - 4(4)(-9)}}{2(4)} = \frac{\pm\sqrt{144}}{8} = \pm\frac{12}{8} = \pm\frac{3}{2}$$

So the two answers are $x = 3/2$ and $x = -3/2$ (or, $x = 1.5$ and $x = -1.5$).

DIFFERENTIATION AND DERIVATIVES

The Slope of a Graph and the Derivative

The derivative of a function at a particular point is most easily viewed as the slope of the function at that point. A slope is really just a rate of change of one variable relative to another. For a straight line, the slope, being the rate of change of y relative to x ($\Delta y/\Delta x$), is everywhere the same and is easy to compute, as shown in Figure B-1.

For functions with nonlinear graphs, determining the slope at a particular point on the curve is slightly more involved. Assume, for the moment, that the function of interest is the parabola, $y = x^2$, with the graph shown in Figure B-2. The general problem of finding the slope of the parabola at any particular point (x, y) is to find the slope of the *tangent* line that just touches the parabola at that point.

The general features of the problem are shown in Figure B-2. To determine the slope of the parabola at the point (x, y), we can first pass a test line through a second, arbitrary point $(x+\Delta x, y+\Delta y)$. As Δx becomes smaller, so will Δy, and eventually the test line should *become* the tangent line as Δx approaches 0. Because the points (x, y) and $(x+\Delta x, y+\Delta y)$ both lie on the parabola $y=x^2$, we can write the two points as (x, x^2) and $(x+\Delta x, (x+\Delta x)^2)$,

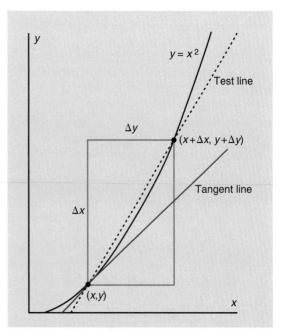

FIGURE B-2 ■ Graph of the parabola $y = x^2$. Two lines are also shown: (1) a tangent line (*solid*) that passes through one point, (x, y), on the parabola; and (2) a test line (*dashed*) that passes through (x, y), as well as a second point, $(x+\Delta x, y+\Delta y)$. As Δx (and thus Δy) grows ever smaller, the test line approaches the tangent line ever more closely. In the limit of infinitesimally small Δx (i.e., as $\Delta x \rightarrow 0$), the test line *becomes* the tangent line. At the point (x, y) the derivative of the function has a value equal to the slope of the tangent line.

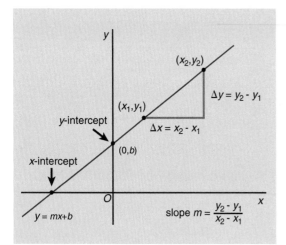

FIGURE B-1 ■ Graph of a line represented by the equation $y = mx + b$. The slope of the line is m, and b is the y-intercept; that is, the line crosses the y-axis at the point $(0, b)$. If two points lying on the line, (x_1, y_1) and (x_2, y_2), are known, the slope can be calculated: $m = \Delta y/\Delta x = (y_2 - y_1)/(x_2 - x_1)$.

respectively. The slope of the line connecting these points must be

$$\text{slope} = \frac{\Delta y}{\Delta x} = \frac{y_2 - y_1}{x_2 - x_1} = \frac{(x + \Delta x)^2 - x^2}{(x + \Delta x) - x}$$

$$= \frac{(x^2 + 2x\Delta x + \Delta x^2) - x^2}{\Delta x} = 2x + \Delta x$$

As Δx approaches 0, the slope, $\Delta y / \Delta x$, approaches $2x$. This means that at *any* arbitrary point on our parabola, the slope of the curve must be $2x$. This slope of the curve (which is equal to the slope of the tangent line touching the curve) is the *derivative* of the curve and is given the symbol $\frac{dy}{dx}$. To summarize:

$$\text{as } \Delta x \to 0, \quad \frac{\Delta y}{\Delta x} \to \frac{dy}{dx}$$

Therefore the slope at *any* point on the parabola $y = x^2$ is given by its derivative

$$\frac{dy}{dx} = 2x$$

Thus, at the point (2,4) on the parabola, the slope is $2(2) = 4$; at the point (5,25), the slope is $2(5) = 10$.

The same approach can be used to find the derivative of any function. For functions that involve only x raised to some power ($y = x^n$), it is relatively easy to show that

$$\frac{dy}{dx} = nx^{n-1} \quad \text{for} \quad y = x^n \qquad [14]$$

The general process whereby we find the derivative of a function is called *differentiation*. A few examples of derivatives that follow from formula [14] are

$$\frac{d}{dx} x^7 = 7x^{(7-1)} = 7x^6$$

$$\frac{d}{dx} x^{-3} = -3x^{(-3-1)} = -3x^{-4}$$

or written in alternative form:

$$\frac{d}{dx}\left(\frac{1}{x^3}\right) = \frac{-3}{x^4}$$

$$\frac{d}{dx}(x^{0.5}) = 0.5\, x^{(0.5-1)} = 0.5\, x^{-0.5}$$

or written in alternative form:

$$\frac{d}{dx}(\sqrt{x}) = \frac{1}{2}\frac{1}{\sqrt{x}}$$

A simple extension of formula [14] (which can be demonstrated by the methods we have already used) expands our ability to differentiate functions:

$$\frac{d}{dx}(ax^n) = a\frac{d}{dx}x^n = anx^{n-1}$$

In other words, multiplying an exponential function by a constant number a means that the derivative is also multiplied by the same factor. In general, for any function $f(x)$, if $\frac{df}{dx}$ is the derivative of the function, then

$$\frac{d}{dx}(a \times f) = a\frac{df}{dx} \quad \text{for} \quad a = \text{constant number} \qquad [15]$$

Derivative of a Constant Number

A constant number, c, can be viewed as the function $y = c$, which is a straight horizontal line intersecting the y-axis at c. The derivative of this function should just be the slope of the horizontal line. But a horizontal line has slope 0 (i.e., regardless of the size of Δx, Δy is always 0, which means that the slope $\Delta y / \Delta x = 0$). This result means that the derivative of any constant number is 0:

$$\frac{d}{dx}c = 0 \quad \text{for} \quad c = \text{constant number} \qquad [16]$$

Differentiating the Sum or Difference of Functions

Applying the "shrinking Δx" definition of the derivative gives us the following rules governing the differentiation of the sum or the difference of functions. Let $f(x)$ and $g(x)$ be different functions of x. If $y = f(x) + g(x)$,

$$\frac{dy}{dx} = \frac{d}{dx}[f(x) + g(x)] = \frac{d}{dx}f(x) + \frac{d}{dx}g(x) \qquad [17a]$$

And if $y = f(x) - g(x)$,

$$\frac{dy}{dx} = \frac{d}{dx}[f(x) - g(x)] = \frac{d}{dx}f(x) - \frac{d}{dx}g(x) \qquad [17b]$$

*An Example Illustrating the Computation
of a Derivative*

To illustrate all the computational rules we have derived
so far, consider differentiating the function

$$y = 7x^3 + \sqrt{x} - \frac{4}{x^3} + 19$$

This problem becomes easier if we rewrite the function using regular exponential notation to give

$$y = 7x^3 + x^{\frac{1}{2}} - 4x^{-3} + 19$$

We see that y is the sum and difference of four functions. The derivative can be obtained by differentiating each of the four parts in turn:

$$\frac{d}{dx}(7x^3) + \frac{d}{dx}(x^{\frac{1}{2}}) - \frac{d}{dx}(4x^{-3}) + \frac{d}{dx}(19)$$

$$= 7(3x^{(3-1)}) + \frac{1}{2}x^{\left(\frac{1}{2}-1\right)} - 4(-3x^{(-3-1)}) + 0$$

to give the result

$$\frac{dy}{dx} = 21x^2 + \frac{1}{2}x^{-\frac{1}{2}} + 12x^{-4}$$

*What Makes the "Natural" Exponential Function
Natural?*

Knowing how to find the derivative of a function, we can now investigate the natural exponential function $y = e^x$. What is the derivative of the natural exponential function? Proceeding as we had done for the parabola, the slope of the exponential function at some x is given by

$$\text{slope} = \frac{\Delta y}{\Delta x} = \frac{e^{(x+\Delta x)} - e^x}{\Delta x}$$

$$= \frac{e^x e^{\Delta x} - e^x}{\Delta x} = \frac{e^x(e^{\Delta x} - 1)}{\Delta x} \qquad [18]$$

As before, we try to obtain the derivative dy/dx by allowing Δx to approach 0. As Δx approaches 0, $e^{\Delta x}$ approaches $e^0 = 1$, and the term $(e^{\Delta x} - 1)$ approaches 0. Apparently, this would make the slope 0/0, which is meaningless. This approach clearly gives no real information about the behavior of the slope as Δx shrinks. A better way to examine the behavior of the slope as

$\Delta x \rightarrow 0$ is needed. First, we look closely at the behavior of $e^{\Delta x}$ when Δx becomes small. Table B-1 summarizes our numerical investigation.

Table B-1 shows that numerically, as Δx gets progressively closer to 0, $e^{\Delta x}$ approaches $(\Delta x + 1)$. The slope expression (Equation [18]) can now be written as

$$\text{slope} = \frac{\Delta y}{\Delta x} = \frac{e^x(e^{\Delta x} - 1)}{\Delta x}$$

$$= \frac{e^x[(\Delta x + 1) - 1]}{\Delta x} = \frac{e^x \Delta x}{\Delta x} = e^x$$

when Δx becomes very small. So, as $\Delta x \rightarrow 0$, the slope expression no longer contains Δx, and becomes the derivative dy/dx:

$$\frac{d}{dx}e^x = e^x \qquad [19]$$

The essence of the natural exponential function is that the value of its derivative at any point is the same as the value of the exponential function itself. Alternative ways of saying this are (1) the natural exponential function is its own derivative or (2) differentiation of the natural exponential function gives back the natural exponential function. This property of the natural exponential function makes it "natural."

Differentiating Composite Functions: the Chain Rule

Consider the function $y = x^6 - 1$, whose derivative we know to be $dy/dx = 6x^5$. One potentially different way of looking at this function is to think of it

TABLE B-1		
Behavior of $e^{\Delta x}$ as $\Delta x \rightarrow 0$		
Δx	$e^{\Delta x}$	$e^{\Delta x} - \Delta x$
0.1	1.1051709 . . .	1.0051709 . . .
0.05	1.0512710 . . .	1.0012711 . . .
0.01	1.0100501 . . .	1.0000501 . . .
0.005	1.0050125 . . .	1.0000125 . . .
0.001	1.0010005 . . .	1.0000005 . . .
0.0005	1.0005001 . . .	1.0000001 . . .
0.0001	1.0001000 . . .	1.0000000 . . .
0.00005	1.0000500 . . .	1.0000000 . . .
0.00001	1.0000100 . . .	1.0000000 . . .

as $y = (x^2)^3 - 1$. In other words, we think of $y = u^3 - 1$, where $u = x^2$ (y is a function of u, but u is, in turn, a function of x). In this context, y is called a *composite* function. We can think of the derivative of the composite function y as having two parts: dy/du and du/dx:

$$y = u^3 - 1, \quad \frac{dy}{du} = 3u^2; \quad u = x^2, \quad \frac{du}{dx} = 2x$$

Multiplying the two parts together gives

$$\frac{dy}{du} \cdot \frac{du}{dx} = (3u^2)(2x) = (3(x^2)^2)(2x)$$
$$= (3x^4)(2x) = 6x^5$$

which is what we had already determined at the outset. We thus infer the general property that when y is a function of u, and u is, in turn, a function of x,

$$\frac{dy}{dx} = \frac{dy}{du} \cdot \frac{du}{dx} \qquad [20]$$

This relationship is called the *chain rule* for the derivative of a composite function. It frequently occurs in the natural sciences that one function may be viewed as a function of another function. A less trivial example is

$$y = e^{-x^2}.$$

In this case the function y is a composite function: $y = e^u$, $u = -x^2$. The derivative is easily found by recalling that the derivative of e^u is e^u and applying the chain rule:

$$\frac{dy}{dx} = \frac{dy}{du} \cdot \frac{du}{dx} = e^u(-2x) = e^{-x^2}(-2x) = -2xe^{-x^2}$$

Derivative of the Natural Logarithm Function

In the section on logarithms we stated that the logarithm can be considered the *inverse* of the exponentiation process. We now wish to find the derivative of a logarithmic function. We can do this by using the idea of the *inverse* of a function. The definition is logical; for example, if $y = e^x$, the inverse would be $x = e^y$. If we take the natural logarithm of both sides of the inverse function, we find

$$\ln x = \ln(e^y) = y(\ln e) = y(1) = y \quad \text{or} \quad y = \ln x$$

Thus the inverse of the natural exponential function is indeed the natural logarithm function. Differentiating the inverse function means

$$\frac{d}{dx} x = \frac{d}{dx} e^y$$

But

$$\frac{d}{dx} x = 1 \quad \text{and} \quad \frac{d}{dx} e^y = e^y \frac{dy}{dx} = e^{\ln x} \frac{dy}{dx} = x \frac{dy}{dx}$$

(because $y = \ln x$ and $e^{\ln x} = x$), so

$$1 = x \frac{dy}{dx}$$

which means

$$\frac{dy}{dx} = \frac{1}{x}$$

or more explicitly

$$\frac{d}{dx} \ln x = \frac{1}{x} \qquad [21]$$

With the techniques developed here, any function that has a derivative can be differentiated. Box B-3 summarizes basic knowledge of differentiation and derivatives.

INTEGRATION: THE ANTIDERIVATIVE AND THE DEFINITE INTEGRAL

Indefinite Integral (Also Known as the Antiderivative)

The term integration has two meanings. First, it is a process whereby for a function $g(x)$, we find a function $f(x)$ such that

$$g(x) = \frac{d}{dx} f(x)$$

In other words, we find a function $f(x)$ whose derivative is the function $g(x)$. Because in this sense, integration is just the reverse of differentiation, the integration process is sometimes referred to as *antidifferentiation*. The function $f(x)$ is called either the *antiderivative* or the *indefinite integral* of $g(x)$. Symbolically, we represent the antidifferentiation process as

$$\int g(x)dx = f(x)$$

BOX B-3

COMMON DERIVATIVES
AND THEIR PROPERTIES

$$\frac{d}{dx}c = 0 \quad \text{for} \quad c = \text{constant number}$$

$$\frac{dy}{dx} = nx^{n-1} \quad \text{for} \quad y = x^n$$

$$\frac{d}{dx}e^x = e^x$$

$$\frac{d}{dx}\ln x = \frac{1}{x}$$

If a = constant number, $\dfrac{d}{dx}[a \times f(x)] = a\dfrac{d}{dx}f(x)$

$$\frac{d}{dx}[f(x) + g(x)] = \frac{d}{dx}f(x) + \frac{d}{dx}g(x) \quad \text{and}$$

$$\frac{d}{dx}[f(x) - g(x)] = \frac{d}{dx}f(x) - \frac{d}{dx}g(x)$$

$$\frac{dy}{dx} = \frac{dy}{du} \cdot \frac{du}{dx} \quad \text{if} \quad y = f(u) \quad \text{and} \quad u = g(x)$$

BOX B-4

INDEFINITE INTEGRALS
AND THEIR PROPERTIES

$$\int nx^{n-1}dx = x^n + C \quad \text{or} \quad \int x^m dx = \frac{1}{m+1}x^{(m+1)} + C$$

$$\int af(x)dx = a\int f(x)dx \quad \text{for} \quad a = \text{constant number}$$

$$\int e^x\, dx = e^x + C$$

$$\int \frac{1}{x}dx = \ln x + C$$

$$\int [f(x) + g(x)]dx = \int f(x)dx + \int g(x)dx$$

$$\text{and} \int [f(x) - g(x)]dx = \int f(x)dx - \int g(x)dx$$

For example, if we have a function $g(x) = nx^{(n-1)}$, its antiderivative is x^n because we recognize that $f(x) = x^n$ is the function, when differentiated, that gives $g(x)$. However, because the derivative of a constant number (C) is zero, we realize that it is not the *only* function whose derivative is $g(x)$; indeed, any function $f(x) = x^n + C$, when differentiated, will give $g(x) = nx^{(n-1)}$. Therefore

$$\int nx^{n-1}\, dx = x^n + C$$

Because C may take on any value, this expression tells us that really an infinite number of functions is the antiderivative of $g(x) = nx^{(n-1)}$. The constant, C, is referred to as the *constant of integration*. Because there is no *unique* antiderivative for $g(x)$, the antiderivative is also known as the *indefinite* integral. The properties of indefinite integrals corresponding to the properties of derivatives listed in Box B-3 are summarized in Box B-4.

Definite Integral

The second type of integration is conceptually equivalent to finding the area under a curve (Figure B-3).

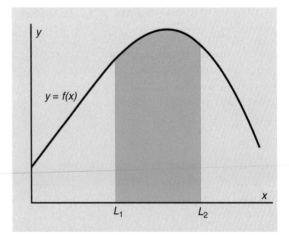

FIGURE B-3 ■ Graphic representation of the definite integral. The graph of the function $y = f(x)$ is shown. The definite integral $\int_{L_1}^{L_2} f(x)dx$ is the *shaded area* under the curve $y = f(x)$ between the limits, L_1 and L_2.

The area under the curve $y = f(x)$ in Figure B-3 is symbolized by the integral

$$\text{Area} = \int_{L1}^{L2} f(x)dx$$

In this context the function $f(x)$ is called the *integrand* and the two numbers L_1 and L_2 are the *lower* and *upper* limits of integration, respectively. To take a concrete example: $y = f(x) = x^2$, and we wish to find out the area under the curve between $x = 3$ and $x = 7$. The definite integral is

$$\int_{x=3}^{x=7} x^2 dx$$

This integral is evaluated in the following way. First we find the antiderivative of the integrand, which in the case of x^2 is $x^3/3$. Next we evaluate the antiderivative at the two integration limits and find the difference between them:

$$\int_{x=3}^{x=7} x^2 dx = \left[\frac{1}{3}x^3\right]_3^7 = \frac{7^3}{3} - \frac{3^3}{3} = \frac{343}{3} - \frac{27}{3} = \frac{316}{3}$$

We note that whereas an indefinite integral is a function, the definite integral is always a number.

DIFFERENTIAL EQUATIONS

First-Order Equations with Separable Variables

Differential equations encountered in cellular physiology are ones wherein the rate of change of a variable is equal to some function:

$$\frac{dy}{dt} = f(t,y)$$

To solve the differential equation is to find the functional form of y. The easiest differential equations to solve are ones in which the independent and dependent variables (t and y, respectively, in the example) can be separated into separate groupings. For example, the equation

$$\frac{dy}{dt} = f(y) \tag{22}$$

can be solved by separating the variables and integrating:

$$y = \int dy = \int f(t)dt \tag{23}$$

A simple physical situation that can be described by a differential equation of this type (Equation [22]) may be a vertical cylinder of constant radius, r, that is being filled so that the volume of water in the cylinder is increasing at the rate of v. Assuming that at time $t = 0$ the cylinder was already filled to a height, h, what is the

function that quantitatively predicts the height of liquid in the cylinder at any time? A volume, v, of water flowing into the cylinder will add a disk-shaped plug of liquid whose height is $v/\pi r^2$; therefore the rate of change of the height, y, is

$$\frac{dy}{dt} = \frac{v}{\pi r^2} \tag{24}$$

This equation is easily solved by direct integration:

$$\int dy = \int \left(\frac{v}{\pi r^2}\right) dt$$

Knowing that the term in parentheses in the right-hand integral is just a constant, we can find the solution from Box B-4:

$$y = \left(\frac{v}{\pi r^2}\right) t + C \tag{25}$$

C, the constant of integration, can be determined from the knowledge that at time $t = 0$, the cylinder was already filled to a height of $y = h$ (knowledge that allows us to determine the value of the constant of integration is referred to as a *boundary condition*). Substituting this information back into Equation [25] gives $C = h$. The exact solution to differential Equation [24] is thus

$$y = \left(\frac{v}{\pi r^2}\right) t + h \tag{26}$$

A graphical representation of the behavior of the liquid column as predicted by Equation [26] is shown in Figure B-4.

Exponential Decay

We now consider a differential equation that is frequently applicable in all natural sciences:

$$\frac{dy}{dt} = -ky \tag{27}$$

This equation basically says that the rate of decrease in some quantity, y, at a particular moment is directly proportional to the magnitude of y at that moment. This condition applies to loss of permeant solute from a compartment enclosed by a semipermeable membrane into a volume of fluid that is much larger than the volume of the compartment: the lower the solute concentration in the compartment, the fewer the

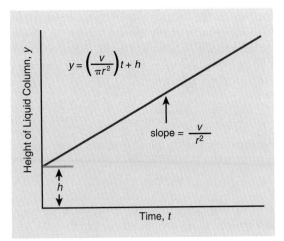

FIGURE B-4 ■ Graph of $y = (v/\pi r^2) \cdot t + h$ (Equation [26]). The equation is that of a straight line with slope $= v/\pi r^2$ and y-intercept $= h$.

number of molecules that can cross the membrane at any given moment, and the more slowly the concentration inside the compartment will change. Similarly, the condition applies in the case of radioactive decay, where the rate of decay (the rate of change of the quantity of radioactive atoms) is directly proportional to the number of radioactive atoms that are left (the fewer radioactive atoms there are, the fewer that can decay at any given moment).

Equation [27] can be solved by separating variables (collecting terms containing t with dt and collecting terms containing y with dy) and then integrating:

$$\int \frac{dy}{y} = \int -k\,dt$$

Again, the answer can be written down with the help of the integrals in Box B-4,

$$\ln y = -kt + C$$

and then rewritten in exponential notation,

$$y = e^{-kt} \cdot e^{C}$$

Using the boundary condition that at time $t = 0$, y had its initial value, y_0, we arrive at the exact solution of Equation [27]:

$$y = y_0 \cdot e^{-kt} \qquad [28]$$

Because the parameter, k, determines how fast y decreases, k is often referred to as the *rate constant*, which has dimensions of 1/time. The inverse of k ($1/k$), with dimensions of time, is known as the *time constant* (symbolized by the Greek letter τ). The larger the rate constant, k (or, the smaller the time constant, τ), the faster y decreases. The behavior predicted by Equation [28] is shown in Figure B-5.

First-Order Linear Differential Equations

The last type of differential equation useful in this book is a first-order linear equation, which can always be put into the form

$$\frac{dy}{dt} + Py = Q \qquad [29]$$

where P and Q are functions of the dependent variable, t. This equation can be transformed into a separable equation if the entire equation is multiplied by a correctly chosen function, which is referred to as an *integrating factor*, φ:

$$\varphi = e^{\int P\,dt} \qquad [30]$$

whose derivative is

$$\frac{d\varphi}{dt} = e^{\int P\,dt} P = \varphi P \qquad [31]$$

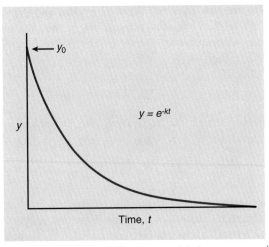

FIGURE B-5 ■ Graph of the exponential decay, $y = y_0 e^{-kt}$. Initially ($t = 0$), $y = y_0$. As t increases, y asymptotically approaches 0.

Multiplying Equation [29] by φ yields

$$\varphi\frac{dy}{dt} + \varphi Py = \varphi Q$$

which is, in light of Equation [31],

$$\varphi\frac{dy}{dt} + y\frac{d\varphi}{dt} = \varphi Q \qquad [32]$$

Consulting the properties of derivatives (see Box B-3), we recognize the left-hand side of Equation [32] as the derivative of a product function:

$$\frac{d}{dt}(\varphi y) = \varphi Q$$

This equation is now separable because the right-hand side, φQ, is a function of t exclusively and can be grouped with dt, whereas $d(\varphi y)$ is by itself. Separation of variables and integration yields

$$\varphi y = \int \varphi Q \, dt$$

which gives the solution to the differential Equation [29] as

$$y = \frac{1}{\varphi}\int \varphi Q \, dt \qquad [33]$$

This solution is useful in a discussion of the electrical behavior of a biological membrane, which can be represented as an R-C circuit (see Appendix D, Box D-3). The relevant differential equation there is

$$\frac{dV_m}{dt} + \frac{1}{RC}\cdot V_m = \frac{I_T}{C} \qquad [34]$$

where V_m, R, and C are, respectively, the membrane voltage, membrane resistance, and membrane capacitance and I_T is the total current passing through the membrane. It is clear that Equation [34] is identical in form to Equation [29]; therefore the solution [33] applies. The integrating factor is

$$\varphi = e^{\int \frac{1}{RC}dt} = e^{\frac{t}{RC}}$$

and the solution is

$$V_m = e^{-\frac{t}{RC}}\int e^{\frac{t}{RC}}\left(\frac{I_T}{C}\right)dt = e^{-\frac{t}{RC}}\left[\frac{I_T}{C}\cdot e^{\frac{t}{RC}}\cdot RC + Const\right]$$

$$= I_T R + Const \cdot e^{-\frac{t}{RC}}$$

where $Const$ is the constant of integration. A boundary condition helps to define the exact solution: at $t = 0$, $V_m = 0$; therefore $Const = -I_T R$. The exact solution is thus

$$V_m = I_T R\left[1 - e^{-\frac{t}{RC}}\right] \qquad [35]$$

It is easy to verify through Equation [35] that (1) at very long times ($t \rightarrow \infty$), the membrane voltage will reach a steady-state value, $V_{m,\infty} = I_T R$, and (2) $R{\times}C$ has the dimensions of time and can be defined as the membrane time constant, τ_m. These two observations allow the solution to be written in a more compact form:

$$V_m = V_{m,\infty}\left[1 - e^{-\frac{t}{\tau_m}}\right] \qquad [36]$$

APPENDIX C

Root-Mean-Squared Displacement of Diffusing Molecules

Initially there are N molecules, all at starting position 0. For any particular molecule (for our purposes, tagged with a label i), regardless of where the molecule is, it can move only either one step to the left (a distance of $-\delta$) or one step to the right (a distance of $+\delta$). This means that the location of any molecule i after n steps must be related to its immediately previous location (at $[n-1]$ steps) by either $+\delta$ or $-\delta$. In other words,

$$x_i(n) = x_i(n-1) \pm \delta \qquad [1]$$

The first question we can ask is, "What is the average position of all of the molecules after all of them have taken n steps?" This is easy to calculate by simply adding up the positions of all the molecules and then dividing by the total number of molecules, N. That is,

$$\text{Average position} =$$
$$\frac{x_1(n) + x_2(n) + x_3(n) + \ldots + x_{N-2}(n) + x_{N-1}(n) + x_N(n)}{N}$$
$$[2]$$

Using the symbol $<x_i(n)>$ to represent the average position, and making use of summation notation, we can write the average position as

$$<x_i(n)> = \frac{\sum_{i=1}^{N} x_i(n)}{N} \qquad [3]$$

Using the relationship between the position after n steps and the position after $(n-1)$ steps (Equation [1]) gives

$$<x_i(n)> = \frac{\sum_{i=1}^{N}[x_i(n-1) \pm \delta]}{N}$$
$$= \left\{\frac{\sum_{i=1}^{N} x_i(n-1)}{N}\right\} + \left\{\frac{\sum_{i=1}^{N}(\pm \delta)}{N}\right\} \qquad [4]$$

In Equation [4] the first term enclosed in braces is just the average position after $(n-1)$ steps, $<x_i(n-1)>$. The second term in braces is the average of all the steps that all the molecules have just taken. Because the molecules moved randomly, essentially half of the steps must have size $+\delta$ and half must have size $-\delta$. This means that all the steps average to 0, so the second term in braces is equal to 0. Therefore Equation [4] simplifies to

$$<x_i(n)> = <x_i(n-1)> \qquad [5]$$

Equation [5] says that the average position of all the molecules remains the same from one step to the next, regardless of how many steps have been taken.

The average position of the molecules is clearly not a very informative parameter. Nonetheless, we know that with time the molecules will progressively spread out in space (see Figure 2-4). How can we make this observation more quantitative? We can decide how to proceed by trying to understand why the average position (i.e., the average displacement from the initial point 0) is so uninformative. The reason is that because, on average, every $+\delta$ step must be balanced by a $-\delta$ step, the arithmetical average must always come out to be 0. One way to get around the fact that there are $+$ and $-$ steps is to square the displacements and *then* average them. Because any positive or negative number squared gives a positive number, averaging the squared displacements guarantees that the result would be non-0.

Squaring the displacements shown in Equation [1] yields

$$x_i^2(n) = [x_i(n-1) \pm \delta]^2$$
$$= [x_i(n-1) \pm \delta] \times [x_i(n-1) \pm \delta] \qquad [6]$$

Actually multiplying out the last two terms gives

$$x_i^2(n) = x_i^2(n-1) \pm 2\delta x_i(n-1) + \delta^2 \qquad [7]$$

287

Summing each term in Equation [7] for all N molecules and then dividing by N gives the average for each term:

$$\frac{\Sigma_{i=1}^{N} x_i^2(n)}{N} = \frac{\Sigma_{i=1}^{N} x_i^2(n-1)}{N} + \frac{\Sigma_{i=1}^{N} \pm 2\delta x_i(n-1)}{N} + \frac{\Sigma_{i=1}^{N} \delta^2}{N}$$

[8]

The term on the left side of the equal sign is the average squared displacement after n steps. The first term on the right is the average squared displacement after $(n - 1)$ steps. By the argument we used earlier, that $+$ and $-$ displacements are equally likely, the second term on the right must be 0. Finally, the last term on the right is just adding δ^2 N times and then dividing by N again; the value must therefore be simply δ^2. After the foregoing simplifications, Equation [8] turns into

$$<x_i^2(n)> = <x_i^2(n-1)> + \delta^2$$

[9]

Knowledge about the distribution of molecules at time $= 0$ allows Equation [9] to be put into a much simpler and more useful form. At time $= 0$ (when 0 steps have been taken), all the molecules are clustered at position 0; the mean squared position or displacement must necessarily be 0. In symbols: $<x_i^2(0)> = 0$. Putting this back into Equation [9] gives

$$<x_i^2(1)> = <x_i^2(0)> + \delta^2 = \delta^2$$

In other words, the mean squared displacement after all the molecules have taken a *single* step ($n = 1$) of size δ is just $1\delta^2$. Using this new result in Equation [9] again gives

$$<x_i^2(2)> = <x_i^2(1)> + \delta^2 = 2\delta^2$$

In other words, the mean squared displacement after all the molecules have taken *two* steps ($n = 2$) of size δ is just $2\delta^2$. In fact, Equation [9] can be used iteratively to generate the mean squared displacement after *any* number of steps. The sequence of results generated in this way is thus

$$<x_i^2(0)> = 0 \quad \text{after 0 steps,}$$

$$<x_i^2(1)> = \delta^2 \quad \text{after 1 step,}$$

$$<x_i^2(2)> = 2\delta^2 \quad \text{after 2 steps,}$$

$$<x_i^2(3)> = 3\delta^2 \quad \text{after 3 steps, etc.}$$

Or, in general,

$$<x_i^2(n)> = n\delta^2 \quad \text{after } n \text{ steps}$$

[10]

This result is expressed in terms of the number of steps the molecules have taken. In reality, it is impossible to monitor all the steps that all the molecules actually take. Therefore it would be much more convenient to express the result in terms of *time* rather than the number of steps. If each step is taken in a time increment of Δt, then the elapsed time, t, after n steps is $t = n \times \Delta t$, which also means that $n = t/\Delta t$. These two facts allow us to put Equation [10] into the following form

$$<x_i^2(t)> = \left(\frac{t}{\Delta t}\right)\delta^2 = \left(\frac{\delta^2}{\Delta t}\right)t$$

[11]

If we define a diffusion coefficient or diffusion constant, $D = (\delta^2/2\Delta t)$,* Equation [11] becomes

$$<x_i^2(t)> = 2Dt$$

[12]

Now, the average, or mean, *squared* displacement has weird dimensions of *length squared*. To obtain something that is more intuitively accessible, we can take the square root of both sides of Equation [12] to get the root-mean-squared (RMS) position or displacement, which again has dimensions of length:

$$\text{RMS displacement} = d_{RMS} = \sqrt{<x_i^2(t)>} = \sqrt{2Dt}$$

[13]

The RMS displacement is a measure of how spread out the initially clustered molecules become after some time has elapsed.

The RMS displacement derived here is for molecules diffusing along a single dimension. What happens in the more common case of molecules diffusing in two dimensions (e.g., along a membrane surface) or three dimensions (e.g., in solution)? Molecules cannot distinguish directions, so their diffusional behavior should be independent of direction. Figure C-1 shows that the Pythagorean theorem gives the answer:

$$(d_{RMS}^{2-D})^2 = (d_x^{1-D})^2 + (d_y^{1-D})^2 \quad \text{or}$$

[14]

$$d_{RMS}^{2-D} = \sqrt{(d_x^{1-D})^2 + (d_y^{1-D})^2}$$

---

*The thing to notice about the diffusion coefficient is that it embodies microscopic molecular properties (δ, the size of the step that a molecule can take, and Δt, the time over which such steps are taken). In addition, the diffusion coefficient has dimensions of length squared over time, as we had deduced from our discussion of Fick's First Law.

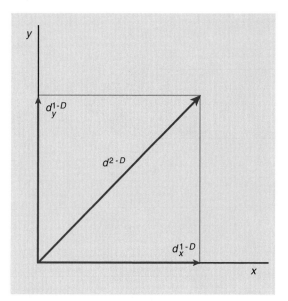

FIGURE C-1 ■

Because $d_{RMS} = \sqrt{2Dt}$. for any single dimension x, y, or z, substitution into Equation [14] gives the result for diffusion in two dimensions (2-D):

$$d^{2-D} = \sqrt{(d_x^{1-D})^2 + (d_y^{1-D})^2}$$
$$= \sqrt{2Dt + 2Dt} = \sqrt{4Dt} \qquad [15]$$

Similarly, for diffusion in three dimensions (3-D), the result is

$$d^{3-D} = \sqrt{6Dt} \qquad [16]$$

BIBLIOGRAPHY

Feynman RP, Leighton RB, Sands ML: *The Feynman lectures on physics*, vol 1, Reading, MA, 1965, Addison-Wesley.

Summary of Elementary Circuit Theory

CELL MEMBRANES ARE MODELED WITH ELECTRICAL CIRCUITS

The plasma membrane can be modeled as an electrical circuit. Figure D-1A is a schematic view of the structure of a biological membrane with a single open K^+–selective ion channel. This physical entity is electrically equivalent to the circuit shown in Figure D-1B. The circuit consists of a resistor in series with a battery, and this combination is in parallel with a capacitor. In Figure D-1, we follow the convention that the positive pole of a battery is represented by a longer bar. The electrical behavior of most resting biological membranes is, in fact, indistinguishable from a circuit similar to that shown in Figure D-1B. Therefore we can use such equivalent circuits to obtain quantitative descriptors of membrane electrical behavior, such as the length constant and the membrane time constant (see Chapter 6). The equivalent circuit is also used to describe current flow across membranes and the effect of current flow on membrane potential (V_m), such as how the action potential is generated and propagated along an elongated structure such as an axon or a skeletal muscle cell (see Chapter 7).

DEFINITIONS OF ELECTRICAL PARAMETERS

Electrical Potential and Potential Difference

The potential difference between two points is the amount of work done per unit charge to move charge from one point to the other. The unit of measure of potential difference is the volt. Potential difference (often called voltage) reflects the electrostatic force exerted on a charge. We use the symbol E to represent

theoretical potentials, such as the equilibrium potential of an ion (see Chapter 4), and the symbol V for actual voltages, such as the membrane potential (V_m).

Current

When a potential difference exists across space, positive charges will move toward the region with the more negative potential and negative charges will move toward the region with the more positive potential. This movement of charge is a *current (I)*. The current at a point in a circuit is defined as the net movement of positive charge past that point per unit time. The movement of 1 coulomb of charge per second is a current of 1 ampere (A). In wires and electronic devices the current is carried by electrons. In biological systems, however, current is carried by ions (e.g., Na^+, K^+, Cl^-, Ca^{2+}) moving in an aqueous environment.

By convention, when an arrow is used to indicate current flow, it shows the direction of the *net* movement of *positive* charge. The physical reality corresponding to such a representation could be either positive charges moving in the direction of the arrow or negative charges moving in the opposite direction.

Resistance and Conductance

The current that flows through a conductive material is proportional to the potential difference, or voltage (V), across the material. We refer to such a material as either a conductor or a resistor. In electronic circuits resistors are made of carbon or some other low-conductivity material and are characterized by their resistance (R). Intuition tells us that the current through the resistor should increase as the driving

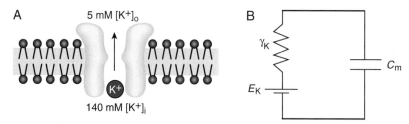

FIGURE D-1 ■ **A,** Schematic drawing of a lipid bilayer membrane containing a single open K^+ channel. A K^+ concentration gradient exists across the membrane. **B,** The electrical equivalent of the membrane shown in **A**. The open K^+ channel is modeled as a conductor (or equivalently, a resistor) with a conductance, γ_K, equal to that of the open channel. The resistor is in series with a battery that represents the K^+ concentration gradient, and it has a voltage equal to E_K. The capacitor, with capacitance C_m, represents the ability of the lipid bilayer to separate charge and is parallel to the channel.

force (V) across the resistor increases. This relationship is quantitatively given by Ohm's Law, which states that the current through a resistor is directly proportional to the voltage across it:

$$V = I \times R \qquad [1]$$

Resistance is measured in ohms (Ω): a current of 1 ampere will flow through a 1-ohm resistor with a driving force of 1 volt across it. Open ion channels behave electrically like conductors (or resistors): the membrane potential drives ion movement through open channels. The ability of a channel to pass current is characterized as its conductance (g or γ), which is the reciprocal of resistance: $g = 1/R$. The unit of conductance is the siemens (S): $1\,S = 1/\Omega = \Omega^{-1}$.

Capacitance

A capacitor is a device that can store, or separate, charges of opposite sign. A parallel plate capacitor has two parallel conducting plates (e.g., made of metal foil) separated by an insulator (e.g., mica, Mylar, glass, air). Biological membranes have capacitive properties because the lipid bilayer is an effective electrical insulator that allows soluble ions to be separated across the membrane. The amount of charge stored on a capacitor is directly proportional to the voltage across the capacitor:

$$q = C \times V \qquad [2]$$

where q is the charge in coulombs, V is the voltage in volts, and C is the capacitance of the device in farads. Note that $+q$ coulombs are stored on one plate

and $-q$ coulombs on the other plate. The capacitance of a given device, or a given area of membrane, is a constant, so Equation [3] shows that if q increases, V across the capacitor must increase.

By taking the time derivative of Equation [2], we obtain

$$\frac{dq}{dt} = C \times \frac{dV}{dt} \quad \text{or} \quad I_c = C \times \frac{dV}{dt} \qquad [3]$$

The quantity dq/dt (charge per unit time, or coulombs per second) is by definition a current (I_c, the capacitive current). Thus, in contrast to a resistor, where the current is proportional to the voltage, in a capacitor the current is proportional to the *rate of change* of the voltage. This important relationship shows that if the voltage is changing (i.e., $dV/dt \neq 0$), capacitive current is flowing. Conversely, $I_c = 0$ when the voltage is constant (i.e., when $dV/dt = 0$).

CURRENT FLOW IN SIMPLE CIRCUITS

The circuits in the following sections are analyzed by using the rules described in Box D-1.

A Battery and Resistor in Parallel

The circuit in Figure D-2A, shows a battery with a voltage of V_b volts connected by a switch to a resistance of R ohms. With the switch open no current flows, so the voltage across R is zero. When the switch is closed, a current (I) flows from the positive pole

BOX D-1

KIRCHHOFF'S LAWS ARE USED TO ANALYZE CIRCUITS

Two relationships, known as Kirchhoff's Laws, are useful in circuit analysis.

Kirchhoff's Current Law

Current is the movement of charge in a circuit, and when an arrow is used to indicate the direction of current flow, it shows the direction of *net positive charge* movement. Thus, when current flows through a resistor, the end the current enters is at a positive voltage relative to the end the current leaves.

Charge can never accumulate at any point in a circuit. Thus Kirchhoff's First Law states that the sum of the currents flowing into a point in a circuit must equal the sum of the currents flowing out of that point (i.e., there is conservation of charge). Another way to state this law is that the sum of all currents entering and leaving a point in a circuit is zero: we adopt the convention that currents entering are positive and currents leaving are negative. Using Kirchhoff's Current Law for the point labeled 1 in Figure Box D-1 gives the result

$$I_A = I_B + I_C \qquad \text{[B1]}$$

We will come back to this example later in Box D-2. From Kirchhoff's Current Law we can also deduce that in a series circuit (where elements are connected end to end), the current must be the same everywhere.

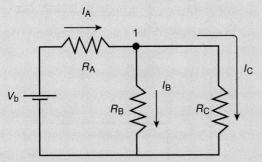

FIGURE BOX D-1 ■ A battery of voltage V_b is connected to three resistors, R_A, R_B, and R_C, in the manner shown. A current, I_A, flows from the positive terminal of the battery through R_A. At the point labeled 1, some of the current, I_B, flows through R_B and the rest of the current, I_C, flows through R_C.

Kirchhoff's Voltage Law

Elements connected in parallel in a circuit have the same voltage across them. Kirchhoff's Voltage Law states that the algebraic sum of all voltage drops around a closed loop is zero. The following sign conventions should be used: (1) the voltage drop across a resistor is positive in the direction of current flow and (2) the voltage drop across a battery is taken as positive if we meet the + pole of the battery going around the loop and negative if we meet the − pole. In the circuit in Figure Box D-1 the voltage drops across resistors R_A and R_B are $I_A \times R_A$ and $I_B \times R_B$, respectively. If we apply Kirchhoff's Voltage Law to the loop with battery V_b and the resistors R_A and R_B, we would write

$$I_A \times R_A + I_B \times R_B - V_b = 0 \qquad \text{[B2]}$$

For the loop with R_B and R_C the voltage law gives

$$I_C \times R_C - I_B \times R_B = 0 \quad \text{or} \quad I_C \times R_C = I_B \times R_B \quad \text{[B3]}$$

Using Equations [B1], [B2], and [B3], we can solve for I_A, I_B, and I_C as follows. Rearranging Equation [B2] gives

$$I_A = \frac{V_b - I_B \times R_B}{R_A} \qquad \text{[B4]}$$

Substitution of Equations [B3] and [B4] into equation [B1] gives the following equation, which can be solved for I_B:

$$\frac{V_b - I_B \times R_B}{R_A} - I_B - I_B \times \frac{R_B}{R_A} = 0$$

and

$$I_B = \frac{V_b}{R_A + R_B + \dfrac{R_A \times R_B}{R_C}}$$

If we let V_b = 10 volts, R_A = 10 ohms, R_B = 10 ohms, and R_C = 20 ohms, these equations give 0.6, 0.4, and 0.2 ampere for I_A, I_B, and I_C, respectively.

A

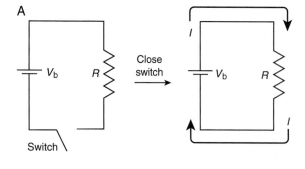

B

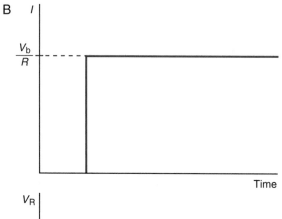

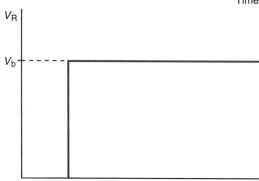

FIGURE D-2 ■ **A,** A battery, with voltage V_b, is connected to a resistor of R ohms through a switch. When the switch is closed, a current *(I)* flows from the positive terminal of the battery, through the resistor, and back to the battery. **B,** At the instant the switch is closed, the voltage drop across the resistor, V_R, is equal to V_b, and a current of magnitude V_b/R flows through the resistor.

of the battery through the resistor and back to the negative pole of the battery. The current instantaneously takes on a value of V_b/R amperes (from Ohm's Law), as shown in Figure D-2B. When a capacitor is added to this circuit, time delays are introduced

(i.e., voltage changes are no longer instantaneous), as shown in a later section.

A hydraulic version of this purely resistive circuit is shown in Figure D-3. A pump that delivers a constant hydrostatic pressure (ΔP) is connected by a valve to a small-diameter, water-filled tube. With the valve closed the hydrostatic pressure difference between the two ends of the tube is zero, so no water flows. When the valve is opened (equivalent to closing a switch in an electrical circuit), water immediately flows out the tube at a constant rate (equivalent to the current flow in an electrical circuit), driven by pressure ΔP (equivalent to the voltage).

A Resistor and Capacitor in Parallel

Next we consider the effect of placing a capacitor in parallel with a resistor (Figure D-4A). Current flow through this circuit is shown with a constant current generator connected to the circuit through a switch. When the switch is closed, a constant current (I_T) is passed through the circuit. At the instant the switch is closed (Figure D-4B), all this current flows through the capacitor (Box D-2). This capacitive current increases the charge separation across the capacitor, and

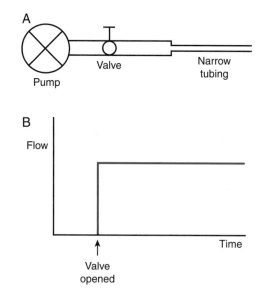

FIGURE D-3 ■ **A,** A hydraulic analogy of the pure resistive electrical circuit. A constant flow pump is connected through a valve to a length of small-diameter tubing. **B,** At the instant the valve is opened, water flows out of the tube at a rate equal to the flow from the pump.

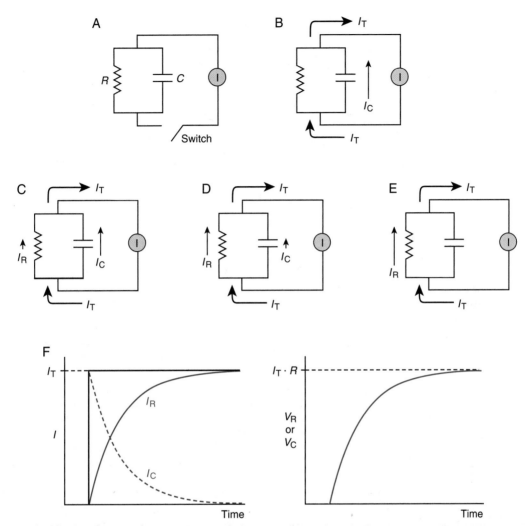

FIGURE D-4 ■ **A,** A parallel combination of a resistor and capacitor is connected to a constant current source through a switch. **B,** At the instant the switch is closed, all the current from the source (I_T) flows through the capacitor (I_C). **C,** As soon as the voltage across the circuit changes (as the result of I_C), some of the applied current begins to flow through the resistor (I_R). **D,** As the voltage continues to change, more current flows through the resistor and correspondingly less through the capacitor. **E,** In the final steady state, all the applied current flows through the resistor. **F,** Graphs showing that in a parallel R-C circuit, in response to a constant current step, the changes in I_R, I_C, and V all follow an exponential time course.

a potential difference, V_C, develops across the capacitor. Because R and C are arranged in parallel, the same potential difference develops across R: $V_C = V_R$. This causes current to flow through the resistor. As the resistive current develops, the capacitive current must decrease because the total current through the circuit, I_T, is constant (Figure D-4C). Eventually the voltage across the circuit will stop changing, a new steady state will be reached, and all the current will flow through the resistor (Figure D-4E). A mathematical analysis of this circuit shows that I_C, I_R, and V_R all follow exponential time courses (Box D-2), and they are plotted as functions of time in Figure D-4F. Comparing Figure D-2B, and Figure D-4 shows that the addition of a capacitor, in parallel with a resistor, causes a delay in the voltage change across the circuit.

BOX D-2

A PARALLEL *R-C* CIRCUIT PREDICTS
AN EXPONENTIAL CHANGE IN MEMBRANE POTENTIAL

The predicted time course of the change in V_m in response to a step of constant current for a parallel R-C circuit (Figure D-4A) can be derived as follows. The total membrane current must be the sum of an ionic current and a capacitive current, so we can write

$$I_T = I_R + I_C = \frac{V_R}{R} + C\frac{dV_C}{dt} \qquad [B1]$$

where I_R is the current through the resistor and I_C is the capacitive current. The resistor and capacitor are in parallel, so $V_R = V_C$. If we let $V_m = V_R = V_C$, we get

$$I_T = \frac{V_m}{R} + C\frac{dV_m}{dt} \qquad [B2]$$

This linear first-order differential equation can be solved with the methods described in Appendix B to give the following relationship:

$$V_m(t) = V_{m,\infty}\left[1 - e^{-\frac{t}{\tau_m}}\right] \qquad [B3]$$

where $\tau_m = R \times C$ and $V_{m,\infty} = I_T \times R$. We can then readily obtain the following equations for $I_R(t)$ and $I_C(t)$.

For the resistive, or ionic current, we get

$$I_R(t) = \frac{V_m(t)}{R} = \frac{V_{m,\infty}}{R}\left[1 - e^{-\frac{t}{\tau_m}}\right] = I_T\left[1 - e^{-\frac{t}{\tau_m}}\right]$$

For the capacitive current:

$$I_C(t) = C\frac{dV_m(t)}{dt} \qquad [B5]$$

By taking the time derivative of Equation [B3], we get

$$\frac{dV_m(t)}{dt} = V_{m,\infty}\frac{d}{dt}\left[1 - e^{-\frac{t}{\tau_m}}\right]$$

$$= \frac{V_{m,\infty}}{\tau_m}\cdot e^{-\frac{t}{\tau_m}} \qquad [B6]$$

Combining Equations [B5] and [B6] gives

$$I_C(t) = \frac{CV_{m,\infty}}{\tau_m}e^{-\frac{t}{\tau_m}} = \frac{CI_TR}{CR}e^{-\frac{t}{\tau_m}}$$

$$= I_T\,e^{-\frac{t}{\tau_m}}$$

Using these equations, we can show that at $t = 0$, $I_R = 0$ and $I_C = I_T$, but at $t = \infty$, $I_R = I_T$ and $I_C = 0$. Thus, at the instant the current step, I_T, is applied, all the current is capacitive (I_C). I_C then decays to zero following an exponential time course. The resistive (or ionic) current (I_R) is zero initially, but increases following an exponential time course. When a new steady state is reached, all the applied current is ionic current ($I_R = I_T$).

In a parallel R-C circuit the voltage does not change instantaneously in response to an applied current, as it does in a purely resistive circuit. Instead, the voltage change follows an exponential time course.

The time necessary to add, or store, charges on a capacitor may be easier to understand in terms of the hydrostatic pressure analogy. A "hydraulic" version of the parallel R-C circuit is shown in Figure D-5. A constant flow pump (the constant current generator) and a length of small-diameter tubing (the resistor) are connected to the base of a large-diameter, cylindrical tank (the capacitor). When the pump is first turned on

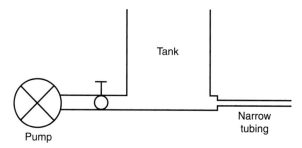

FIGURE D-5 ■ A hydraulic analogy of the electrical R-C circuit. The constant flow pump, storage tank, and small-diameter tubing are analogous to a constant current generator, a capacitor, and a resistor, respectively.

(with the tank empty), all the water flows into the tank; none flows out the tube because no driving force, or pressure difference, is present to force water out through the restricted tubing. As the water level in the tank increases, the hydrostatic pressure increases and water begins to flow out the tube, thus leaving less water available to fill the tank. The hydrostatic pressure in the tank and the outflow increase until the outflow is equal to the flow delivered by the pump and a new, stable, steady state is reached. The hydrostatic pressure increases along an exponential time course. The hydrostatic pressure (or voltage) increases slowly because it takes time to fill the tank (or store charge on the capacitor).

Answers to Study Problems

CHAPTER 2

1. The relationship between the distance and time of diffusion is $d \propto \sqrt{t}$; that is, the distance diffused is proportional to the square root of time (e.g., for three-dimensional diffusion, $d = \sqrt{6Dt}$). One way to solve the problem is to plug numbers in: first, $5 \mu m = \sqrt{6 \cdot D \cdot 1 \sec}$; second, $10 \mu m = \sqrt{6 \cdot D \cdot t}$. The first equation allows us to solve for D: squaring the first equation gives $25 \mu m^2 = 6D \sec$, which means that $D = 25/6 \mu m^2/\sec$. Substituting D into the second equation gives $10 \mu m = \sqrt{6 \cdot (25/6 \mu m^2 \sec^{-1}) \cdot t} = \sqrt{(25 \mu m^2 \sec^{-1}) \cdot t}$. Squaring both sides gives $100 \mu m^2 = (25 \mu m^2 \sec^{-1}) \cdot t$, which means that $t = 4 \sec$. We see that, because of the square root dependence on time, to go two times the distance by diffusion takes four (2^2) times as long.

2. The expressions should be $J_{inward} = P_K \times [K^+]_o$ and $J_{outward} = P_K \times [K^+]_i$. A net flux is positive when it brings material *into* the cell. Therefore the correct expression for the net flux of K^+ is $J_K = J_{inward} - J_{outward} = P_K ([K^+]_o - [K^+]_i)$.

CHAPTER 3

1. In solving all these problems, "very large volume of plasma" implies the "infinite bath" condition. In other words, the extracellular fluid volume is so large that a bit of solute or water entering or leaving the cell has essentially *no effect* on either the extracellular volume or the concentrations of solutes in the extracellular fluid.

   a. Answer: Water will move out of the cell. The final cell volume will be one half the initial volume. Explanation: The cell was initially equilibrated, which must mean that the intracellular concentration of permeant solute was

$C_P = 300$ mM and the intracellular concentration of impermeant solute was $C_{NP} = 10$ mM. When considering the situation at equilibrium, we need not worry about the permeant solute because it can cross the cell membrane (and the intracellular and extracellular permeant solute concentrations will always become equal automatically). When the extracellular impermeant solute concentration is increased to 20 mM, an osmotic imbalance occurs, with the inside of the cell having a deficit of impermeant solute. Because the impermeant solute concentration in the cell cannot be changed by solute movement (impermeant solute cannot cross the cell membrane), the only way the cell can cope with the osmotic imbalance is to *lose water and shrink its volume*. The reduction in cell volume would increase the impermeant solute concentration inside until the intracellular and extracellular impermeant solute concentrations become equal. To calculate the volume change, we note that the initial equilibrium condition is $C_{NP, Cell, Initial} = C_{NP, Bath, Initial}$ (impermeant solutes in the cell and bath are balanced) and the final equilibrium condition is $C_{NP, Cell, Final} = C_{NP, Bath, Final}$ (impermeant solutes in the cell and bath are once again balanced). Concentration is just the number of moles per volume of solution; therefore the two equalities may be written as

$$\frac{n_{NP, Cell}}{V_{Cell, Initial}} = C_{NP, Bath, Initial} = 10 \text{ mM} \quad \text{and}$$

$$\frac{n_{NP, Cell, Final}}{V_{Cell, Final}} = C_{NP, Bath, Final} = 20 \text{ mM}$$

We know that the *number of moles* of impermeant solute inside the cell must remain constant because any impermeant solute originally inside the cell must remain there at all times. In other words, $n_{NP, \text{Cell, Initial}} = n_{NP, \text{Cell, Final}} = n_{NP,\text{cell}}$. Substituting this back into the two foregoing equations and rearranging gives

$$V_{\text{Cell, Initial}} = \frac{n_{NP, \text{Cell}}}{10 \text{ mM}} \quad \text{and} \quad V_{\text{Cell, Final}} = \frac{n_{NP, \text{Cell}}}{20 \text{ mM}}$$

Solving these two equations (e.g., by dividing the second equation by the first) gives

$$\frac{V_{\text{Cell, Final}}}{V_{\text{Cell, Initial}}} = \frac{1}{2}$$

In other words, the final cell volume will be half the initial volume.

b. Answer: Water initially moves out of the cell (so the cell volume decreases), but then the volume gradually returns to the initial volume as equilibrium is reestablished (i.e., the final cell volume is the same as the initial cell volume). Explanation: When the extracellular permeant solute concentration is increased to 400 mM, initially an osmotic imbalance will occur. Because the membrane is more permeable to water than solutes, water will respond to the osmotic imbalance first by leaving the cell, so the cell should start to shrink. Even as water leaves the cell, the permeant solute, driven by its concentration gradient across the cell membrane, gradually permeates into the cell. Water follows osmotically, causing the cell volume to grow. Osmotic equilibrium can be reestablished only when the *impermeant* solutes are balanced across the cell membrane. Because the extracellular impermeant solute concentration is constant at 10 mM, equilibrium requires that the intracellular impermeant solute concentration return to 10 mM, which was its initial value. This can happen only if the cell swells back to its initial volume.

c. Answer: Water initially enters the cell and causes an increase in cell volume. Then, in response to permeant solute movement, water leaves the cell and cell volume decreases. The final cell volume is one half the initial cell volume.

Explanation: As soon as the extracellular solute concentrations change, an osmotic imbalance occurs; initially the total solute concentration inside is greater than the total solute concentration outside. Because water is the most permeant species, it crosses the cell membrane the fastest and is the first to respond to the osmotic imbalance by moving into the cell. This causes the cell to swell initially. Even as water enters the cell, the permeant solute, driven by its concentration gradient across the cell membrane, gradually permeates out of the cell. Water follows osmotically, causing the cell to shrink. As before, when a new osmotic equilibrium is established, the final volume of the cell is determined only by the presence of *impermeant* solutes. Because the outside impermeant solute concentration was changed from 10 to 20 mM, the situation concerning impermeant solutes is identical to that in part *a* of this problem—the final cell volume is one half the initial cell volume.

CHAPTER 4

1. E_{Na} is defined by the Nernst equation:

$$E_{Na} = \frac{RT}{(+1)F} \ln \frac{[Na^+]_o}{[Na^+]_i}$$

After substituting in $RT/F = 26.7$ mV, and $[Na^+]_i = 5$ mM, we can solve for $[Na^+]_o = 10.6$ mM.

2. a. V_m can be calculated through the GHK equation:

$$V_m = 26.7 \ln \frac{0.8(3)+1.0(145)+0.5(5)}{0.8(140)+1.0(15)+0.5(105)} = -4.8 \text{ mV}$$

b. Use the Nernst equation:

$$E_{Cl} = \frac{RT}{(-1)F} \ln \frac{[Cl^-]_o}{[Cl^-]_i} = \frac{26.7}{-1} \ln \frac{105}{5} = -81.3 \text{ mV}$$

c. If Cl^- is allowed to move, it will move so as to drive V_m toward E_{Cl}. Because $V_m = -4.8$ mV, Cl^- must move *into the cell* to drive V_m toward $E_{Cl} = -81.3$ mV. By definition, an inward flux is a *positive* flux (see Box 4-5).

d. The Cl^- flux carries negative charges into the cell, which has the same electrical effect as bringing positive charges out of the cell. A positive current is defined as the flow of positive charges out of the cell. Therefore the Cl^- current is a *positive* (or *outward*) *current*.

3. The Na^+ pump normally counteracts the osmotic consequences of the Donnan effect and thus helps maintain cell volume. When the Na^+ pump is inhibited, the osmotic imbalance arising from the Donnan effect would tend to cause the cell to be hyperosmotic relative to the extracellular fluid. This means that water would enter and cause the cell to swell.

CHAPTER 5

1. The primary function of gated ion channels is to increase the permeability of the membrane to a specific ion. To increase the membrane K^+ permeability, the β-cell could alter the K^+ channel properties in the following ways:

- The number of *open* channels could be increased or decreased. The overall permeability of the membrane will be directly related to the number of open channels.
- The amount of time the channel is opened could be varied: the longer the channels are open, the higher the membrane permeability.
- The permeability (or conductance) of a single open channel could be increased. Typically, cellular mechanisms exist to regulate the number of open channels and the amount of time the channel is open. The permeability of a single open channel is usually constant.

2. The most likely location of this amino acid would be in the segment of the P region that forms the selectivity filter of the Na^+ channel. The selectivity filter is a narrow region in the permeation pathway that determines which ionic species are allowed to pass. The selectivity filter works, in part, by making it energetically more favorable for a specific ion to enter the pore.

A single amino acid substitution in the selectivity filter could significantly alter its structure or chemical properties. For example, the amino acid substitution described in this question could make it energetically more favorable for the divalent Ca^{2+} ion to enter the pore.

CHAPTER 6

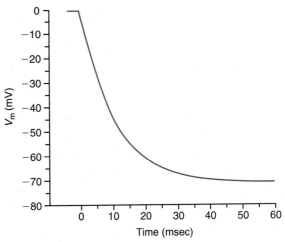

FIGURE E-1 ■ When the Cl^- channel opens (at time = 0 in this graph), V_m goes from an initial value of 0 mV to a final value of –70.5 mV along an exponential time course. The time constant of the exponential (the membrane time constant, τ_m) is 10^{-2} seconds.

1. a. The concentration gradient for Cl^- causes Cl^- ions to enter the cell through the open channel. Thus a slight excess of negative charges builds up inside the cell and moves V_m in the negative direction. The developing V_m progressively impedes Cl^- entry until $V_m = E_{Cl} = -70.5$ mV, at which point net Cl^- flux becomes zero and V_m stops changing.

b. The V_m follows a single-exponential time course (Figure E-1) with a time constant, τ_m,

$$\tau_m = R_m \times C_m$$

$$R_m = \frac{1}{\gamma_{Cl}} = \frac{1}{10^{-11}S} = 10^{11} \text{ ohms}$$

$$C_m = 10^{-6} \text{ F/cm}^2 \times 10^{-7} \text{ cm}^2 = 10^{-13} \text{ F}$$

$$\tau_m = 10^{11} \text{ ohms} \times 10^{-13} \text{ F} = 10^{-2} \text{ sec}$$

2. a. The change in V_m will decrease as a function of distance away from the site of current injection according to the following equation:

$$\Delta V_m(x) = \Delta V_0 e^{\left(\frac{-x}{\lambda}\right)}$$

In this problem $\Delta V_0 = -30$ mV, and the length constant, λ, can be calculated:

$$\lambda = \sqrt{\frac{r_m}{r_i}} = \sqrt{\frac{2.5 \times 10^4 \, \text{ohm} \cdot \text{cm}}{1 \times 10^5 \, \text{ohm} \cdot \text{cm}^{-1}}} = \sqrt{0.25 \, \text{cm}^2} = 0.5 \, \text{cm}$$

The change in V_m at $x = 4$ mm (0.4 cm) can then be calculated:

$$\Delta V_m(0.4) = (-30 \text{ mV}) \times e^{\frac{-0.4 \text{ cm}}{0.5 \text{ cm}}}$$

$$= -30 \times 0.45 = -13.5 \text{ mV}$$

Thus V_m is –83.5 mV at a distance of 4 mm from the site of current injection.

b. If the Cl^- channels are blocked, the membrane resistance (r_m) would increase. This would increase the length constant. With an increase in length constant, V_m would change more slowly as a function of distance away from the site of current injection. Thus V_m would be *more negative* at 4 mm from the site of current injection because V_m would not have decreased as much because of the longer length constant.

3. a. The membrane time constant, τ_m ($= r_m \times c_m$), provides a quantitative measure of the rate of decay of the EPSP. Given the values in the table, dendrites 1 and 2 have the same value of τ_m (4×10^{-4} sec). Thus the EPSP decay rates would be the same.

b. The length constant,

$$\lambda = \sqrt{\frac{r_m}{r_i}}$$

provides a quantitative measure of the decay of subthreshold signals with distance: with a shorter length constant, a change in membrane potential will decay faster as a function of distance. The length constants of the two dendrites can be determined as follows:

$$\lambda_{\text{Dendrite 1}} = \sqrt{\frac{2 \times 10^6 \, \text{ohm} \cdot \text{cm}}{1 \times 10^{10} \, \text{ohm} \cdot \text{cm}^{-1}}}$$

$$= \sqrt{2 \times 10^{-4} \, \text{cm}^2} = 0.014 \, \text{cm}$$

$$\lambda_{\text{Dendrite 2}} = \sqrt{\frac{2 \times 10^4 \, \text{ohm} \cdot \text{cm}}{1 \times 10^6 \, \text{ohm} \cdot \text{cm}^{-1}}}$$

$$= \sqrt{2 \times 10^{-2} \, \text{cm}^2} = 0.14 \, \text{cm}$$

Thus the size of the EPSP would decrease more with distance in dendrite 1 because dendrite 1 has a shorter length constant.

A subthreshold EPSP would not be detected in the cell body because at a distance of 2 cm from the synapse the amplitude of the subthreshold EPSP would decline essentially to zero, even in the case of dendrite 2, which has the longer length constant. This is shown by the following calculation:

$$\Delta V_m(2 \text{ cm}) = \Delta V_0 \cdot e^{\frac{-2 \text{ cm}}{0.14 \text{ cm}}} = \Delta V_0 \cdot 6.2 \times 10^{-7} \approx 0$$

CHAPTER 7

1. a. The Na^+ conductance can be calculated using Ohm's Law. The maximum Na^+ conductance occurs when all the Na^+ channels are open (at +20 mV in this problem). Ohm's Law can be used to calculate $g_{Na,max}$ at +20 mV:

$$g_{Na} = \frac{I_{Na}}{V_m - E_{Na}}$$

$$g_{Na,max} = \frac{-2.0 \times 10^{-3} \, \text{A/cm}^2}{(0.02V - 0.06V)} = 50 \times 10^{-3} \, \text{S/cm}^2$$

b. When all the Na^+ channels are open, that is, when the channel open probability is 1 ($p_o = 1$), the Na^+ conductance is maximal, or $g_{Na} = g_{Na,max}$.

c. The relationship between g_{Na} and channel open probability is given by $g_{Na} = N_T \times p_o \times \gamma_{Na}$, where N_T is the total number of Na^+ channels, p_o is the probability that a single channel is open, and γ_{Na} is the conductance of a single Na^+ channel. Because $g_{Na,max} = N_T \times \gamma_{Na}$, then $g_{Na} = p_o \times g_{Na,max}$ or

$$p_o = \frac{g_{Na}}{g_{Na,max}}$$

At -20 mV,

$$g_{Na} = \frac{-2.0 \times 10^{-3} \, \text{A/cm}^2}{(-0.02V - 0.06V)} = 25 \times 10^{-3} \, \text{S/cm}^2$$

Thus

$$P_o = \frac{25 \times 10^{-3}\ S \cdot cm^{-2}}{50 \times 10^{-3}\ S \cdot cm^{-2}} = 0.5$$

d. These calculations illustrate one of the most important properties of voltage-gated ion channels: the open probability of the channel changes with V_m.

2. a. The depolarization opens voltage-gated Na^+ channels. The Na^+ electrochemical gradient causes Na^+ ions to enter the cell through the open channels, thereby generating an inward I_{Na}. The inward I_{Na} deposits some positive charges on the inside surface of the membrane (i.e., it produces an outward I_C), causing V_m to move in the positive direction. This depolarization opens more voltage-gated Na^+ channels, and this positive feedback of V_m on the opening of Na^+ channels causes V_m to approach E_{Na} rapidly. Next, two processes begin at about the same time. The Na^+ channels close as a result of inactivation, and voltage-gated K^+ channels start to open. The K^+ electrochemical gradient causes K^+ to exit the cell. This outward ionic current (I_K) deposits positive charges on the outside surface of the membrane (or, it generates an inward I_C), which makes V_m move in the negative direction.

b. The initial outward current is an outward capacitive current that depolarizes the membrane toward threshold. The inward current that follows is I_{Na} that flows during the upstroke of the AP. The final phase of outward current is I_K that flows during the repolarization of the AP.

c. If a neuron is stimulated at a point along the axon, an AP would propagate away in both directions. Propagation is normally unidirectional because neurons are stimulated at only one end, near the cell body in α motor neurons and at distal terminals in sensory neurons. The AP then propagates along the axon to the other end of the cell, and the membrane behind the AP becomes refractory.

3. ■ Length constant. The conduction velocity increases with an increase in the length constant. During a propagated AP the upstroke of the AP results from an inward Na^+ current. Some of this inward current flows down the axon ahead of the AP and provides a stimulus to bring V_m from the resting level toward the AP threshold. In an axon with a longer length constant the membrane would be brought to threshold farther from the point where the AP upstroke is occurring.

■ Time constant. The conduction velocity increases with a decrease in the time constant. With a shorter time constant, V_m changes more rapidly in response to current flow and the AP rises and declines more rapidly. Thus the AP moves more quickly from one patch of membrane to the next and consequently increases the conduction velocity.

■ Na^+ channel density. The conduction velocity increases with an increase in the Na^+ channel density. If the Na^+ channel density increases, the magnitude of the Na^+ current also increases. The rate of rise of the AP (dV_m/dt) is directly proportional to the capacitive current, I_C. Because I_C increases with I_{Na}, depolarization occurs more rapidly with a higher Na^+ channel density.

CHAPTER 8

1. Even the small influx of Ca^{2+} ions that occurs during an AP can significantly increase $[Ca^{2+}]_i$ because the normal resting $[Ca^{2+}]_i$ is very low (~100 nM). The following calculation illustrates the point. In cardiac ventricular myocytes an average I_{Ca} of approximately 20×10^{-12} A flows for approximately 300 msec (roughly comparable to the duration of a cardiac AP). The amount of charge, in coulombs, that enters the cell is equal to the product of the current, in coulombs/second, and time, in seconds:

$$g_{Ca} = I_{Ca} \cdot t = 20 \times 10^{-12}\ coul \cdot sec^{-1} \cdot 0.3\ sec$$
$$= 6 \times 10^{-12}\ coul$$

Charge can be converted to moles to obtain the number of moles of Ca^{2+} that entered the cell:

$$n_{Ca} = \frac{q_{Ca}}{zF} = \frac{6 \times 10^{-12}\ coul}{2 \times 96,485\ coul \cdot mole^{-1}} = 3 \times 10^{-17}\ moles$$

The volume of the spherical cell is

$$\text{Volume} = \frac{4\pi r^3}{3} = \frac{4\pi (5 \times 10^{-4}\text{cm})^3}{3}$$

$$= 5.2 \times 10^{-10}\text{ cm}^3 = 5.2 \times 10^{-13}\text{ liter}$$

This is the average volume of a ventricular myocyte. Neglecting any Ca^{2+} buffering that would occur in the cell, the I_{Ca} would produce a change in $[Ca^{2+}]_i$ of

$$\frac{3 \times 10^{-17}\text{ moles}}{5.2 \times 10^{-13}\text{ liters}} = 58 \times 10^{-6}\text{ M (58 }\mu\text{M)}$$

Such a change in concentration would be relatively insignificant for Na^+ or K^+ because their concentrations are in the millimolar range. However, because resting $[Ca^{2+}]_i$ is very low ($\sim$100 nM; see Chapter 11), not much Ca^{2+} need enter the cell to raise $[Ca^{2+}]_i$ significantly. In actuality, the rise in $[Ca^{2+}]_i$ is never as large as this calculation suggests because most cells have an abundance of Ca^{2+}-buffering proteins that rapidly bind much of the Ca^{2+} entering the cells.

2. The depolarization of the resting potential would increase the number of K_A channels that are inactivated. Thus fewer K_A channels would be available to contribute to the cell's electrical activity. When V_m depolarizes following one of the APs in the burst, less outward I_A is activated to oppose the depolarization. As a result, threshold is reached faster, resulting in a higher AP frequency in the burst.

3. Insulin secretion from pancreatic β-cells is activated by Ca^{2+} influx through L-type Ca^{2+} channels that open during APs. The Ca^{2+} influx could be enhanced in several ways:
   a. A lengthening of the duration of the burst of APs would increase Ca^{2+} influx because the L-type Ca^{2+} channels open more frequently. The burst duration could be prolonged by blocking BK_{Ca} or K_{ATP} channels.
   b. A change in the number, or open probability, of L-type Ca^{2+} channels could also be used to alter Ca^{2+} influx. The activity of the L-type Ca^{2+} channel in pancreatic β-cells is potentiated by protein kinases A and C. Cellular mechanisms that activate protein kinase A or C in the β-cell

therefore enhance Ca^{2+} influx, which, in turn, triggers increased insulin secretion.

CHAPTER 9

1. a. Glucose is electrically neutral so it has no electrical component for its potential energy. That leaves the chemical component: $\mu_{G,in} = \mu^0_G + RT\ln[G]_i$.
   b. Ca^{2+} is electrically charged, so we expect that V_m will influence its potential energy. For Ca^{2+} inside the cell, $\mu_{Ca,in} = \mu^0_{Ca} + RT\ln[Ca^{2+}]_i + 2FV_m$. Recall that V_m is defined as the inside potential relative to the outside potential, and we define the outside potential to be 0 mV. Therefore, irrespective of charge, the electrical component of the potential energy is 0 for any extracellular species. For Ca^{2+} outside the cell, the electrochemical potential energy is thus $\mu_{Ca,out} = \mu^0_{Ca} + RT\ln[Ca^{2+}]_o$.

CHAPTER 10

1. The cotransport systems normally mediate large net movements of solutes into cells. This results in an osmotic burden that should cause cells to swell. Some exceptions are the Na^+-dependent neurotransmitter reuptake mechanisms, but here the extracellular concentrations are very low, so only small amounts of solute are involved. Thus even a large (100- to 1000-fold) concentration gradient does not impose a large osmotic burden on the neurons or glia. Epithelial cells avoid osmotic catastrophe by expressing transport mechanisms (ion channels or simple carrier systems) that permit the net transfer of solutes *across* the cells so that normally no net buildup of osmotically active solutes (*osmolytes*) takes place within the cells. In contrast, the exchangers, by exchanging one solute for another, normally have negligible effect on the total solute concentration within cells. The Na^+ that enters the cells by Na^+-coupled transport is rapidly extruded by the Na^+ pump (see Chapter 11) so that Na^+ does not accumulate within the cells.

2. $[Ca^{2+}]_i$, at equilibrium, would be close to 100 mM:

$$[Ca^{2+}]_i = [Ca^{2+}]_o \cdot e^{\left(\frac{-zV_mF}{RT}\right)} = 1 \cdot e^{\left(\frac{(-2)\cdot(-60)}{26.7}\right)} = 90\text{ mM.}$$

This very high level of intracellular Ca^{2+} would form $CaCO_3$ and $Ca_3(PO_4)_2$ precipitates. The high $[Ca^{2+}]_i$ also would contribute to high intracellular osmolarity, which would lead to cell swelling.

3. $[Na^+]_i$ must be higher than in most cells (i.e., much greater than ~6 to 10 mM), or V_m must be considerably more positive than in most cells. Indeed, both are true because in these photoreceptor cells there is a large inward (Na^+) depolarizing current in the dark (called the dark current). Therefore, with a 3-Na^+-for-1-Ca^{2+} exchange, the energy in the Na^+ electrochemical gradient is insufficient to maintain $[Ca^{2+}]_i$ close to 100 nM. The extra energy needed to extrude Ca^{2+} from these cells is obtained by coupling of a fourth Na^+ and a K^+.

4. One alternative is an ATP-driven proton pump (see Chapter 11), but the turnover rate of pumps is at least 10-fold slower than that of coupled carriers that mediate countertransport (Table 10-1; the turnover rate of the Na^+/H^+ exchanger may be much greater than that of the Na^+/Ca^{2+} exchanger because only 1 Na^+ needs to bind to the Na^+/H^+ exchanger to initiate the transport cycle). Therefore too many copies of an H^+ pump may be needed to transfer H^+ at a sufficiently rapid rate to keep up with cellular demands. *Note*: Whether the transport system uses ATP directly is not a consideration because the same amount of energy must be used, in the long run (see Chapter 11), to extrude H^+.

5. The 1-Na^+-to-1-glucose cotransporter (e.g., SGLT-2) should be able to concentrate glucose about 100-fold across the apical membrane of the proximal tubule epithelial cell. Because glucose can exit the cell by facilitated diffusion across the basolateral membrane, the intracellular glucose concentration will approximate that in the plasma (5 mM). The SGLT-2 Na^+-glucose cotransporter should *theoretically* be able to remove sufficient glucose from the tubular fluid to lower the glucose concentration in the renal tubular fluid to approximately 0.05 mM. However, this assumes that the cotransporter operates at 100% efficiency *and* that it reaches equilibrium, both of which seem unlikely. On these assumptions, in the 60 liters of fluid/day that would leave the proximal tubules, the total amount of glucose remaining in the urine would be a *minimum* of 3 millimoles/day or 540 mg/day (which appears to be negligible). The 1-Na^+-to-1-glucose cotransporter, however, is unlikely to remove the glucose in the filtered load fully. This is because efficiency and transport rate fall off as the glucose concentration in tubular fluid declines to less than the carrier's K_m for glucose ($\approx$ 2 mM, the concentration at which transport velocity is half-maximal). A second cotransporter, with a 2-Na^+-to-1-glucose coupling ratio and higher affinity for glucose ($K_m \approx$ 0.4 mM) and for Na^+ ($K_m \approx$ 3 versus 100 mM), in the distal part of the proximal tubule, reduces glucose in the urine to a negligible level.

CHAPTER 11

1. By stimulating the Na^+ pump, the α-adrenergic agonists lower $[Na^+]_i$ and transiently hyperpolarize the smooth muscle cells (because the pump is electrogenic). The lower $[Na^+]_i$ (and local $[Na^+]_o/[Na^+]_i$ gradient; see Box 11-2 and Figure 11-3) tends to drive more Ca^{2+} out of the cells through the NCX and thus lowers $[Ca^{2+}]_i$ and removes Ca^{2+} from the contractile apparatus.

2. The Na^+ electrochemical gradient generated by the Na^+ pump serves as a storage battery with a large capacity. No other ion electrochemical gradient in the cell is comparable in this respect. (This results from the fact that $[Na^+]_o >> [Na^+]_i$ and the extracellular fluid volume is very large.) Moreover, the secondary active transport systems all have faster turnover rates than ATP-driven pumps.

3. The same transport systems in different cells, or even in different places in a single cell, may have different functions and may need to be regulated differently. One example given in this chapter is the different distribution and function of the Na^+ pump catalytic (α) subunit isoforms within individual cells. Another example is that the SERCA in cardiac muscle, but not the SERCA in twitch-type skeletal muscle, is regulated by a small inhibitory protein, phospholamban. When phosphorylated, phospholamban does not inhibit SERCA, so cardiac relaxation is accelerated. The main point is that these numerous, functionally different transport systems are required to maintain exquisite

control of the intracellular environment in cells with different needs.

CHAPTER 12

1. a. The mean number of quanta released, the quantal content, is the product of p, the probability of quantal release, and n, the number of release sites. Thus, the quantal content is $0.06 \times 50 = 3$. The mean MEPP (quantal) amplitude is 0.5 mV, so the mean EPP amplitude is $3 \times 0.5 = 1.5$ mV at low $[Ca^{2+}]_o$.

   b. The probability of a failure (i.e., the release of 0 quanta) can be calculated from the binomial distribution as follows:

   $$P(0:50) = \frac{50!}{0! \times 50!} \times 0.06^0 \times 0.94^{50} = 0.045$$

   Thus 4.5% of the nerve stimuli will result in a failure.

2. a. Stimulate the motor nerve to fire an AP while recording intracellularly (current clamp or voltage clamp) from a muscle fiber. Compare the size of the EPPs in diseased muscle fibers with the size of EPPs in muscle fibers from healthy individuals.

   b. 1. Fewer vesicles of transmitter (ACh) could be released by the AP either because the probability of release is low or because there are fewer active zones.

   2. There could be fewer ACh receptors on the muscle fiber.

   3. The ACh receptors could have a lower single-channel conductance.

   4. The ACh receptors could open with a lower probability.

   5. The ACh receptors could have an altered permeability ratio, such that they would be more permeable to K^+ than to Na^+ ions.

   6. Less ACh could be packaged in, and therefore released from, each vesicle.

   7. The muscle resting potential could be unusually depolarized, which would decrease the driving force for the neurotransmitter-activated current.

   8. The nerve AP could be reduced in amplitude because of changes in its Na^+ or K^+ channels.

9. The Ca^{2+} channels at the nerve terminal could be altered so that they admitted less Ca^{2+} when the AP invaded the terminal (there could be fewer Ca^{2+} channels, or they could have altered voltage-dependence or permeability to Ca^{2+}, or be less favorably situated near the release sites.)

CHAPTER 13

1. The synaptic cleft may be visualized as a flat, pancake-like volume (the depth or thickness of the cleft is 20 to 40 nm, whereas the linear dimension is <0.5 μm, or a few hundred nanometers wide, so the width is much greater than the thickness). Neurotransmitter molecules should therefore spread toward the edges of the cleft by two-dimensional diffusion. The distance, d, over which molecules can spread by two-dimensional diffusion in a time, t, is $d = \sqrt{4Dt}$, which can be rearranged to $t = d^2/4D$. For $d \approx 0.5$ μm (equivalent to 0.5×10^{-4} cm), the time scale should be $t = (0.5 \times 10^{-4}$ cm$)^2/4(5 \times 10^{-6}$ cm^2/sec$) = 1.25 \times 10^{-4}$ sec. This is the time scale for the molecules to spread to the edge of the synapse. Clearance would be expected to take several-fold longer, but the time scale would still of the order of a millisecond or so. Therefore, diffusion is an effective means for removing transmitter from the synapse.

2. Most behaviors are complex phenomena that are mediated by several different neurotransmitters. A single agent may modify the dominant symptoms but is unlikely to be effective therapy for all symptoms of a disorder. Furthermore, most neurotransmitters act at diverse synapses in different parts of the CNS and thus have multiple actions, so that pharmacological intervention may also have significant undesired side effects resulting from actions of the agents at synapses that are not involved in the targeted behavior.

3. a. None of the ions listed in the table has an equilibrium potential close to −10 mV, so the channel must allow more than one type of ion to permeate. The channel could be permeable to both K^+ and Na^+ or to both K^+ and Ca^{2+}.

b. The channel selectivity can be determined by changing the concentration of one of the ions. For example, if the extracellular Na^+ is decreased to 10 mM, E_{Na} would be 0 mV instead of +71.5 mV. If this change made the reversal potential more negative, this would indicate that Na^+ ions move through this channel. If Ca^{2+} ions permeate the channel, a decrease in extracellular Ca^{2+} should move the reversal potential in the negative direction. Finally, an increase in extracellular K^+, which makes E_K more positive, should move the reversal potential in the positive direction if K^+ ions permeate the channel.

4. a. Under high-frequency stimulation, activated AMPARs mediate inward currents that persistently depolarize the postsynaptic neuron. The depolarization relieves the voltage-dependent blockade of NMDARs by Mg^{2+} and thus enables Ca^{2+} influx through NMDARs to activate Ca^{2+}-dependent signaling processes that underlie LTP. When AMPARs are blocked pharmacologically, depolarization of the postsynaptic neuron by voltage clamp would still relieve Mg^{2+} blockade of NMDARs to enable Ca^{2+} influx. Therefore LTP should still be induced.

b. Because a rise in $[Ca^{2+}]_i$ is essential for activating Ca^{2+}-dependent signaling processes that underlie LTP, blocking $[Ca^{2+}]_i$ rises by Ca^{2+} buffers in the postsynaptic neuron would prevent LTP induction.

CHAPTER 14

1. Contraction of skeletal and cardiac muscles is activated by a "switch" in the thin (actin) filaments. When the cytosolic Ca^{2+} level is elevated, Ca^{2+} binds to TnC. This induces a conformational change in the troponin complex so that tropomyosin rotates out of the way and thereby permits myosin heads to interact with the actin. In smooth muscles, in contrast, the Ca^{2+} binds to CaM. The Ca^{2+}/CaM complex then binds to and activates MLCK. The activated MLCK then phosphorylates the 20-kD myosin RLC. This enables the myosin heads to interact with actin to form cross-bridges. Thus the Ca^{2+}-sensitive "switch" that activates contraction in skeletal and cardiac muscles is located on the thin filaments

(TnC), whereas in smooth muscles it operates through MLCK and the thick filaments (RLC).

2. ■ Skeletal muscle cells are attached to the skeleton by tendons. Most skeletal muscles are under voluntary control. A primary function of many skeletal muscles is to shorten and generate force in order to produce movement of skeletal levers.

■ Cardiac myocytes make up the wall of each heart chamber and must contract synchronously so that the lumen volume is reduced and the chamber can eject its contents. In heart muscle the AP spreads rapidly from cell to cell through gap junctions between the cells. This is important in the heart, where the muscle cells must contract synchronously to reduce the lumen volume.

■ Smooth muscle cells are the type of muscle cells contained within the walls of hollow organs such as the gastrointestinal tract, arteries and veins, urinary bladder, bronchi, and ureters. The contractile fiber bundles in smooth muscle cells are oriented obliquely to the long axis of the cell, so that some contractile force is transmitted laterally. Thus in a tubular organ, such as an artery or the intestine, smooth muscle cells that are oriented nearly end to end around the lumen constrict the lumen when they contract. In contrast, contraction of longitudinally oriented smooth muscle cells, as in the longitudinal muscle layer of the intestine, causes the organ to shorten. In saccular organs such as the urinary bladder, lateral as well as longitudinal forces cause the wall of the organ to contract relatively uniformly, leading to a reduction lumen volume.

3. Many smooth muscles must remain contracted for long periods. To maintain $[Ca^{2+}]_i$ substantially higher than the contraction threshold for a long time, the transport systems that move Ca^{2+} into and out of the cytosol would have to be reset to minimize large Ca^{2+} fluxes and a large energy expenditure that is normally required to resequester or extrude the Ca^{2+}. An alternative is to reset (increase in this case) the sensitivity of the contractile machinery to Ca^{2+}. In this way the contraction can be maintained at a relatively low $[Ca^{2+}]_i$. This is important because sustained elevation of $[Ca^{2+}]_i$ may turn on too many other, extraneous Ca^{2+}-dependent processes.

CHAPTER 15

1. In both skeletal and cardiac muscle, E-C coupling is initiated by a depolarization of the surface membrane that propagates into the T-tubules and activates DHPRs. which are functionally coupled to elements of the SR. These functional interactions occur at triads in skeletal muscle and at dyads and peripheral couplings in cardiac muscle. At these sites, DHPRs in the surface membrane lie opposite Ca^{2+} release channels (i.e., RyRs) in the SR. Activation of the DHPRs causes RyR channels to open, allowing Ca^{2+} to be released from the SR.

   Despite similarities in E-C coupling between skeletal and cardiac muscle, several important differences exist. In skeletal muscle an RyR is opened as a result of a mechanical link between the DHPR and the RyR; Ca^{2+} entry is unnecessary. In cardiac muscle, the RyRs are opened by CICR: Ca^{2+} enters the cell through L-type Ca^{2+} channels, binds to the RyRs, and opens them. Thus extracellular Ca^{2+} is essential for E-C coupling in cardiac muscle but is not required in skeletal muscle. All of the Ca^{2+} required for contraction in skeletal muscle comes from the SR. Another important difference involves the recycling of Ca^{2+}. In skeletal muscle virtually all of the Ca^{2+} released during a contraction is returned to the SR by SERCA. In cardiac muscle a significant amount of Ca^{2+} enters during each AP. Therefore, to maintain $[Ca^{2+}]_i$ in the steady state, NCX (primarily) and PMCA must transport some Ca^{2+} out of the cell after each contraction.

2. Perhaps the Ca^{2+} release channels that are not opened by a mechanical link to a DHPR are opened by CICR. That is, when the RyRs that are mechanically coupled to the voltage-sensitive DHPRs open and release Ca^{2+} from the SR, the released Ca^{2+} may, in turn, activate adjacent RyRs that are *not* directly opposite the DHPRs.

3. a. Mechanisms that influence the availability of Ca^{2+} for smooth muscle contraction include (1) those that mediate or regulate Ca^{2+} entry (e.g., VGCCs, K_{Ca} channels, ROCs, SOCs, the Na^+ pump, and the NCX), (2) those that mediate or regulate Ca^{2+} extrusion from the myocytes (PMCA, the Na^+ pump, and the NCX), and (3) those that mediate or regulate Ca^{2+} sequestration in the SR (SERCA, the Na^+ pump, and the NCX).

   b. Mechanisms that influence smooth muscle contraction by altering the sensitivity of the contractile apparatus to Ca^{2+} (see Chapter 14) include phosphorylation of MLCK and MLCP.

   c. The numerous mechanisms that contribute to the regulation of smooth muscle contraction are apparently needed to vary contraction in a graded fashion, to maintain tonic contraction, and to respond to the multiple types of neuronal and hormonal signals that influence contraction in different smooth muscles (e.g., in circular and longitudinal smooth muscles in the intestinal wall).

   d. The numerous mechanisms involved in the regulation of smooth muscle contraction provide many opportunities for therapeutic intervention with a variety of different agents that can act selectively on the different mechanisms.

CHAPTER 16

1. On stimulation, the muscle first develops force (tension) at constant length (Figure E-2A) between points *a* and *b*. This is a period of isometric force development. When the muscle force exceeds the attached load, the muscle begins to shorten isotonically (i.e., at constant force; from *b* to *c*). At point *c* the muscle begins to relax (lengthen) at constant force. This isotonic relaxation ends at *d* when the muscle returns to its initial length. Muscle force continues to decline at constant length until the muscle is fully relaxed (at *e*). These phases of the contraction (isometric contraction [*a* to *b*], isotonic shortening [*b* to *c*], isotonic relaxation [*c* to *d*], and isometric relaxation [*d* to *e*]) are also illustrated on the length-tension curve in Figure E-2B.

2. The increase in contractility would result in more isotonic shortening and thus a greater length change in individual ventricular myocytes. Because the ventricular myocytes are interconnected to form a sac (the ventricle), the larger length change (i.e., increased shortening) in the individual ventricular myocytes would produce a smaller final ventricular volume.

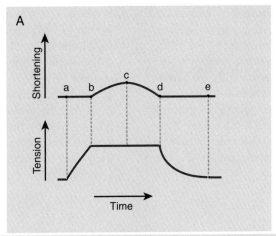

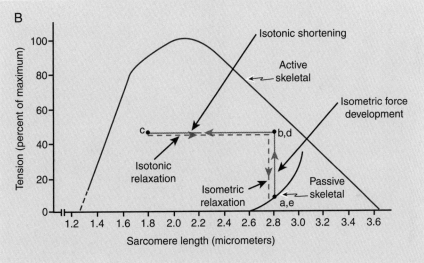

FIGURE E-2 ■ A, During this partly isometric, partly isotonic contraction, the muscle first undergoes isometric force development (from *a* to *b*). This is followed by isotonic shortening (*b* to *c*), isotonic relaxation (*c* to *d*), and, finally, isometric relaxation (*d* to *e*). **B,** The four phases of the contraction are illustrated on the length-tension curve. The muscle begins at rest, at an initial sarcomere length of 2.8 μm (point *a*). After stimulation the muscle develops force at constant length (isometric force). When the muscle force exceeds the attached load (at *b*), the muscle begins to shorten (isotonic shortening). At point *c*, the muscle begins to relax (lengthen) at constant force (isotonic relaxation), until the muscle returns to its initial length (at *d*). Muscle force then declines at constant length (isometric relaxation) until the muscle is fully relaxed (at *e*).

3. The force generated by whole skeletal muscle is altered physiologically in two ways. First, the number and type of motor units activated at the same time can be varied. Second, the frequency of stimulation of the motor unit can be varied. As a result the force can vary from as little as that of a single twitch to as much as that of a fused tetanus.

In ventricular muscle the primary physiological mechanism for altering force involves regulating the level of $[Ca^{2+}]_i$: mechanisms that raise peak $[Ca^{2+}]_i$ or increase the duration of the elevated

$[Ca^{2+}]_i$ will increase force. Conversely, mechanisms that lower peak $[Ca^{2+}]_i$ or reduce the duration of the elevated $[Ca^{2+}]_i$ will decrease force. For example, stimulation of β-adrenergic receptors with norepinephrine increases contractility in cardiac muscle by increasing Ca^{2+} influx through voltage-gated Ca^{2+} channels. As in skeletal muscle, the force developed by cardiac muscle depends on the initial length of the muscle fiber. In ventricular muscle the initial fiber length will be determined by the ventricular end-diastolic volume.

4. When the dephosphorylation of myosin RLC is slow, more cross-bridges can remain attached simultaneously. Moreover, if myosin RLC becomes dephosphorylated while the cross-bridges are still attached, detachment will be slowed. These factors contribute to the tonic contractions that are maintained with very little hydrolysis of ATP by the myosin ATPase. In contrast, skeletal and cardiac muscles could not perform their normal functions if they were unable to relax rapidly.

APPENDIX F
Review Examination

1. Which of the following statements about diffusion is **FALSE?**
   a. Molecules diffuse from regions of high concentration to regions of low concentration.
   b. Diffusion flux (J) is directly proportional to the concentration gradient $(\Delta C/\Delta x)$.
   c. Diffusion tends to even out concentration differences.
   d. Diffusion is the result of the random motion of molecules.
   e. The distance molecules diffuse is directly proportional to time (i.e., $d \propto t$).

2. If a collection of molecules is found to diffuse a distance d in 9 seconds, how long will it take the same molecules to diffuse twice as far?
   a. 9 seconds.
   b. 18 seconds.
   c. 36 seconds.
   d. 81 seconds.
   e. None of the above.

3. Which of the following statements about membrane permeability is **FALSE?**
   a. The cell membrane is more permeable to small, nonpolar solutes than to polar solutes because nonpolar solutes dissolve better in the membrane.
   b. A difference in solute concentration on the two sides of the membrane can drive a diffusive flux of molecules through the membrane.
   c. Increasing the thickness of a membrane reduces the diffusive flux through the membrane.
   d. The net solute flux through a cell membrane is the result of balancing influx and efflux.
   e. Increasing the diffusion coefficient of a solute molecule in the membrane does not change the permeability of the membrane to that solute.

4. If the extracellular and intracellular Na^+ and K^+ concentrations are symbolized as $[Na^+]_o$, $[Na^+]_i$, $[K^+]_o$, and $[K^+]_i$, and the respective membrane permeabilities are symbolized as P_{Na} and P_K, which of the following statements about unidirectional fluxes is correct?
   a. $J_{Na, \, in \rightarrow out} = P_{Na}[Na^+]_i$
   b. $J_{K, \, out \rightarrow in} = P_K[K^+]_o$
   c. $J_{K, \, in \rightarrow out} = P_K[K^+]_i$
   d. All of the above are correct.
   e. All of the above are incorrect.

5. A cell containing 300 mM nonpermeant solute is initially equilibrated with an extremely large volume of plasma that also contains 300 mM nonpermeant solute. The initial cell volume is 1 picoliter. The solute concentration of the plasma is then suddenly increased by the addition of 100 mM of a *permeant* solute. Which of the following is **FALSE?**
   a. After new osmotic equilibrium is reached, the cell volume will be 1 picoliter.
   b. The final concentration of permeant solute in the cell will be 100 mM.
   c. After new osmotic equilibrium is reached, the cell volume will be 0.75 picoliter.
   d. The final concentration of nonpermeant solute inside the cell will be 300 mM.
   e. The cell will initially shrink slightly and then recover to its original volume.

6. Which of the following statements is **FALSE?**
   a. Injecting, into the circulation, a substance that cannot cross the blood-brain barrier tends to reduce brain edema.
   b. Adding a permeant solute to the plasma will cause red blood cells to shrink permanently.

c. Inflammatory mediators released by mast cells increase the permeability of capillary walls, thus leading to edema.

d. Capillary hydrostatic pressure drives fluid filtration, whereas capillary colloid osmotic pressure drives fluid reabsorption.

e. Proteins that cannot readily permeate the capillary wall are the principal cause of the osmotic pressure difference between the plasma and the interstitial tissue fluid.

7. The membrane potential of a cell is $V_m = -60$ mV, and the intracellular and extracellular Cl^- concentrations are $[Cl^-]_i = 7$ mM and $[Cl^-]_o = 105$ mM, respectively. Which of the following statements is **FALSE?**
   a. The Cl^- *current* is inward.
   b. The Cl^- *flux* is inward.
   c. The Cl^- equilibrium potential is $E_{Cl} = -72$ mV.
   d. There is an outward *driving force* on Cl^-.
   e. If $V_m = E_{Cl}$, there will be no net flux of Cl^- ions across the cell membrane.

8. Which of the following is (are) essential for generating a stable, nonzero membrane potential?
   a. The cell membrane must be permeable to at least one type of ion.
   b. The net current through the plasma membrane must be zero.
   c. The net flux of ions through the plasma membrane must be zero.
   d. a and b are correct.
   e. a, b, and c are correct.

9. If the cell's Na^+ pumps are inhibited by ouabain, all of the following will happen **EXCEPT:**
   a. The Donnan effect will cause osmotic imbalance.
   b. $[K^+]$ will fall in the cell.
   c. The membrane potential will hyperpolarize.
   d. The cell volume will increase.
   e. $[Na^+]$ will rise in the cell.

10. For a neuron, the equilibrium potentials for several common ions were found to be $E_H = -12$ mV, $E_{Na} = +50$ mV, $E_K = -90$ mV, $E_{Ca} = +135$ mV, and $E_{Cl} = -50$ mV. If the membrane potential of the neuron is at $V_m = -60$ mV, which of the following statements about ion fluxes is **FALSE?**
   a. Cl^- flux is inward and H^+ flux is inward.
   b. The K^+ and Na^+ fluxes flow in opposite directions.

c. Ca^{2+} flux is inward and Na^+ flux is inward.
d. There is net H^+ influx, which would tend to make the inside of the cell more acidic.
e. If more Cl^- channels are opened, the resulting change in Cl^- flux will tend to depolarize the cell.

11. Intracellular and extracellular concentrations for the three common monovalent ions are given below for a cell.

Ion	In (mM)	Out (mM)
K^+	140	5
Na^+	5	145
Cl^-	5	105

If the relative permeabilities for the three ions are 1:10:0.2 (K^+:Na^+:Cl^-), what is the expected V_m of the cell (rounding to two significant digits)? (**Note:** At 37°C, $RT/F = 26.7$ mV and $2.303RT/F = 61.5$ mV.)
   a. −89 mV.
   b. −53 mV.
   c. −33 mV.
   d. +52 mV.
   e. +90 mV.

12. Which of the following statements is **FALSE?**
   a. Chemical potential energy is stored in concentration gradients.
   b. The electrochemical potential for an ion is zero if the membrane potential is at $V_m = 0$ mV.
   c. If the electrochemical potential of an ion is the same on the inside and outside of the cell membrane, that ion is at equilibrium.
   d. Diffusion is an example of a transport process driven by a gradient in electrochemical potential.
   e. A gradient in electrochemical potential drives the transport of ionic substances.

13. After demyelination, the axonal membrane in the region between two nodes of Ranvier:
   a. Depolarizes at approximately the same rate during a propagated action potential as it did before demyelination.
   b. Has a smaller membrane time constant (τ_m) because the membrane resistance is lower and the membrane capacitance is unchanged.

c. Has a lower membrane resistance and more charge separation across the membrane at the resting potential (−70 mV).

d. Has a larger membrane time constant (τ_m) because the membrane capacitance is larger and the membrane resistance is unchanged.

e. None of the above.

14. Two nonmyelinated axons have the same resistance and capacitance *per unit area* of membrane but are different in diameter. The passive cable properties of these axons were investigated, and the following values of resistance and capacitance per unit length of axon were determined (assume that $r_o = 0$):

	r_m	r_i	c_m
Axon 1	3×10^4 Ω·cm	1.5×10^6 Ω /cm	3×10^{-8} F/cm
Axon 2	3×10^6 Ω·cm	1.5×10^{10} Ω /cm	3×10^{-10} F/cm

Which of the following is **TRUE?**

a. The membrane time constant in axon 1 is smaller than in axon 2.

b. Axon 2 is larger in diameter than axon 1.

c. The effect of a subthreshold stimulus can be seen over a longer distance in axon 2 than in axon 1.

d. The action potential would propagate faster in axon 1.

e. The length constant is the same in both axons.

15. All of the following contribute to the muscle paralysis found in hyperkalemic periodic paralysis **EXCEPT:**

a. Some Na^+ channels are open in the steady state, thus causing membrane depolarization.

b. Some Na^+ channels do not inactivate completely.

c. An increase in extracellular K^+ produces membrane depolarization.

d. Membrane depolarization increases Na^+ efflux through the Na^+ pump and results in a reduction in the amount of Ca^{2+} stored in the sarcoplasmic reticulum.

e. Membrane depolarization causes skeletal muscle to become mechanically inactivated.

16. All of the following are true of voltage-gated L-type Ca^{2+} channels **EXCEPT:**

a. Their open probability increases as V_m moves in the positive direction.

b. They can be opened by phosphorylation.

c. During an action potential generated by voltage-gated Ca^{2+} channels, the amount of Ca^{2+} entering the cell can significantly increase $[Ca^{2+}]_i$.

d. The number of activatable Ca^{2+} channels in cardiac cell membranes can be altered by increasing the concentration of intracellular cAMP.

e. They are blocked by dihydropyridines (e.g., nifedipine).

17. Which of the following statements about current flow across an "isolated patch" of membrane (e.g., a small spherical cell) is **FALSE?**

a. An inward ionic current will produce a depolarization only if it causes an outward capacitive current.

b. Outward capacitive current always causes V_m to become more positive.

c. Inward capacitive current never causes V_m to become more positive.

d. An outward ionic current can never cause V_m to become more positive.

e. While current is being passed across the membrane through an external current source, the capacitive current is equal in magnitude, and opposite in direction, to the ionic current.

18. Assume that a Na^+ channel stays open and allows Na^+ ions to flow through at all times. Which of the following statements about the current through this single *open* Na^+ channel is **FALSE?**

a. The current has a constant amplitude if the membrane potential is held at a constant level.

b. The current would decrease in amplitude as V_m changed from the resting potential to the peak of the action potential.

c. The current would increase and then decrease while V_m was voltage clamped at 0 mV.

d. The current amplitude can be calculated from Ohm's Law.

e. The net current through the channel is zero if V_m is at E_{Na}.

19. Which of the following would be the Cl^- current flowing through 100 open Cl^- channels if V_m is −70 mV, E_{Cl} is −45 mV, and the conductance of a

single open chloride channel is 20 pS (pS = 10^{-12} S; 1 S = 1 A/V).
a. 50 pA.
b. −50 pA.
c. 0.5 pA.
d. −0.5 pA.
e. −250 pA.

20. Which of the following statements about BK_{Ca} channels is **FALSE?**
a. In neurons that generate a burst of action potentials, outward current through BK_{Ca} channels helps to terminate the burst.
b. BK_{Ca} channels could contribute outward current to help repolarize individual action potentials.
c. BK_{Ca} channels always carry outward current under physiological conditions.
d. At V_m = −80 mV, and $[Ca^{2+}]_i$ <100 nM, BK_{Ca} channels are mostly closed.
e. Ca^{2+} ions bind to a site on the extracellular side of the BK_{Ca} channel to activate the channel.

21. When the plasma glucose concentration rises to more than approximately 5 mM, pancreatic β-cells secrete insulin in response to a rise in $[Ca^{2+}]_i$. $[Ca^{2+}]_i$ increases because:
a. Glucose blocks K_{ATP} channels, thus leading to membrane depolarization and the opening of voltage-gated Ca^{2+} channels.
b. Glucose binds to a receptor on the voltage-gated Ca^{2+} channel and increases the channel open probability.
c. Glucose metabolism increases $[ATP]_i$ and ATP directly opens voltage-gated Ca^{2+} channels.
d. Glucose metabolism increases $[ATP]_i$ and ATP blocks K_{ATP} channels, thus leading to membrane depolarization and the opening of voltage-gated Ca^{2+} channels.
e. None of the above.

22. All of the following events are involved in the increase in cardiac contractility caused by epinephrine **EXCEPT:**
a. The cytosolic cAMP concentration increases.
b. Ca^{2+} entry through voltage-gated L-type Ca^{2+} channels is enhanced.
c. Na/Ca exchangers in the plasma membrane are inhibited.
d. cAMP-dependent protein kinase phosphorylates L-type Ca^{2+} channels.
e. Epinephrine activates adenylyl cyclase by binding to β-adrenergic receptors.

23. Which of the following statements is **FALSE?**
a. Voltage-gated K^+ channels in squid axons and A-type K^+ channels are opened by membrane depolarization, but the open probability of Ca^{2+}-activated K^+ channels decreases with depolarization.
b. Under physiological conditions a current flowing through any type of K^+ channel tends to oppose membrane depolarization.
c. A-type K^+ channels inactivate during maintained depolarization.
d. Ca^{2+}-activated K^+ channels can be opened by an increase in $[Ca^{2+}]_i$.
e. All K^+ channels normally pass outward current under physiological conditions.

24. Which of the following statements concerning the voltage-gated macroscopic Na^+ (g_{Na}) and K^+ (g_K) conductances of the squid giant axon is **FALSE?**
a. In a voltage clamp experiment, if the V_m is stepped to 0 mV, g_{Na} would activate faster than g_K.
b. Both g_{Na} and g_K are voltage dependent and time dependent.
c. Both g_{Na} and g_K have an activation and an inactivation phase during a depolarizing voltage clamp step.
d. g_{Na} is proportional to the number of open Na^+ channels.
e. g_K is proportional to the probability that a K^+ channel is open.

25. Which of the following events does **NOT** occur during a propagated action potential along a myelinated nerve axon?
a. Propagation occurs by local circuit current flow.
b. Inward Na^+ current at one node of Ranvier causes inward current to flow at an adjacent node, which depolarizes that node toward threshold.
c. Na^+ channels become inactivated.

d. The action potential rapidly propagates from one node of Ranvier to the next.

e. If E_{Cl} is always more negative than the V_m, an inward Cl^- flux is flowing.

26. All of the following statements are true of membrane capacitance (C_m) **EXCEPT:**

a. The decrease in C_m that results from demyelination is one of the factors contributing to slowing or blocking of action potential propagation.

b. Biological membranes typically have a C_m of approximately 1×10^{-6} F/cm^2.

c. The rate of change of V_m in a small spherical cell, in response to a constant current stimulus, would decrease if C_m increased while R_m did not change.

d. C_m is equal to the amount of charge separated across the membrane, in coulombs, per volt of membrane potential.

e. If the V_m is changing, the amount of charge separated across the membrane is changing and a capacitive current is flowing.

27. All of the following statements about voltage-gated Na^+ channels are true **EXCEPT:**

a. They pass only inward current under normal physiological conditions.

b. K^+ ions interfere with Na^+ channel inactivation in skeletal muscle cells from patients with hyperkalemic periodic paralysis.

c. They have an open probability that depends on time and V_m.

d. They inactivate during the action potential, causing the membrane to be refractory to additional stimuli for a short period of time.

e. Inward current through Na^+ channels in axons generates the upstroke of the action potential.

28. Which of the following is **NOT** a property of *all* ion channels?

a. Ion channels increase the permeability of the membrane to ions.

b. Ion channels are integral membrane proteins that extend across the lipid bilayer.

c. An ion channel is a pore that is not open at all times.

d. Ion channels exhibit selectivity by allowing only certain ions to flow through the channel.

e. All of the above are properties of all ion channels.

29. Which of the following statements about ion channel structure is **FALSE?**

a. Voltage-gated Na^+, K^+, and Ca^{2+} channels contain structural similarities indicating that they are members of a gene superfamily.

b. Voltage-gated Na^+, K^+, and Ca^{2+} channels contain a P region, or loop, that lines the pore of the channel.

c. Inward rectifier K^+ channels and voltage-gated K^+ channels are composed of subunits with six membrane-spanning helical segments.

d. The selectivity filter in the bacterial KcsA K^+ channel is formed by several main-chain carbonyl oxygen atoms from amino acids in the P region.

e. Voltage-gated Na^+, K^+, and Ca^{2+} channels all contain a membrane-spanning segment (S4) having several positive charges that act as a voltage sensor.

30. All of the following are true of membrane transport **EXCEPT:**

a. Aquaporins are responsible for high water permeability across some cell membranes.

b. Single carrier molecules, such as the simple glucose carrier (GLUT), can transport substrate at a rate of 1000 to 5000 molecules/second.

c. Insulin modulates the glucose carriers (GLUTs) in skeletal muscle and thereby speeds the rate of glucose uptake into these cells, but it does not affect the intracellular glucose concentration in the cells under steady-state conditions.

d. The diffusion of nonpolar compounds across cell membranes is directly proportional to their solubility in water.

e. Ions normally cross the plasma membrane at a very slow rate unless they are transported by special integral membrane proteins (e.g., carriers, channels, or pumps).

31. All of the following are characteristic of transport across cell membranes **EXCEPT:**

a. Carriers behave like channels that are open to only one side of the membrane at a time.

b. Carriers that mediate facilitated diffusion of a single solute species cannot maintain a concentration gradient for that solute in the steady state.

c. Carriers that mediate facilitated diffusion of a single solute species can "cycle" only when the solute is bound.

d. Carrier-mediated transport can be modulated by phosphorylation.

e. Carrier-mediated transport can be regulated by cycling carriers into and out of the plasma membrane.

32. Glucose absorption in the intestinal tract and reabsorption in the renal tubules are dependent on Na^+-glucose cotransporters (SGLTs) in the brush border membranes of intestinal and renal epithelial cells. Assume that:

The glucose concentration in the blood plasma = 4 mM.

The Na^+ concentrations are 145 mM in blood plasma, 10 mM in the cytoplasm, and 135 mM in kidney tubular fluid.

V_m is –62 mV (cytosol with respect to tubule lumen) and –58 mV (cytosol with respect to blood plasma).

$RT/F = 26.7$ mV.

What is the theoretical limit to which an Na^+-glucose cotransporter, with a 1 Na^+:1 glucose coupling ratio (SGLT-2), could reduce the glucose concentration in the tubule lumen?

a. 0.4 mM.
b. 0.3 mM.
c. 0.2 mM.
d. 0.04 mM.
e. 0.03 mM.

33. All of the following are true of glucose transport in human cells **EXCEPT:**

a. All cells express Na^+-glucose cotransporters (SGLT).

b. Some cells express Na^+-glucose cotransporters with a coupling ratio of 1 Na^+:1 glucose.

c. Some cells express Na^+-glucose cotransporters with a coupling ratio of 2 Na^+:1 glucose.

d. All cells that express Na^+-glucose cotransporters (SGLT) must also express (simple) glucose carriers (GLUT).

e. The simple glucose carrier (GLUT) activity in some cell types is modulated by insulin.

34. The Na^+/Ca^{2+} exchanger in vertebrate photoreceptors (rod and cone cells in the retina) mediates the exchange of 4 Na^+ for 1 Ca^{2+} plus 1 K^+. A rise in the extracellular K^+ concentration can be expected to:

a. Decrease the exchanger-mediated transport of Ca^{2+} from the cytoplasm to the extracellular fluid.

b. Increase the exchanger-mediated transport of Ca^{2+} from the cytoplasm to the extracellular fluid because of a reduction in the K^+ concentration gradient.

c. Increase the exchanger-mediated transport of Ca^{2+} from the cytoplasm to the extracellular fluid because of a membrane depolarization.

d. Increase the exchanger-mediated entry of Na^+ into the cells.

e. b, c, and d are all correct.

35. The following are all characteristics of Na^+ pumps **EXCEPT:**

a. Na^+ pumps in the kidney and in the brain consume more than half of the ATP hydrolyzed in those organs.

b. The hormone aldosterone regulates Na^+ pumps by promoting the synthesis and insertion of new Na^+ pumps into the plasma membrane of certain epithelial cells.

c. Na^+ pumps can be regulated by phosphorylation at a site other than the catalytic site.

d. Na^+ pumps are electrogenic: they generate a current and normally contribute a few (e.g., 1 to 3) millivolts to the resting V_m.

e. Most cells have two different Na^+ pumps that are the products of different genes: one is located in the plasma membrane, and the other in the endoplasmic reticulum.

36. All of the following are true of transport ATPases **EXCEPT:**

a. The Na^+ pump (Na^+,K^+-ATPase) ultimately provides the energy for most secondary active transport mechanisms.

b. The Na^+,K^+-ATPase normally generates an inward electric current.

c. The gastric H^+,K^+-ATPase can extrude protons from an intracellular environment with pH of 7.2 to a very acidic extracellular environment (pH 2 to 3).

   d. Wilson's disease and Menkes' disease are associated with genetic defects in two different copper-transporting ATPases.

   e. The multidrug resistance (MDR) transport proteins are ATPases that actively transport anticancer agents such as doxorubicin (Adriamycin) out of cells.

37. Inhibition of the Na^+ pump by ouabain (with no change in plasma Na^+ concentration) will lead to all of the following **EXCEPT:**

   a. Swelling of skeletal muscle cells.

   b. Reduction of neurotransmitter (e.g., dopamine) reuptake at nerve terminals.

   c. Augmentation of Ca^{2+}-dependent secretion of catecholamine by adrenal medullary cells.

   d. Augmentation of glucose and amino acid (e.g., alanine) absorption from the intestinal lumen.

   e. Cell acidification.

38. The Na^+ pump does all of the following **EXCEPT:**

   a. Helps regulate the glucose concentration in skeletal muscle.

   b. Contributes directly to the resting V_m.

   c. Contributes indirectly to the resting V_m by regulating $[K^+]_i$.

   d. Helps regulate Ca^{2+} stores in the sarcoplasmic and endoplasmic reticulum and Ca^{2+} signaling.

   e. Helps regulate cell volume.

39. The multidrug resistance (MDR) proteins are all of the following **EXCEPT:**

   a. P-type transport ATPases.

   b. Transport proteins that may be upregulated in the presence of their substrates.

   c. Primary active transporters.

   d. Transport proteins that actively concentrate many types of drugs in cells.

   e. Transport proteins that underlie drug resistance to many anticancer agents.

40. Which *one* of the following statements about regulation of solute transport is **TRUE:**

   a. Insulin stimulates glucose uptake into skeletal muscle cells by increasing the number of Na^+-glucose cotransporter molecules in the plasma membrane.

   b. Histamine stimulates the secretion of acid into the stomach lumen by increasing the number of H^+,K^+-ATPase molecules in the parietal cell basolateral membrane.

   c. Histamine stimulates the secretion of acid into the stomach lumen by increasing the number of Na^+/H^+ exchanger molecules in the oxyntic cell apical membrane.

   d. Dopamine can modulate the activity of some Na^+ pumps by promoting phosphorylation of the pump at a site *different* from the phosphorylation site required for activation of Na^+ extrusion and pump cycling.

   e. The Na^+/H^+ exchanger requires a 2 Na^+:1 H^+ coupling ratio to lower intracellular pH to 7.4 when extracellular pH = 7.4.

41. All of the following statements about Ca^{2+} homeostasis are true **EXCEPT:**

   a. The free Ca^{2+} concentration in the ER/SR ($[Ca^{2+}]_{ER/SR}$), of quiescent (resting) cells is normally greater than 10^{-4} M.

   b. Most of the Ca^{2+} in the lumen of the ER or SR is free Ca^{2+} (i.e., not bound to proteins).

   c. $[Ca^{2+}]_i$ in quiescent cells is normally approximately 10^{-7} M.

   d. Partial Na^+ pump inhibition can induce a large increase in $[Ca^{2+}]_{ER/SR}$ despite a very small increase in $[Na^+]_i$.

   e. The plasma membrane Ca^{2+} pump (PMCA) plays an important role in maintaining $[Ca^{2+}]_i$ in quiescent cells.

42. A 25-year-old man comes to your office with a chief complaint that his muscles are stiff after he exercises. On physical examination you find nothing remarkable. However, when you ask him to flex and extend his arms rapidly and repeatedly, you notice that he soon slows down and his muscles seem to remain tense and contracted longer and longer. He reports that his older brother has a similar problem. The impaired skeletal muscle relaxation could be explained by a defect causing *reduced activity* of which one of the following skeletal muscle channels/transporters?

   a. Ryanodine receptors (SR Ca^{2+} release channels).

   b. Dihydropyridine receptors (L-type Ca^{2+} channels).

c. Plasma membrane Ca^{2+} pump (PMCA).

d. SR Ca^{2+} pump (SERCA).

e. Na^+/Ca^{2+} exchanger.

43. All of the following statements about transepithelial cell transport are true **EXCEPT:**

   a. The main driving force for Cl^- reabsorption in the small intestine and the renal proximal tubule cells is the -3 to -5 mV (lumen negative) transepithelial electrical potential gradient.

   b. Most of the Cl^- reabsorbed in the small intestine and the renal proximal tubules is transported through the cells (i.e., by the transcellular pathway).

   c. Pancreatic exocrine secretions are a major source of the Na^+ that is used to drive Na^+-coupled solute transport across intestinal cell apical membranes.

   d. Net transport of many solutes across epithelia is ultimately driven by the Na^+ pump in the basolateral membrane.

   e. Net solute transport across epithelia depends on the presence of different transporters (carriers, pumps, or channels) in the apical and basolateral membranes.

44. All of the following statements about water (re)absorption across epithelia are true **EXCEPT:**

   a. Net water transport may occur by active transport of water molecules.

   b. Net water transport is always coupled to solute transport.

   c. Net water transport across leaky epithelia is always driven by an osmotic gradient.

   d. Water absorbed across tight epithelia normally passes through the epithelial cells (i.e., it occurs through the transcellular route).

   e. Antidiuretic hormone (ADH) promotes water transport across some tight epithelia by promoting the insertion of aquaporins (water channels) into the apical membrane of the epithelial cells.

45. Secretory diarrhea is dependent on which one of the following cAMP-mediated mechanisms:

   a. Increased activation of Cl^- channels in the apical membrane of intestinal epithelial cells.

   b. Increased activation of Cl^- channels in the basolateral membrane of intestinal epithelial cells.

   c. Increased insertion of aquaporins (water channels) into the apical membrane of intestinal epithelial cells.

   d. Increased insertion of aquaporins into the basolateral membrane of intestinal epithelial cells.

   e. Inhibition of the Na^+ pump in the basolateral membrane of intestinal epithelial cells.

46. All of the following statements about gap junction channels are true **EXCEPT:**

   a. They are formed by two hemichannels.

   b. They connect two neurons at an electrical synapse.

   c. They allow current to flow directly from one cell to another.

   d. They allow the passage of monovalent ions, but not other molecules.

   e. They allow rapid communication because of a negligible synaptic delay.

47. The size of the EPP at the NMJ can be reduced by all of the following **EXCEPT:**

   a. Decreasing the concentration of ACh in synaptic vesicles.

   b. Inhibiting the enzyme acetylcholinesterase.

   c. Decreasing the Na^+ permeability of nAChRs.

   d. Decreasing $[Ca^{2+}]_o$.

   e. Blocking nAChRs with curare.

48. All of the following events contribute to the synaptic delay at chemical synapses **EXCEPT:**

   a. The EPP causing voltage-gated Na^+ channels to open.

   b. The presynaptic AP causing voltage-gated Ca^{2+} channels to open.

   c. ACh diffusing across the synaptic cleft.

   d. ACh opening nAChR channels in the postsynaptic membrane.

   e. Ca^{2+} entering the presynaptic cell and triggering transmitter release.

49. All of the following contribute to the size of MEPPs at the NMJ **EXCEPT:**

   a. The amount of ACh stored in a synaptic vesicle.

   b. $[Ca^{2+}]_o$.

   c. The activity of acetylcholinesterase.

d. The conductance of nAChR channels in the postsynaptic membrane.

e. The open probability of nAChR channels in the postsynaptic membrane.

50. All of the following statements about chemical synapses are true **EXCEPT**:

a. High concentrations of neurotransmitters are stored in synaptic vesicles in the presynaptic nerve terminal.

b. At an active zone, synaptic vesicles must be "docked" before neurotransmitters can be released.

c. Neurotransmitters are released by fusion of synaptic vesicles with the cell membrane at the nerve terminal.

d. Transporters are the predominant mechanism for removing neurotransmitters from the synaptic cleft.

e. Released neurotransmitters can trigger the opening of ion channels to activate the postsynaptic neuron.

51. All of the following statements about transmitter release are true **EXCEPT**:

a. Synaptotagmin is a protein on the synaptic vesicle that is required for neurotransmitter release.

b. Ca^{2+} ions trigger neurotransmitter release by activating synaptotagmin.

c. Toxins such as α-latrotoxin and botulinum toxin (BoTox) block synaptic transmission by inhibiting neurotransmitter release.

d. SNAP and SNARE proteins are essential components of the neurotransmitter release machinery in the nerve terminal.

e. After neurotransmitter release, the synaptic vesicle membrane is immediately retrieved back into the nerve terminal.

52. All of the following statements about short-term synaptic plasticity are true **EXCEPT**:

a. Synaptic depression, synaptic facilitation, and post-tetanic potentiation are all forms of short-term plasticity.

b. Synaptic depression is fully explained by depletion of releasable synaptic vesicles.

c. Ca^{2+} ions are essential for induction of synaptic facilitation.

d. In synaptic depression, the second of a pair of stimuli evokes a smaller postsynaptic response.

e. Post-tetanic potentiation requires a rise of $[Ca^{2+}]_i$ in the presynaptic terminal.

53. All of the following statements are true about nAChRs **EXCEPT**:

a. nAChR channels are permeable to both Na^+ and K^+.

b. Prolonged exposure of nAChRs to ACh causes desensitization.

c. Opening of nAChR channels at the NMJ produce an EPP.

d. nAChRs can be activated by nicotine.

e. The E_{rev} of nAChR channels is normally close to E_{Na}.

54. The main excitatory neurotransmitter in the brain is:

a. Acetylcholine.

b. Norepinephrine.

c. Glutamate.

d. Glycine.

e. Dopamine.

55. Unconventional neurotransmitters such as endo-cannabinoids:

a. Are stored in small translucent synaptic vesicles.

b. Are stored in large dense-core vesicles.

c. Are stored in small dense-core vesicles.

d. Are released immediately following synthesis.

e. Activate ionotropic receptors.

56. All of the following statements about glutamate receptors are trued **EXCEPT**:

a. AMPARs mediate fast excitatory synaptic transmission.

b. NMDAR activation requires a coagonist.

c. KARs are monovalent cation-selective channels.

d. NMDAR channels are blocked by benzodiazepines.

e. Activation of mGluRs located on presynaptic terminals may modulate neurotransmitter release.

57. All of the following statements about ionotropic glutamate receptors are true **EXCEPT**:

a. All ionotropic glutamate receptors are channels permeable to both Na^+ and K^+.

b. NMDAR channels are blocked by extracellular Mg^{2+} at negative V_m.

c. In response to glutamate, NMDAR channels open more rapidly than AMPAR channels.

d. NMDAR channels are permeable to Ca^{2+}.

e. NMDAR channels conduct outward current better than inward current.

58. All of the following statements about inhibitory synaptic transmission are true **EXCEPT**:

a. GABA receptor activation can increase Cl^- conductance.

b. GABA receptor activation can decrease Ca^{2+} conductance.

c. GABA receptor activation can increase K^+ conductance.

d. Ionotropic GABA receptors are targets of benzodiazepines.

e. A genetic defect in GABA receptors is responsible for human startle disease.

59. All of the following statements about neurotransmission are true **EXCEPT**:

a. ADP is a cotransmitter that is released with norepinephrine.

b. Adenosine activates P1 receptors and indirectly opens inward-rectifying K^+ channels.

c. All P1 purinergic receptors are metabotropic, and are members of the G-protein-coupled receptor family.

d. The endocannabinoid 2-arachidonoylglycerol (2-AG) is released by activation of metabotropic glutamate receptors.

e. The endocannabinoid 2-arachidonoylglycerol (2-AG) is a retrograde transmitter that inhibits transmitter release at glutamatergic synapses.

60. All of the following statements about long-term synaptic plasticity are true **EXCEPT**:

a. Long-term potentiation (LTP) and long-term depression (LTD) are persistent changes in the efficiency of neurotransmission at a synapse.

b. Once induced by synaptic stimulation, LTP and LTD can persist for at least days to weeks.

c. A rise in $[Ca^{2+}]_i$ is necessary for induction of all forms of LTP and LTD.

d. Ca^{2+} influx through NMDAR channels can activate LTP or LTD.

e. The frequency of stimulation determines whether LTP or LTD is induced at a synapse.

61. Which of the following statements about the structure of skeletal muscle is **FALSE?**

a. Thick filaments interdigitate with thin filaments, so that each thick filament is surrounded by a hexagonal array of thin filaments.

b. Bundles of thick and thin filaments run obliquely across the cell.

c. Thick filaments are composed mostly of myosin.

d. Thin filaments are composed of actin, tropomyosin, troponin, and nebulin.

e. Surrounding each myofibril is an extensive membrane-enclosed intracellular compartment, the sarcoplasmic reticulum.

62. In the presence of high $[Ca^{2+}]_i$ all of the following steps occur in the cross-bridge cycle in skeletal muscle **EXCEPT**:

a. ATP binds to myosin and thus causes the cross-bridge to detach.

b. With ADP and Pi bound to the myosin head, myosin binds to actin.

c. When ATP is hydrolyzed to ADP and Pi, the angle between the myosin head and the thin filament changes from 45 to 90 degrees.

d. The release of ADP generates the power stroke.

e. Force is generated when the angle between the myosin head and the thin filament changes from 90 to 45 degrees.

63. All of the following are features of both skeletal and cardiac muscle **EXCEPT**:

a. Individual muscle cells are electrically connected through gap junction channels.

b. Ca^{2+} ions initiate contraction by binding to troponin C.

c. The contractile elements are organized in sarcomeres.

d. Force is generated by a sliding filament mechanism.

e. In resting muscle cells, cross-bridges cannot form because tropomyosin covers the myosin binding sites on actin.

64. In skeletal muscle the activation of contraction by Ca^{2+} involves the regulatory proteins tropomyosin and troponin in which of the following ways?
  a. When $[Ca^{2+}]_i$ rises into the micromolar range, Ca^{2+} binds to troponin C and weakens the bond between troponin I and actin.
  b. Troponin T links the troponin complex to tropomyosin.
  c. Tropomyosin moves on the thin filament to either cover or expose myosin binding sites on actin.
  d. a, b, and c are all true.
  e. Only b and c are true.

65. All of the following are true of both skeletal and cardiac muscle contraction **EXCEPT:**
  a. Contraction is initiated by a depolarization of the surface membrane.
  b. Almost all of the Ca^{2+} that activates the contraction is pumped back into the SR by SERCA.
  c. L-type Ca^{2+} channels (DHPRs) in the surface membrane are activated by depolarization.
  d. Ca^{2+} is released from the SR into the cytoplasm through open Ca^{2+}-release channels (RyRs) in the SR.
  e. Ca^{2+} release channels in the SR are opened at specialized junctions between SR and T-tubule membranes.

66. The activation of Ca^{2+} release from the SR during a contraction in skeletal muscle:
  a. Occurs by the same mechanism as in cardiac muscle.
  b. Is caused by Ca^{2+} entering the cell through L-type Ca^{2+} channels in the surface membrane.
  c. Occurs through Ca^{2+}-release channels opened by IP_3 binding.
  d. Is independent of V_m.
  e. None of the above.

67. Which of the following is **FALSE?**
  a. The opening of RyRs in cardiac muscle involves Ca^{2+}-induced Ca^{2+} release.
  b. The activity of both the Na^+/Ca^{2+} exchanger and SR Ca^{2+} pump (SERCA) decreases $[Ca^{2+}]_i$ and causes the relaxation of cardiac muscle.

  c. Extracellular Ca^{2+} is required for E-C coupling in cardiac muscle.
  d. All of the Ca^{2+} that activates cardiac muscle contraction comes from the SR.
  e. The Ca^{2+} release through Ca^{2+} release channels can be visualized as Ca^{2+} "sparks."

68. Which of the following statements about FF (fast, fatigable), FR (fast, fatigue-resistant), and S (slow) motor units in skeletal muscle is **FALSE?**
  a. S motor units generate the smallest forces and do not fatigue.
  b. Whole muscle often contains mixtures of FF, FR, and S motor units.
  c. A motor unit is a single skeletal muscle cell and the α motor neuron that innervates it.
  d. FR motor units are more resistant to fatigue than FF motor units.
  e. FF motor units generate the largest tetanic forces and are usually recruited after S and FR motor units.

69. The force generated by cardiac muscle can be increased by all of the following **EXCEPT:**
  a. Increasing the duration of the cardiac action potential.
  b. Inhibiting Na^+/Ca^{2+} exchangers in the plasma membrane.
  c. Activating β-adrenergic receptors in cardiac muscle.
  d. Increasing the end-diastolic volume of the heart.
  e. Increasing the frequency of stimulation of the heart.

70. In a pure isometric contraction the force generated by whole skeletal muscle can be increased by all of the following **EXCEPT:**
  a. Increasing the number of motor units that are activated.
  b. Increasing the frequency of stimulation of the muscle.
  c. Increasing the sarcomere length from 1.4 to 1.6 μm.
  d. Increasing the sarcomere length from 2.4 to 3.0 μm.
  e. Decreasing the rate of Ca^{2+} uptake by SERCA into the SR.

71. All of the following are true of the relationship between isotonic force and velocity of shortening in skeletal muscle **EXCEPT:**
    a. As the isotonic force generated by the muscle increases, the velocity of shortening decreases.
    b. The maximum velocity of shortening (V_0) occurs when the muscle shortens with no attached load.
    c. V_0 a reflection of the maximum rate of cross-bridge cycling.
    d. V_0 is inversely proportional to myosin ATPase activity.
    e. If the load attached to the muscle is greater than the maximum force the muscle can generate, the velocity of shortening is zero.

72. The maximum force in a twitch contraction in skeletal muscle is smaller than the maximum tetanic force because:
    a. The amount of Ca^{2+} released during a twitch is not enough to saturate all the troponin C binding sites.
    b. During a twitch there is not enough time to fully stretch series elastic elements in the muscle.
    c. $[Ca^{2+}]_i$ remains elevated for a longer time during a tetanus.
    d. a, b, and c are all true.
    e. Only b and c are true.

73. All of the following statements about malignant hyperthermia (MH) are true **EXCEPT:**
    a. MH is triggered by exposure to volatile, halogenated anesthetics.
    b. Dihydropyridines (e.g., nifedipine) are used to prevent attacks of MH.
    c. An episode of MH is characterized by a sustained increase in $[Ca^{2+}]_i$ in skeletal muscle.
    d. Most cases of MH are associated with a point mutation in the gene that encodes the skeletal muscle RyR.
    e. In an episode of MH, continuous cross-bridge cycling and SERCA activity produce an increase in body temperature that can be fatal.

74. Which one of the following statements is **FALSE?**
    a. In an isotonic contraction the velocity of shortening is constant.
    b. In an isometric contraction the muscle generates force and then relaxes at constant length.
    c. In an isotonic contraction the muscle shortens against a constant load.
    d. Under physiological conditions most skeletal muscle contractions are partly isotonic and partly isometric.
    e. When skeletal muscle is activated with an attached load, the first phase of the contraction is isometric force development.

75. All of the following are true **EXCEPT:**
    a. Smooth muscles do not exhibit the striation patterns observed in skeletal muscle because the thick and thin filament bundles are not in register.
    b. Smooth muscle, like skeletal muscle, requires troponin C (TnC) for contractile activation.
    c. The dense bodies of smooth muscles are analogous to the Z lines of skeletal muscle because both are composed primarily of α-actinin.
    d. The dense bodies of smooth muscles are analogous to the Z lines of skeletal muscle because the thin filaments are inserted into the dense bodies.
    e. In saccular organs such as the urinary bladder, both longitudinal and lateral forces cause relatively uniform contraction of the organ wall.

76. Smooth muscle contraction can be activated by all of the following **EXCEPT:**
    a. Ca^{2+} binding to calmodulin.
    b. Ca^{2+} binding to SERCA.
    c. Phosphorylation of myosin regulatory light chain (RLC).
    d. IP_3.
    e. Ca^{2+} binding to RyRs.

77. Smooth muscle relaxation is promoted by all of the following **EXCEPT:**
    a. Dephosphorylation of myosin regulatory light chain (RLC).
    b. Elevating the concentration of cAMP in the smooth muscle cytoplasm.
    c. Decreasing the activity of myosin RLC phosphatase.
    d. Phosphorylation of myosin light chain kinase.
    e. Opening K_{Ca} channels.

78. Mechanisms involved in the activation of smooth muscles may include all of the following **EXCEPT:**
    a. Depolarization initiated by some neurotransmitters such as ACh.
    b. ACh-induced release of nitric oxide (NO).
    c. Release of Ca^{2+} from the SR.
    d. Pharmacomechanical coupling that involves little or no change in V_m.
    e. Ca^{2+}-independent changes in the sensitivity of the contractile apparatus to Ca^{2+}.

79. Activation of smooth muscle differs from that of skeletal muscles in that:
    a. Smooth muscles are never activated by membrane depolarization.
    b. Smooth muscles are not dependent on SR Ca^{2+} release.
    c. Smooth muscles are activated by a troponin-tropomyosin–regulated mechanism.

    d. Smooth muscles generate much less force per unit cross-sectional area (= "stress").
    e. Smooth muscle contraction velocity is slower.

80. All of the following are true of tonic smooth muscles **EXCEPT:**
    a. The unitary actin-myosin cross-bridge force is much smaller than in skeletal muscle.
    b. Smooth muscle cross-bridge attachments last several times longer than skeletal muscle cross-bridge attachment.
    c. Compared with skeletal muscle, a much larger fraction of the smooth muscle actin-myosin cross-bridges are attached at any time.
    d. Tonic smooth muscles can maintain contraction with little consumption of ATP.
    e. The rate of ADP dissociation from myosin in smooth muscles is very slow when the cross-bridges are attached.

ANSWERS TO REVIEW EXAMINATION

1. e	17. e	33. a	49. b	65. b
2. c	18. c	34. a	50. d	66. e
3. e	19. b	35. e	51. e	67. d
4. d	20. e	36. b	52. b	68. c
5. c	21. d	37. d	53. e	69. e
6. b	22. c	38. a	54. c	70. d
7. a	23. a	39. d	55. d	71. d
8. d	24. c	40. d	56. d	72. e
9. c	25. b	41. b	57. c	73. b
10. a	26. a	42. d	58. e	74. a
11. d	27. b	43. b	59. a	75. b
12. b	28. e	44. a	60. c	76. b
13. c	29. c	45. a	61. b	77. c
14. d	30. d	46. d	62. d	78. b
15. d	31. c	47. b	63. a	79. e
16. b	32. e	48. a	64. d	80. a

INDEX